Lecture Notes in Computer Scie

Edited by G. Goos and J. Hartmanis

G. Cohen S. Litsyn
A. Lobstein G. Zémor (Eds.)

Algebraic Coding

First French-Israeli Workshop
Paris, France, July 19-21, 1993
Proceedings

Springer-Verlag
Berlin Heidelberg New York
London Paris Tokyo
Hong Kong Barcelona
Budapest

Series Editors

Gerhard Goos
Universität Karlsruhe
Postfach 69 80
Vincenz-Priessnitz-Straße 1
D-76131 Karlsruhe, Germany

Juris Hartmanis
Cornell University
Department of Computer Science
4130 Upson Hall
Ithaca, NY 14853, USA

Volume Editors

Gérard Cohen
Antoine Lobstein
Gilles Zémor
École Nationale Supérieure des Télécommunications
46 rue Barrault, F-75634 Paris Cedex, France

Simon Litsyn
Department of Electrical Engineering, Tel Aviv University
Ramat Aviv 69978, Israel

CR Subject Classification (1991): E.3-4, G.2

ISBN 3-540-57843-9 Springer-Verlag Berlin Heidelberg New York
ISBN 0-387-57843-9 Springer-Verlag New York Berlin Heidelberg

CIP data applied for

Typesetting: Camera-ready by author
SPIN: 10131992 45/3140-543210 - Printed on acid-free paper

PREFACE

The first French-Israeli Workshop on Algebraic Coding took place at Ecole Nationale Supérieure des Télécommunications, Paris, July 19-21, 1993. It was a continuation of a French-Soviet Workshop hold in 1991 and edited by the same board (Springer-Verlag LNCS 573).

Convolutional codes and special channels

V. Balakirsky formulates a necessary and sufficient condition for a linear convolutional time-variant encoder to be noncatastrophic.
G. Kaplan, S. Shamai and Y. Kofman address code design and selection rules under power and decoding delay constraints for a slowly-fading channel modeling a mobile communication system.
G. Poltyrev and J. Snyders consider the assignment of codes to users of a multiple access channel, allowing correction of errors and separation of messages, under the hypothesis that the subset of active users is known to the receiver.
V. Blinovsky and M. Pinsker obtain an upper bound on the number of codewords which must be stored in order to achieve capacity when applying list decoding to an arbitrarily varying channel.

Covering codes

I. Honkala studies the problem of lowerbounding the minimum cardinality of a code with given length n and covering radius 1, in the case when the code is binary and n is congruent to 5 (mod 6).
E. Kolev and I. Landgev address the same problem for mixed binary/ternary codes with length up to 8 and covering radius up to 3.
A. Lobstein and V. Pless present a new table for the smallest length of a binary linear code with given codimension m and covering radius r, for $m \leq 24$ and $r \leq 12$.
I. Bocharova and B. Kudryashov give a new upper bound for the covering radius of convolutional codes, by means of random coding.

Cryptography

O. Delos and J.J. Quisquater describe a multi-signature scheme involving cooperating entities, with no interaction needed between the cosigners.
D. Naccache and D. M'Raïhi present an efficient alternative approach to Montgomery's algorithm for modular operations.
C. Blundo, A. De Santis, L. Gargano and U. Vaccaro consider the problem of designing efficient secret sharing schemes with veto capabilities from qualified minorities.
J.P. Tillich and G. Zémor study weaknesses and strengths of group-theoretic hash functions based on computations in arithmetic groups $SL_2(\mathbf{F}_p)$.

Sequences

S. Bitan and T. Etzion present new constructions for optimal optical orthogonal codes, i.e. families of w-sets of integers modulo n in which no difference is repeated.
A. Gavish and A. Lempel consider so-called complementary pairs of sequences over the alphabet $\{-1, 0, 1\}$. A pair of words is complementary if for a given nonzero shift their aperiodic autocorrelation functions sum up to zero. The authors are interested in minimizing the number of zeros in complementary pairs: they derive some tight bounds on this quantity.
R. Roth considers sequences having rth order spectral null at zero frequency. Upper bounds are derived for the size of the set of all such sequences. Some subsets defined as null spaces of certain submatrices of Hadamard matrices provide codes with codewords belonging to the considered set and large minimal distance.
A. Barg uses the cyclic structure of shortened Kerdock codes to present a family of codes with asymptotically optimal correlation properties.

Graphs and codes

N. Alon and B. Sudakov solve a problem due to Ahlswede *et al.* concerning the maximal number of subsets of constant-weight binary words, such that in each pair of subsets it is possible to find a pair of nonintersecting words, one from each. It turns out that asymptotically, the maximal number of subsets is equal to half of the total number of constant-weight words.
O. Moreno and V. Zinoviev give sufficient conditions for 4-regular graphs to have 3-regular subgraphs by variations on the use of the Chevalley-Warning theorem.
M. Karpovsky, S. Chaudhry, L. Levitin and C. Moraga present bounds and constructions for codes detecting and correcting given error patterns caused by faulty processing elements in multiprocessor systems.

Sphere packings and lattices

A. Bonnecaze and P. Solé construct formally self-dual binary codes and unimodular lattices using quaternary codes. They obtain new constructions of Leech and Gosset lattices.
P. Boyvalenkov and S. Nikova propose a new method for obtaining lower bounds on the size of a spherical t-design, with special emphasis on the cases $t = 9$ or 10.
G. Poltyrev constructs a class of lattices from linear codes using Conway-Sloane construction A, and shows that they achieve capacity for the unrestricted AWGN channel.
P. Loyer and P. Solé generalize the Conway-Sloane lattice decoding algorithm to the L_p norm, computing, in particular, some Voronoi diagrams and covering radii.
O. Amrani, Y. Be'ery and A. Vardy develop new fast algorithms for the soft decoding of the Golay code and the Leech lattice.

Bounds for codes

S. Kovalov derives new conditions that the last element of the distance spectrum of optimal binary codes should satisfy.
S. Litsyn and A. Vardy prove that the code with parameters $(10, 40, 4)$ is unique. This allows them to derive new upper bounds on the number of words in single-error-correcting codes of lengths 10 and 11, namely 78 and 156, respectively.
G. Kabatianski and A. Lobstein provide a new upper bound for binary arithmetic codes, which is asymptotically better than previously known bounds.
G. Cohen, L. Huguet and G. Zémor derive new bounds on the maximum possible dimension of binary codes in terms of their generalized distances.
G. Zémor introduces threshold probabilities θ of linear codes when studying residual error probability; in particular, the residual error probability is shown to always be an exponential function of the minimal distance when the channel error probability is separated from θ.
J. Rifa-Coma gives a decoding algorithm for BCH codes that achieves a little more than the conventional algorithms in the case when the designed and true minimum distances differ.
N. Sendrier analyses the trade-off for error-correcting codes between algorithmic complexity and decoding performance. Low-rate product codes in particular, although of poor minimum distance, possess an efficient natural decoding algorithm.
I. Dumer and P. Farrell study the performance of linear codes on the erasure channel; in particular the case of BCH and concatenated codes is considered.

This meeting was sponsored by the Centre National de la Recherche Scientifique, l'Association Franco-Israelienne de Recherche Scientifique et Technique, le Ministère de l'Enseignement Supérieur et de la Recherche, the French Section of IEEE, and ENST.

We would like to thank the referees,
N. Alon, B. Arazi, A. Barg, G. Battail, R. Calderbank, G. Cohen, J.L. Dornstetter, I. Dumer, S. Eliahou, T. Etzion, M. Girault, I. Honkala, G. Kabatianski, S. Kovalov, J. Lahtonen, A. Lempel, S. Litsyn, A. Lobstein, O. Moreno, V. Pless, G. Poltyrev, S. Qiu, J.J. Quisquater, J. Rifa-Coma, N. Sloane, P. Solé, H. van Tilborg, R. Zamir, G. Zémor.

December 1993 G. Cohen, S. Litsyn, A. Lobstein, G. Zémor

AFTERMATH

As mentioned in the foreword, the French-Soviet workshop took place in july 1991 ; one month later, the USSR collapsed. A few months after the French-Israeli meeting, Jericho and Gaza underwent a dramatic change of status. Understandably, we have become increasingly concerned as to the consequences of the choice of our partner for the next binational event. Indeed, so much so that we decided to switch to a politically invariant characterization of our workshop. We are now preparing the *first Mediterranean Workshop on Algebraic Coding*: this decision was reached only after careful study, and has been approved by earthquake forecast experts.

ECOLE NATIONALE SUP
LES FORCES ELEMENTAIRES POUR LES
TRANSMISSIONS

CONTENTS

Convolutional codes and special channels

Covering codes

Cryptography

Sequences

Graphs and codes

Sphere packings and lattices

Bounds for codes

A Necessary and Sufficient Condition for Time-Variant Convolutional Encoders to be Noncatastrophic *

V.B.Balakirsky
Russia
The author is with the Department of Information Theory, University of Lund, P.O. Box 118, S-221 00 Lund, Sweden on leave from The Data Security Association "Confident", St.-Petersburg, Russia
E-mail: volodya@dit.lth.se

Abstract. Linear convolutional encoders whose rate and generator polynomials periodically vary with time are considered. A system of linear recurrent equations defined by the parameters of the encoder is introduced, and it is shown that the encoder is noncatastrophic if and only if there exists an autoregressive filter realizing these equations. A necessary and sufficient condition for encoders to be noncatastrophic, formulated using this fact, may be simpler than a condition that can be obtained from the known result by Massey and Sain.

1 Introduction.

It is well-known [1] that a time-invariant binary linear convolutional code having rate $R = K/N$ may be defined by the matrix $\mathbf{G}$ of size $K \times N$. The k-th row and n-th column of this matrix contain the polynomial

$$g_{kn} = g_{kn}(z) = \sum_{j=0}^{\nu} g_{kn}^{(j)} z^j, \quad g_{kn}^{(j)} \in GF(2)$$

that is referred to as *the generator polynomial*; $k = 1, ..., K, n = 1, ..., N$, and z is a formal variable. The parameter ν is referred to as *the constraint length* of the encoder. The relationship between the information sequence $u_1^{(0)}, ..., u_K^{(0)}$, $u_1^{(1)}, ..., u_K^{(1)}, ...$ and the coded sequence $x_1^{(0)}, ..., x_K^{(0)}, x_1^{(1)}, ..., x_K^{(1)}, ...$ is established by the following rule :

$$x_n^{(t)} = \sum_{k=1}^{K} \sum_{j=0}^{\nu} g_{kn}^{(j)} u_k^{t-j}, \tag{1.1}$$

where $u_k^{(-1)} = ... = u_k^{(-\nu)} = 0$, $k = 1, ..., K$.
Convolutional encoders transforming any information sequence, having infinite

* Work partially supported by The Data Security Association "Confident", Smolny, St.-Petersburg, Russia and partially by a Scholarship from the Swedish Institute, Stockholm, Sweden.

Hamming weight, to a coded sequence, that also has infinite Hamming weight, are known as *noncatastrophic encoders*. These codes play the most important role when systems of information transmission over memoryless channels are being constructed [1, 2]. Covolutional encoders that do not have this property are referred to as *catastrophic encoders*. For example, the encoder of rate 2/4 defined by the matrix

$$G = \begin{pmatrix} 1+z & 1+z & 0 & 0 \\ 0 & z & 1 & 1+z \end{pmatrix} \tag{1.2}$$

is catastrophic because if $u_1^{(t)} = 1$ and $u_2^{(t)} = 0$, $t = 0, 1, ...$ then $x_0^{(0)} = x_1^{(0)} = 1$ but all other symbols of the coded sequence are equal to zero. A necessary and sufficient condition for convolutional encoders to be noncatastrophic was obtained by Massey and Sain [2] and generalized by Olson [3]. To use this condition it is required to calculate the determinants of all $K \times K$ submatrices of G. Let $C = N!/(K!(N-K)!)$, and let $\Delta^{(n)}$ be the determinant of the n-th submatrix, $n = 1, ..., C$. Besides, let

$$\Delta = g.c.d.\{\Delta^{(1)}, ..., \Delta^{(C)}\}$$

be the greatest common divisor of the polynomials in brackets. *A convolutional code is noncatastrophic if and only if the polynomial Δ is equal to some power of z, i.e.,*

$$\Delta \in \{1, z, z^2, ...\}.$$

In particular, the encoder defined by (1.2) is catastrophic, since

$$\Delta = g.c.d.\{z(1+z), 1+z, (1+z)^2, 1+z, (1+z)^2, 0\} = 1+z.$$

It is well-known [1] that decoding of the received sequence may be described as a motion along the branches of a directed graph, named code trellis. Each level of the trellis contains $2^{K\nu}$ nodes, 2^K branches leave each node, and 2^K branches enter each node located at levels $\nu+1, \nu+2, ...$ The subblocks of length N are assigned to branches.

A decoding procedure is realized in a most simple way for codes of rate $1/N$ because every step is taken after choosing one of only two alternatives. To expand the rate region that may be obtained using these codes, the "punctured" codes were introduced [1,4]. For these codes not all symbols of the coded subblock are transmitted. It is known that both this coding method and time-varying of generator polynomials in accordance with a periodic law [5] can provide good error-correcting codes. We examine a coding method including these approaches as particular cases.

2 Mixed Convolutional Codes.

Suppose there are M convolutional codes of rates $1/N_1$,...,$1/N_M$ and constraint length ν defined by the generator polynomials

$$G_m = (g_{m,11}, \dots g_{m,1N_m}), \quad m = 1, ..., M. \tag{2.1}$$

Suppose also that there is a function φ such that

$$\varphi(t+iT)=\varphi(t)\in\{1,...,M\},\quad i=1,2,...,t=0,...,T-1, \tag{2.2}$$

i.e. T is the minimal period of φ. If the information sequence $u^{(0)},u^{(1)},...$ is encoded by all codes then M code subblocks

$$\mathbf{x}_m^{(t)}=(x_{m,1}^{(t)},...,x_{m,N_m}^{(t)})\in\{0,1\}^{N_m}$$

are generated at time t. These subblocks are obtained by (1.1) after substitution 1 for K, and $x_{m,1n}^{(t)},g_{m,1n}^{(j)},u^{(t-j)}$ for $x_n^{(t)},g_{kn}^{(j)},u_k^{(t-j)}$, respectively; $m=1,...,M$. Let one of these subblocks be sent at the output of the encoder while all the others are lost :

$$\mathbf{x}^{(t)}=\mathbf{x}_{\varphi(t)}^{(t)},\quad t=0,1,..., \tag{2.3}$$

where $\mathbf{x}^{(0)},\mathbf{x}^{(1)},...$ is the coded sequence. Then the rate of the code is equal to $R_\varphi=1/N_\varphi$ bits per channel symbol, where

$$N_\varphi=(N_{\varphi(0)}+...N_{\varphi(T-1)})/T. \tag{2.4}$$

Convolutional codes constructed in accordance with (2.1)-(2.3) will be referred to as *mixed convolutional codes.* The encoder consists of a shift register with $\nu+1$ cells, adders modulo 2 and switches (Fig.1). The code may be described using a code trellis having the same structure as the code trellis of some code of rate $1/N$. The only difference lies in the fact that coded subblocks of various lengths may be assigned to the branches.
The codes under consideration may be employed when one has to construct a system of information transmission over a memoryless channel with unknown parameters. Then several adders can be included into the encoder. The decoder is realized as a device having 2^ν buffers. These buffers are sequentially filled using choices from only two alternatives for any function φ that may be assigned just before operations. If the parameters of the channel are changed then φ may be also changed, and it does not essentially affect the complexity of the encoding and decoding algorithms.
Some noncatastrophic encoders with the constraint length $\nu=4$, that are obtained using four codes of rates $1,1/2,1/3$ and $1/4$, are shown in Table 1. The coefficients of the generator polynomials are given in octal. For example, $G_2=(23,35)$ means that $g_{2,11}=1+z^3+z^4, g_{2,12}=1+z+z^2+z^4$. The function φ is defined by the T-tuple $\varphi=(\varphi(0),...\varphi(T-1))$. *The free distance* of the code is denoted by d_f, and $\hat{d_f}$ denotes the value taken from an upper bound on the free distance that was obtained in [6] as a combination of the Heller's and Griesmer's bounds : if d_f is the free distance, then

$$d_f\leq\min_{t=0,...,T-1}\min_{k\geq1}\frac{2^{k-1}}{2^k-1}\sum_{\tau=t}^{t+\nu+k}N_{\varphi(\tau)}$$

and

$$\sum_{i=1}^{k}d_i\leq\min_{t=0,...,T-1}\sum_{\tau=t}^{t+\nu+k}N_{\varphi(\tau)},\quad k=1,2,...$$

where

$$d_1 = d_f,$$

$$d_i = \begin{cases} d_{i-1}/2 & \text{if } d_{i-1} \text{ is even} \\ (d_{i-1}+1)/2 & \text{if } d_{i-1} \text{ is odd} \end{cases} \quad i > 1.$$

Any code, we are considering, may be represented as a time-invariant code of rate $T/(TN_\varphi)$, where N_φ is defined by (2.4). For example, the coded sequence generated by the encoder given in Fig.1 coincides with the coded sequence belonging to the code defined by (1.2) if $T = 2$, $\varphi = (1,2)$, $u^{(2t)} = u_1^{(t)}$, $u^{(2t+1)} =$ $= u_2^{(t)}, t = 0, 1, \ldots$ .. Therefore, we can use the result of Massey and Sain to determine whether an encoder is catastrophic. However, when $T > 1$ this procedure may be rather difficult.

As it is easy to see from (1.1) and (2.3), the encoder is catastrophic if and only if there exists $t_0 < \infty$ and a nonzero information sequence such that the following equations :

$$\sum_{j=0}^{\nu} g_{\varphi(t),1n}^{(j)} u^{(t-j)} = 0, \quad n = 1, \ldots, N \tag{2.5}$$

are valid for all $t \geq t_0$. Further considerations deal with an analysis of these equations, and the following notation will be used.

Let $a^{(0)}, a^{(1)}, \ldots \in GF(2)$ be the coefficients of a polynomial $a(z)$; $b^{(0)}, b^{(1)}, \ldots \in GF(2)$ be the coefficients of a polynomial $b(z)$, etc. If there are no restrictions on the considered polynomials then dependence on the formal variable z will be omitted. *The degree* and *delay* of a polynomial a are defined as :

$$del(a) = \min_{j:a^{(j)}=1} j, \quad deg(a) = \max_{j:a^{(j)}=1} j.$$

If a and b have finite degree then

$$(a,b) = \sum_{j \geq 0} a^{(j)} b^{(j)}$$

is referred to as *the inner product* of these polynomials. An integer

$$rank(a_1, a_2, \ldots) = r$$

is referred to as *the rank of polynomials* $a_1, a_2, \ldots$ if there are polynomials $a_{i_1}, \ldots, a_{i_r}$ such that from

$$\sum_{j=1}^{r} c_j a_{i_j} = 0, \quad c_1, \ldots, c_r \in GF(2)$$

follows that $c_1 = \ldots = c_r = 0$, while this statement is not valid for any $r+1$ polynomials.

3 Some Properties of Autoregressive Filters.

Let

$$f = f^{(0)} + ... + f^{(\mu-1)}z^{\mu-1} + z^{\mu}$$

be a polynomial and let

$$u^{(t)} = \sum_{j=0}^{\mu-1} f^{(j)} u^{(t-\mu+j)}, \quad t = \mu, \mu+1, ... \tag{3.1}$$

As it is well-known, (3.1) is realized by an autoregressive filter. Let us introduce some complementary notation to express the symbols of the sequence satisfying this equation in terms of polynomial f and symbols $u^{(0)}, ..., u^{(\mu-1)}$.
For all a and b let $R_a[b]$ be a polynomial defined as the remainder of the division of b by a, i.e.,
(i) $deg R_a[b] < deg(a)$;
(ii) there is a polynomial A such that

$$b + R_a[b] = Aa$$

The statements given below are well-known [7].
(R1) Conditions (i) and (ii) uniquely determine $R_a[b]$.
(R2) The following equations :

$$R_a[b] + R_a[c] = R_a[b+c]$$
$$R_a[zb] = R_a[zR_a[b]]$$

are valid for all b and c.
Lemma (f). If a sequence $u^{(0)}, u^{(1)}, ...$ satisfies (3.1) then

$$u^{(t)} = (R_f[z^t], u), \quad t = 0, 1, ..., \tag{3.2}$$

where

$$u = u^{(0)} + ... + u^{(\mu-1)}z^{\mu-1}. \tag{3.3}$$

Proof. It is obvious that (3.2) are valid for $t < \mu$. Let $t \geq \mu$ be fixed and let

$$u^{(t-\mu+j)} = (R_f[z^{t-\mu+j}], u), \quad t = 0, ..., \mu-1. \tag{3.4}$$

Then, using statements (R1), (R2) and properties of the inner product we obtain :

$$u^{(t)} = \sum_{j=0}^{\mu-1} f^{(j)}(R_f[z^{t-\mu+j}], u) =$$
$$= (R_f[f^{(0)}z^{t-\mu} + ... + f^{(\mu-1)}z^{t-1}], u) =$$
$$= (R_f[z^{t-\mu}f + z^t], u) = (R_f[z^t], u),$$

i.e., (3.4) is also valid for $j = \mu$, and (3.2) is proved by induction. // We propose the following generalization of the previous formulas. Suppose, there is an integer T, polynomials $f_0, ..., f_{T-1}$ of degree μ , and

$$\theta(\tau + iT) = \tau, \quad i = 0, 1, ..., \tau = 0, ..., T-1.$$

Let us consider the following recurrent equation :

$$u^{(t)} = \sum_{j=0}^{\mu-1} f^{(j)}_{\theta(t)} u^{(t-\mu+j)}, \quad t = \mu, \mu+1, ... \tag{3.5}$$

that may be realized using time-variant autoregressive filter (Fig.2).
For any polynomials $a_0, ..., a_{T-1}$ of degree μ and any polynomial b, we denote by $R_{\mathbf{a}}[b]$, where $\mathbf{a} = (a_0, ..., a_{T-1})$, the polynomial with the following properties:
(**i**) $deg R_{\mathbf{a}}[b] < \mu$;
(**ii**) there are polynomials $A_0, ..., A_{T-1}$ such that

$$b + R_{\mathbf{a}}[b] = \sum_{i=0}^{T-1} z^i A_i(z^T) a_i.$$

We may prove the statements given below in the same way as (R1), (R2) and Lemma (f).
(**R1**) Conditions (**i**) and (**ii**) uniquely determine $R_{\mathbf{a}}[b]$.
(**R2**) The following equations :

$$\begin{aligned} R_{\mathbf{a}}[b] + R_{\mathbf{a}}[c] &= R_{\mathbf{a}}[b+c] \\ R_{\mathbf{a}}[z^T b] &= R_{\mathbf{a}}[z^T R_{\mathbf{a}}[b]] \end{aligned}$$

are valid for all b and c.
Lemma f. If a sequence $u^{(0)}, u^{(1)}, ...$ satisfies (3.5) then

$$u^{(t)} = (R_{\mathbf{f}}[z^t], u), \quad t = 0, 1, ...,$$

where $\mathbf{f} = (f_0, ..., f_{T-1})$ and the polynomial u is defined in (3.3).

4 A Necessary and Sufficient Condition for Mixed Convolutional Encoders to be Noncatastrophic.

For any nonnegative integer t let $[t]$ be *the residue class modulo* T containing t, i.e. $[t]$ consists of all integers $t + iT$, where $i \geq -t/T$. The integer

$$lea[t] = \min_{t^* \in [t]} t^*$$

will be referred to as *the leader* of the residue class containing t. We will use the encoder defined as follows :

$$M = 2, \quad T = 2, \quad \varphi = (1, 2), \tag{4.1}$$

$$g_{1,11} = 1 + z^3 + z^4, \quad g_{1,12} = z^4 + z^6 + z^8,$$

$$g_{2,11} = z^2 + z^4 + z^6$$

to illustrate further considerations. For this code, equations (2.5) are written below :

$$\begin{cases} u^{(t)} = u^{(t-3)} + u^{(t-4)}, \\ u^{(t-4)} = u^{(t-6)} + u^{(t-8)}, \\ u^{(t-3)} = u^{(t-5)} + u^{(t-7)}, \quad t = 8, 10, \dots . \end{cases} \tag{4.2}$$

Substituting t for $t-4$ in the second equation, and $t-1$ for $t-3$ in the third equation, we obtain that (4.2) generates the following system :

$$\begin{cases} u^{(t)} = u^{(t-3)} + u^{(t-4)}, \\ u^{(t)} = u^{(t-2)} + u^{(t-4)}, \\ u^{(t-1)} = u^{(t-3)} + u^{(t-5)}, \quad t = 8, 10, \dots . \end{cases} \tag{4.3}$$

Adding the first equation in (4.3) to the second one we obtain :

$$\begin{cases} u^{(t-2)} = u^{(t-3)}, \\ u^{(t)} = u^{(t-3)} + u^{(t-4)}, \\ u^{(t-1)} = u^{(t-3)} + u^{(t-5)}, \quad t = 8, 10, \dots . \end{cases} \tag{4.4}$$

and, finally, substituting t for $t-2$ in the first equation of (4.4), we write :

$$\begin{cases} u^{(t)} = u^{(t-1)}, \\ u^{(t)} = u^{(t-3)} + u^{(t-4)}, \\ u^{(t-1)} = u^{(t-3)} + u^{(t-5)}, \quad t = 8, 10, \dots . \end{cases} \tag{4.5}$$

If there exist nonzero symbols $u^{(3)}, u^{(4)}, u^{(5)}, u^{(6)}$, that determine a sequence satisfying (4.5), then this sequence also satisfies (4.2) when $t \geq 12$. The solution is as follows :

$$u^{(4)} = u^{(3)}, \quad u^{(6)} = u^{(5)}, \quad (u^{(3)}, u^{(5)}) \neq (0,0),$$

and the encoder is catastrophic.

Suppose there are nonzero generator polynomials $G_1, \dots, G_M$ and a function φ of period T. Let us introduce polynomials g_j, $j = 1, \dots, TN$, where N_φ is defined in (2.4), setting

$$g_j = z^l g_{\varphi(t),1n}, \tag{4.6}$$

where

$$\begin{aligned} l &= -del(g_{\varphi(t),1n}) + lea[t + del(g_{\varphi(t),1n})], \\ j &= N_{\varphi(0)} + \dots + N_{\varphi(t-1)} + n, \\ & n = 1, \dots, N_{\varphi(t)}, \quad t = 0, \dots, T-1. \end{aligned}$$

For all $i = 0, \dots, \nu$, let us do the following operations : if there is s such that

$$del(g_s) = i, \tag{4.7}$$

then set $s_0 = s, g = g_s$, and for all $s \neq s_0$ satisfying (4.7) set $g_s = g_s + g$.
Let

$$\mu = \max_s\{deg(g_s) - del(g_s)\}$$

and

$$F_t = \{z^{\mu+t}g_s(z^{-1}) \text{ for all } s : del(g_s) = t\},$$

where $t = 0, ..., T$. If F_t is empty, we set $F_t = \{1\}$.
For example, considering the encoder defined by (4.1), we obtain :

$$g_1 = 1 + z, \quad g_2 = 1 + z + z^4, \quad g_3 = z + z^3 + z^5,$$

$$\mu = 4,$$

$$F_0 = \{z^3 + z^4, 1 + z + z^4\}, \quad F_1 = \{1 + z^2 + z^4\}.$$

Theorem. A convolutional encoder defined by the generator polynomials G and a function φ of period T is noncatastrophic if and only if

$$rank(R_{\mathbf{f}}[z^t] + R_{\mathbf{f}'}[z^t], \ \mathbf{f}, \mathbf{f}' \in \mathbf{F}, \ t = \mu, \mu + 1, ...) = \mu, \tag{4.8}$$

where $\mathbf{F} = F_0 \times F_1 \times ... \times F_{T-1}$.
In particular, as it is easy to see, the only calculations of the expression at the left hand side of (4.8) should be done for

$$\mathbf{f} = (z^3 + z^4, 1 + z^2 + z^4), \quad \mathbf{f}' = (1 + z + z^4, 1 + z^2 + z^4),$$

when we consider the encoder (4.1). Then

$$(1 + z^4)(1 + z + z^4) + z^4(z^3 + z^4) = z^3(1 + z^2 + z^4)$$

and

$$\begin{aligned}
R_{\mathbf{f}}[z^{5+2i}] &= R_{\mathbf{f}'}[z^{5+2i}], \quad i = 0, 1, ..., \\
R_{\mathbf{f}}[z^4] + R_{\mathbf{f}'}[z^4] &= (z^3) + (1 + z) = 1 + z + z^3, \\
R_{\mathbf{f}}[z^6] + R_{\mathbf{f}'}[z^6] &= (z + z^3) + (z^2 + z^3) = z + z^2, \\
R_{\mathbf{f}}[z^{4+4i}] + R_{\mathbf{f}'}[z^{4+4i}] &= R_{\mathbf{f}}[z^4] + R_{\mathbf{f}'}[z^4], \\
R_{\mathbf{f}}[z^{6+4i}] + R_{\mathbf{f}'}[z^{6+4i}] &= R_{\mathbf{f}}[z^6] + R_{\mathbf{f}'}[z^6], \\
i &= 0, 1, ... \ .
\end{aligned}$$

Thus, the left hand side of (4.8) is equal to 2, and the encoder is catastrophic.

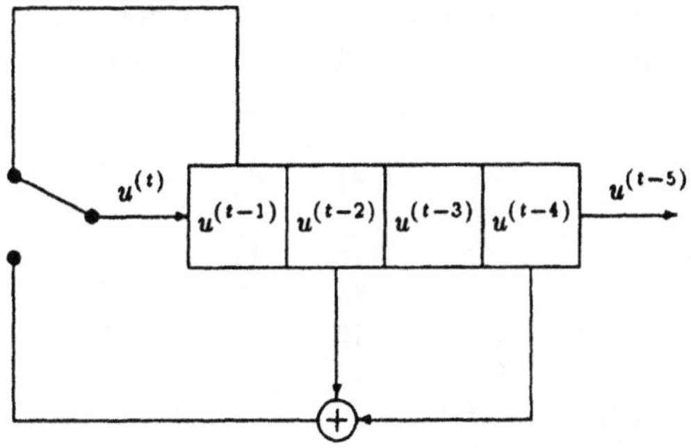

Fig. 1. A time-variant autoregressive filter; $\mu = 4$, $f_0 = z^3 + z^4$, $f_1 = 1 + z^2 + z^4$.

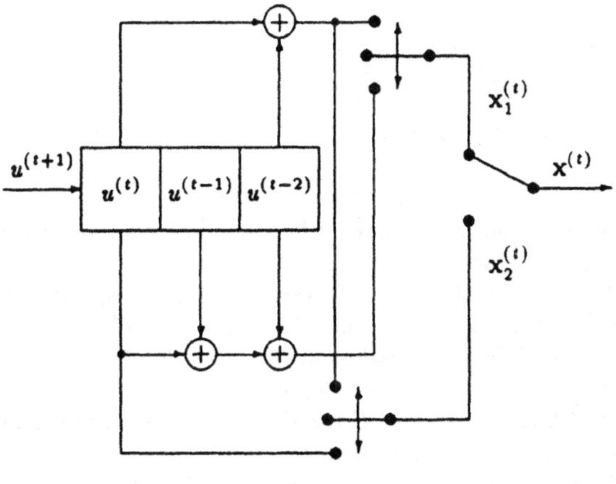

Fig. 2. An encoder of the mixed code with $G_1 = (1 + z^2, 1 + z + z^2)$, $G_2 = (1, 1 + z^2)$.

Table 1. *Parameters of convolutional encoders based on four codes with* $\nu = 4$ *and* $G_1 = (23), G_2 = (23,35), G_3 = (25,33,37), G_4 = (25,27,33,37)$.

T	φ	R_φ	d_f	$\hat{d}_f$
1	4	0.25	16	16
2	3,4	0.286	13	13
1	3	0.333	12	12
4	2,3,3,3	0.364	10	10
3	2,3,3	0.375	9	10
2	2,3	0.4	9	9
3	2,2,3	0.429	8	8
4	2,2,2,3	0.444	7	8
1	2	0.5	7	8
4	1,2,2,2	0.571	5	6
3	1,2,2	0.6	5	6
2	1,2	0.667	5	5
3	1,1,2	0.75	3	4
4	1,1,1,2	0.8	3	4

References

[1] A.J.Viterbi and J.K.Omura, *Principles of Digital Communication and Coding.* NY: McGraw, 1979.

[2] J.L.Massey and M.K.Sain, "Inverses of linear sequential circuits," *IEEE Trans.Computers,* C-17, no. 4, 1968, pp.330-337.

[3] R.R.Olson, "Note on feedforward inverses for linear sequential circuits," *IEEE Trans.Computers,* C-19, no. 12, 1970, pp.1216-1221.

[4] G.C.Clark and J.B.Cain, *Error-Correction Coding for Digital Communication.* NY : Plenum Press, 1981.

[5] M.Mooser, "Some periodic convolutional codes better than any fixed code", *IEEE Trans.Inform.Theory,* IT-29, no. 5, 1983, pp.750-751.

[6] V.B.Balakirsky, "Spectrum characteristics of rate-variant convolutional codes," *Proceedings of NIIR,* M.: Radio i svyaz, 1988, no. 2, pp.28-33.

[7] R.Lidl and H.Niderreiter, *Finite Fields.* NY : Addison Wesley, 1983.

On the Design and Selection of Convolutional Codes for a Bursty Rician Channel

Gideon Kaplan, Shlomo Shamai (Shitz)[1] and Yosef Kofman[2]

[1] Department of Electrical Engineering
Technion - Israel Institute of Technology, Haifa 32000, Israel
[2] Teledata Communications Ltd., Herzlia, Israel.

Abstract. This work addresses code design and code selection rules under power and decoding delay constraints for antipodal (BPSK) modulated and convolutionally encoded communication system. The system operates over a slowly-fading AWGN channel, described here by the 'block fading' model. We specialize to coherent detection and maximum likelihood decoding with ideal channel information (the instantaneous fading values). The dominant design criterion in this scenario is the code diversity level in terms of blocks while the standard Hamming distance plays a secondary role. A code design procedure is presented along with a code-search algorithm. Performance results of a selected code are assessed via simulation and compared to those achieved by a Reed-Solomon code with erasure and error decoding.

1 Introduction

Consider the following model of a slowly-fading channel. The fading process is realized in blocks, the duration of which is, say, η transmitted symbols, that is: the fading level remains constant over the block, and furthermore, the fading values are assumed independent from block to block. Such a channel is referred to as the block-fading Gaussian channel in [1]; it is a useful model for a mobile communication system employing Frequency Hopping with η symbols/hop [2] and can serve as a simplified model for correlated fading channels. The fading values are random variables governed by the Rician distribution [3]. A coded communication system operating under both (average) power and stringent decoding delay constraints is addressed. The latter presumably prohibits the use of ideal interleaving-deinterleaving; moreover, the delay in terms of fading blocks is limited to a relatively modest number [1].

In this work we focus on code design rules and on code selection for a rate b/n, convolutionally-encoded Binary Phase-Shift Keying signal operating over this channel, with a receiver employing coherent detection, soft-decision maximum likelihood (M.L.) decoding, and assisted by perfect side information (S.I.) on the fading level of each block. Recall that for a slowly-fading channel, pilot-aided phase and fading level estimation are quite useful, thus the above simplifying assumptions are not overly idealistic.

In Section II, following the system description, code selection criteria are examined. As might be expected, the code diversity in terms of blocks, or the minimum effective length denoted by L_{eff}, is the *dominant* factor for code design when high values of the average signal to noise ratio prevail. A secondary code design criterion is concerned with the binary Hamming weights of the blocks spanning each codeword. A similar result was found for a trellis-coded MPSK system over an *interleaved* fading channel [5]. Our code design problem has a close relation to the design of q-ary convolutional codes for a q-ary orthogonal modulation system, addressed in [7][8]. There, the q-ary code weight is the dominant design criterion, regardless of the operating signal to noise ratio (SNR), while here this is not exactly the case. Results from [7][8] are adapted and these works are extensively referenced. The channel model has also a close connection to the bursty erasure channel, examined, e.g. in [4] and lately in [6],[9].

Section II further examines a bound on the L_{eff}, while Section III discusses binary, rate k/η code constructions based on maximum distance separable (MDS) block codes. The presented results are extensions of known results for 1-binary to 1 q-ary convolutional codes [7][8]. There are several other known methods for constructing q-ary convolutional codes with a large weight [10]-[12], which are *mainly* directed to low-rate or long constraint lengths. In view of the fact that q-ary codes are often poor in terms of their *binary* Hamming distances, we comment in Section 3.2 on the improvement of the Hamming weights by concatenation. Section IV is devoted to presentation of a code search algorithm and performance evaluations for selected codes. Performance results computed by simulation, demonstrate that performance degrades considerably when the burst length increases, in contrast to the bursty erasure channel described in [4]. The results are also compared to those of erasure- and error decoding of a Reed-Solomon (RS) code (following [13]).

2 Code Design Criteria

2.1 System Description

A coded, BPSK modulated signal is transmitted over the non-selective, Rician block-fading channel with a rate of R bits/symbol. Denote the average received energy per bit by E_b, and per symbol by $E_s = RE_b$. Then, $E_s = E_p + E_r$, where E_r designates the average scattered or diffuse component, and E_p the specular, or direct-path component. For a Rayleigh fading channel, $E_r = E_s$ [3]. The received signal is accompanied by additive white Gaussian noise (AWGN) with a one-sided power spectral density of N_o. Assume first, that a binary rate R block code C with M codewords $\underline{x}_1, \ldots \underline{x}_M$ is employed, where each codeword comprises N symbols; hereafter it is assumed that η, the fading block length, divides N. For a convolutional code, we assume throughout that the fading 'burst' includes an integral number of branches in the trellis diagram of the code used (a 'synchronous' bursty channel).

To evaluate the diversity of the code on the block-fading channel, to be addressed later on, one is motivated to transform the coding system to a one in which each burst comprises exactly one branch of the convolutional code.

Suppose a rate $\frac{1}{n}$ code is employed over the channel with a fade length of J *branches* (where $J = \eta/n$, is assumed an integer throughout); then it is useful to examine the equivalent (in terms of bit error probability) $\frac{J}{Jn}$ convolutional code, resulting by lumping of contiguous J branches of the original code (that is, branches 1 to J, then $J+1$ to $2J$, etc.) into a single branch of the equivalent code. The transition matrix of the resultant code is essentially that of the original one taken to the J-th power. Note that the maximum-likelihood Viterbi decoder, employing also perfect S.I., operates over the 'original' rate $\frac{1}{n}$ code. The above discussion readily generalizes to rate b/n coding over the channel.

2.2 An Upper Bound on the Error Probability

For any transmitted codeword, say $\underline{x}_m$, denote the received vector by $\underline{y}$, the corresponding fading levels sequence by $\underline{z}$, a vector of length $B \triangleq N/\eta > 1$, and the conditional sequence transition probability by $P_N(\underline{y}|\underline{x}_m, \underline{z})$. The average pairwise error probability (p.e.p) can be obtained in a closed-form expression (see, e.g. [3]). To get insight on the governing parameters, we prefer to invoke the Bhattacharyya upper bound [14] on the conditional p.e.p and then average over the fading Probability Density Function, to obtain:

$$\begin{aligned} P(\underline{x}_m \rightarrow \underline{x}_{m'}) &= E_{z_1,\ldots,z_B}\, P(\underline{x}_m \rightarrow \underline{x}_{m'}|\underline{z}) \\ &\leq \prod_{i=1}^{B} \frac{1}{1+d_i E_r/N_o} \cdot \exp\left\{-\frac{d_i E_p/N_o}{1+d_i E_r/N_o}\right\} \triangleq \prod_{i=1}^{B} D_i^{-1} \quad . \end{aligned} \tag{1}$$

where z_i, $i \in \{1,\ldots,B\}$, are i.i.d random variables designating the block fading levels, and d_i is the binary Hamming distance between the i-th 'blocks' of $\underline{x}_m$ and $\underline{x}_{m'}$. For the binary-input, output symmetric channel under investigation, one may take the all-zero word ($\underline{0}$) as the transmitted one, with no loss in generality as far as error probability evaluation is concerned [14].

In the following, two definitions are introduced; the first will be used for code search, while the second, suitable for high average SNR's, will be invoked for code design.

Definition 1. The equivalent free distance of a linear block or convolutional code operating over the block-fading channel for a *given* average E_b/N_o is defined by

$$D_{eq} \triangleq \min_{\substack{\underline{x}' \neq \underline{x} \\ \underline{x},\underline{x}' \in C}} \prod_{i=1}^{B} D_i \quad . \tag{2}$$

For $E_s/N_o \gg 1$ and for small to medium values of $\gamma \triangleq E_p/E_r$ — the Rician parameter, which is the useful range for a mobile communication channel (e.g. [1],[13]), then E_r/N_o has a value $\gg 1$ and it follows that:

$$\begin{aligned} D_i^{-1} &\cong \tfrac{1}{d_i E_r/N_o} \cdot \exp(-\gamma)\,, \quad &\text{if } d_i \neq 0 \\ &= 1 &\text{if } d_i = 0 \end{aligned} \quad . \tag{3}$$

Substituting the right-hand side (RHS) of (3) into (2), yields,

$$P(\underline{x}_m \rightarrow \underline{x}_{m'})\Big|_{\frac{E_r}{N_o} \gg 1} \tilde{<} \left(\frac{E_r}{N_o}\right)^{-L_{\underline{x}_m,\underline{x}_{m'}}} \cdot e^{-\gamma \cdot L_{\underline{x}_m,\underline{x}_{m'}}} \cdot \prod_{i=1}^{B} (d_i)^{-1} \qquad (4)$$

where $L_{\underline{x}_m,\underline{x}_{m'}}$ denotes the number of different blocks (of η symbols) in the two codewords (that is, number of blocks for which $d_i \neq 0$). It is clear that under high average SNR's, $L_{\underline{x}_m,\underline{x}_{m'}}$, depending only on the algebraic structure of the code, is the dominant factor governing eq. (4). This provides thus the motivation to introduce (similarly to [5])

Definition 2. The minimum effective length of a linear code is defined by

$$L_{eff} \triangleq \min_{C/\underline{0}} \ \{\textit{number of blocks with } w_i \neq 0\},$$

where w_i designates the binary Hamming weight of the i-th block, and $C/\underline{0}$ designates the ensemble of all codewords except the all-zero one, or all paths of the convolutional code diverging from and remerging to the all-zero state [14].

The average bit error probability for convolutional codes with M.L. decoding, is upper bounded, for a given E_s/N_o, by (e.g. [8])

$$P_b \leq \sum_{L=L_{eff}}^{\infty} \sum_{\{w_i\}} N_I[L,\{w_i\}] \cdot P_{\underline{x}\rightarrow\underline{x}'}[L,\{w_i\}] \qquad (5)$$

where $N_I[L,\{w_i\}]$ designates the total number of nonzero information bits attached to paths with L non-zero blocks and with a set of binary Hamming block weights of $\{w_i\}$.
Although the whole weight spectrum of a code (in terms of (5)) affects the performance, a code search algorithm may first find codes maximizing D_{eq} (and preferably those that have the minimum number of information errors associated with the worst paths). This is indeed the criterion employed in our code search algorithm. Further refinements can be made by considering more coefficients of the weight spectrum or via a bit error probability simulation program. A simplified methodology will be used here for code design – that is, a design suitable for high E_s/N_o values, which is based on maximizing L_{eff} of the code for a given rate and constraint length. A secondary consideration should be given to the product of the block Hamming weights. Suppose that $\underline{x}_{m'}$ has a total Hamming weight of d, and includes B blocks with $d_i \neq 0$, where $d = \sum_{i=1}^{B} d_i$ (see (4)). By standard Lagrangian methods it is found that the maximal product of d_i's results when $d_i = d/B$, independent of i ('two-valued' codes – with $d_i = 0$ or $d_i =$ constant on any path). Possible means for increasing the w_i's of the blocks of a selected code will be also discussed.

Returning to our *main* code design rule, it is clearly equivalent to maximizing the code *diversity* (or number of degrees of freedom). It is noted that a similar criterion evolves for the bursty erasure channel [4] as might be expected, and also in a scenario of coded q-ary orthogonal signaling [7][8]. Suppose each branch of the code represents one q-ary symbol, then, the dominant code design rule is to maximize the minimum q-ary Hamming distance, designated d_q, between any two q-ary paths, where $d_q = 0$ if the corresponding q-ary symbols agree, and $d_q = 1$ if they differ.

In our application, suppose a rate $\frac{b}{n}$ binary coding is employed on a channel with fade length $\eta = J \cdot n$, then the equivalent code, as previously discussed, is a rate $\frac{Jb}{Jn} = \frac{k}{\eta}$ code. For this code, if the η binary symbols of each branch (spanning exactly *one* fading burst) are interpreted as a single q-ary symbol, where $q = 2^\eta$, then the effective length (number of non-zero blocks) on each path is equal to its q-ary Hamming weight. Thus, an extensive use of [7][8] is made here. Note, however, that in our case the code selection criterion does depend on the operating average E_s/N_o.

2.3 A Bound on L_{eff}

Trumpis considered [7] the minimum q-ary Hamming distance for a "rate 1", that is 1-binary input to 1 q-ary output convolutional codes, and obtained upper and lower bounds on this quantity for a given constraint length, v. The latter is assumed to equal the total number of memory elements of the encoder. Here the upper bound is extended to rate $\frac{k}{\eta}$ codes.

Lemma 3. *An upper bound on L_{eff} for rate k/η codes is expressed by*

$$L_{eff} \leq \min\left\{\lfloor\frac{v}{k}\rfloor + 1,\ \min_{tk\geq\eta} \frac{2^\eta - 1}{2^\eta} \cdot \frac{2^{tk}}{2^{tk}-1} \cdot \left(\lfloor\frac{v}{k}\rfloor + t\right)\right\} \tag{6}$$

where $\lfloor x \rfloor$ designates the largest integer lower or equal to x.

Proof. The proof of the first term is based on examining the shortest path diverging and remerging with the all-zero state. The 2nd part is proved by an extension of a Heller-type bound, examining paths with $k \cdot t$ arbitrary bits $+v$ "tail bits". □

Example 1. A rate $\frac{1}{2}$ coding is required for communicating over a block fading channel with a block length of $\eta = 8$ symbols. The equivalent convolutional code in terms of error performance is therefore a rate $\frac{k}{\eta} = \frac{4}{8}$ code. Assuming a specific rate $\frac{1}{2}$ code with a constraint length of v was selected, the upper limit on the L_{eff} is $\frac{v}{4}+1$ (here $v/4$ is assumed an integer). It is clear that in order to increase the actual code diversity by one, the constraint length of the rate $\frac{1}{2}$ code has to be increased by k, or 4 in our case. The performance deterioration due to a (large) fading block length is thus apparent as also discussed in [1] and further shown in Section 4.2.

3 Code Construction

3.1 Construction of rate $\frac{k}{\eta}$ codes achieving the maximal L_{eff}

Here a construction of rate k/η codes which achieves the maximal L_{eff} is outlined, as an extension to "rate 1" and rate $\frac{1}{n}$ q-ary codes of [7][8]. The encoder, depicted in Fig. 1, is composed of a single shift register with $k \cdot v$ binary cells (see [3]). In each time unit, a new binary k-tuple is pushed into the encoder shift register, and the output is a binary η-tuple, as follows. The input symbols, represented as k-tuples, are a subset of $GF(q = 2^\eta)$ whose elements are represented as binary η-tuples. We adopt the convention throughout this section that they represent η-tuples with $\eta - k$ zeros in the $\eta - k$ most significant bits (msb's) followed by the k input bits. This subset of $GF(q)$ is denoted here as the input (k-tuple) set. If k divides η, then the encoder input and output ensembles are isomorphic of $F_1 = GF(2^k)$ and $F_2 = GF(2^\eta)$, respectively, where F_2 is an extension field of F_1 [8]. The v 'sets' of k binary cells in the encoder, combined with the current input k-tuple (all represented as q-ary elements, as detailed above) are multiplied by tap gains g_i, $i \in \{0, \ldots, v\}$, which are elements of $GF(q)$, and added up to produce the output η-tuple. The structure is equivalent to k shift registers of v binary cells each.

Lemma 4. *A k-binary to one q-ary convolutional code ($k < \eta$) with a generator polynomial $G(D)$, the coefficients of which are elements of $GF(q = 2^\eta)$, is catastrophic if there exists a non-monomial factor of $G(D)$, denoted $A(D)$, such that $A^{-1}(D)$ is an infinite-weight polynomial where the weight is given by the number of nonzero coefficients from the input set.*

The *proof* is evident, based on the standard definition of a catastrophic code [16][8], that is an infinite Hamming weight input binary information sequence produces a finite Hamming weight (over $GF(q)$) output coded sequence.
If the output field is an extension of the input field, that is when k divides η, the condition for a code to be non-catastrophic becomes an iff condition, simplifies considerably and is given by [8, Thm. 1] for rate $\frac{1}{n}$ 'F_1 to F_2' codes. □

In the following, the notation $p(x) \in K[x]$ means that the coefficients of the polynomial $p(x)$ belong to the input set. The construction of a rate k/η convolutional code achieving the maximal L_{eff} is based on the following theorem.

Theorem 5. *Let*

$$g(D) = g_o + g_1 D + \cdots + g_v D^v \tag{7}$$

be a generator polynomial of a q-ary, MDS and cyclic [20] block code, $(N, N-v)$, with $d_{\min} = v + 1$. A convolutional code of rate k/η and $L_{eff} = v + 1$, can be constructed by the encoder of Fig. 1, using an encoder polynomial $G(D) = g(D)$, provided that:

(a) $h(D) \notin K[D]$, where $h[D]$ is the generator polynomial of the dual code.
(b) If $a(D) \cdot h(D)$ represents a legitimate input sequence, then $W_H\{a(D) \cdot (D^N + 1)\} \geq d_{\min}$.

Proof. The proof uses standard arguments, e.g. [18], [8, Lemma 2, Thm 3]. The main idea is to invoke the 'weight-retaining' property [17] to prove that given that the polynomial $g(x)$ generates a q-ary cyclic code with a minimum Hamming distance of d, then for any polynomial $U(x)$ with coefficients from the input set, representing the input sequence, one has $W_H\{U(x) \cdot g(x)\} \geq d$, provided that $U(x)$ is not a multiple of $h(x)$; otherwise, condition (b) guarantees a weight $\geq d_{\min}$. □

The outlined method allows for designing of k/η codes with constraint lengths between 2 and $q-2$ which achieve the maximal L_{eff} of $v+1$, using an encoder register of $k \cdot v$ cells. One concludes that this method is adequate for designing codes with relatively *modest* lengths, which is commensurate with delay-constrained systems as described in section 1 (see also [1]). A drawback of the outlined method, though, is that it requires the decoder to operate on a k/η code even if the required coding rate is $\frac{1}{n}$, i.e. $k = J$, thus requiring a complexity penalty on the decoder side. Another drawback is concerned with the block Hamming weights of such codes. Another construction is based on interlacing of k binary convolutional encoders (when k divides η).

3.2 Improving the Binary-Tuple Hamming Weight

A distinct feature of codes over $GF(q)$, even those achieving a large L_{eff}, is that their minimum Hamming weights when viewed as *binary* codes, are definitely modest. As an example, a (N, K) RS code has a minimum binary Hamming weight of (only) $D = N - K + 1$, the same as its minimum value of d_q. Hence – the motivation to increase the w_i's or d_i's (see (1)). The basic idea is to use *concatenation*. Justesen suggested [18] a particular form of 'time-varying' rate $\frac{1}{2}$ inner code, combined with an RS outer code. However these codes are known to be effective in the *asymptotic* sense, that is for very large outer code lengths. In [6], a two-valued codebook was suggested for a bursty erasure channel. Such a code has branches with weights of either zero or $\frac{n}{2}$, for rate $\frac{1}{n}$ coding. To construct the code, a (matrix) mapper is connected to the output of the encoder. In order to obtain a rate $\frac{1}{n}$ code, the designer first chooses a rate $\frac{1}{s}$ code over $GF(q' = 2^s)$ which achieves the maximal L_{eff} for the given constraint length; then, a $\frac{1}{n}$ $(n > s)$ code is obtained by applying a Hadamard matrix on the basic encoder output. An improved and more versatile construction is based on concatenation of a rate $\frac{k}{s}$ code, achieving the maximal L_{eff}, with a Reed-Muller (RM) code as an inner code to form a rate $\frac{k}{n}$ code.
The above mentioned two-valued 'Hadamard' codes are very useful for another application, that is a coded system where joint noncoherent detection of blocks is applied (see [15]), for there the *optimum* weight of each branch of n symbols is $\frac{n}{2}$.

4 Selected Codes and Performance Results

4.1 Code Search

A convolutional code search program for rate $\frac{1}{n}$ codes operating over a bursty fading channel with $J \geq 1$ branches per burst, using an encoder with a constraint length of v, was written. The criterion for code selection is based on (1), (2) or on maximizing D_{eq}, representing the code by its equivalent $\frac{J}{Jn}$ code. The algorithm indicates the value of D_{eq}, the number of paths with the same D_{eq}, and the total number of information bit errors associated with these paths (see (5)). This procedure is performed for a given E_b/N_o. Codes with rates of $\frac{1}{2}$ and $\frac{1}{3}$, practical constraint lengths of 5 to 9, and $1 \leq J \leq 5$ were examined.

The generators of the best non-catastrophic codes in terms of the largest D_{eq} and the minimal number of total bit errors on the paths with D_{eq}, for a Rayleigh channel with fading block lengths of $J = 1$ to 3 branches, and rate $\frac{1}{2}$ or $\frac{1}{3}$ coding, for practical v values of 4 to 6, and $E_s/N_o = 6$dB (as an example) are listed in Table 1. As previously mentioned, one should also examine paths with $L_{eff}+1$, $L_{eff}+2$, etc., or use simulation, in order to choose the best code. It is shown that in some cases the optimal code on the AWGN channel is not the optimal code for the block fading channel. As an example, the optimum rate $\frac{1}{2}$, $v = 6$ code over the AWGN – the (133, 171) code – is not the optimal code for the cases $J = 1$ or 2. It has $D_{eq}^{-1} = 6.9 \cdot 10^{-6}$, for the $J = 1$ case (with $E_s/N_o = 6$dB) while the optimum code is the (125, 91) code, achieving $D_{eq}^{-1} = 3.95 \cdot 10^{-6}$; the difference in performance will be more emphasized at higher SNR's. However, based on the code search results it can be *conjectured* that for a $J \gg 1$, it is reasonable to use the optimum codes for the AWGN channel for the block-fading application.

4.2 Simulation Results for Selected Codes

A simulation program for a soft-decision Viterbi decoder, operating over the block-fading channel, is employed to evaluate the average bit error probability. Fig. 2 depicts the results for the standard (133, 171) code, under Rayleigh fading with $\eta = 1, 6$, and 10; the $\eta = 1$ case represents the ideally-interleaved slowly fading channel. It is noted that not only is there a noticeable degradation for a large value of η, the *slope* of the bit error probability vs. E_b/N_o curve is severely affected. This results by the reduced diversity level with the increase of η and was also observed for the correlated Rayleigh channel (see, e.g. [19]). Fig. 2 also depicts the results for decoding with no side information on the fading levels; the resultant degradation is modest.

We compare the performance results to an often-used block code on bursty channels, that is an RS code over $GF(q = 2^\eta)$. The 'basic' code length is $N = 2^\eta - 1$ (η-tuples), however the code can be shortened and keep its MDS property. The main advantage of an RS code is in its large diversity, when the burst (fade) length matches the symbol's. However, it has three drawbacks: (a) the Hamming distances as a *binary* code are only modest, (b) the number of adversaries is large – e.g., the number of codewords with weight D is $(q-1)\binom{N}{D}$ [20], (c) a soft decision ML decoding mechanism is quite difficult to accomplish. One can, however, use erasure and error decoding (following [13]). Fig. 2 also depicts a *lower* bound on the bit error probability (based on [13]) vs. average E_b/N_o for the (16, 8) RS code over a Rayleigh channel with a burst length of $\eta = 6$. Note that the delay of the encoder-decoder pair for the (16,8) code is 96 information bits, which is comparable to that of existing realizations for the $v = 6$ code. The RS (16, 8) code is outperformed by the $v = 6$, rate $\frac{1}{2}$ convolutional code for $E_b/N_o \leq$ 16dB. The merits of convolutional codes for the block-erasure channel were lately demonstrated analytically in [9]. In our case, one may attribute the superiority of the convolutional code (under a given decoding delay) to the deficiencies of the RS codes as mentioned above. However, refined decoding algorithms for the latter [2] should also be considered.

References

1. Ozarow, L.H., Shamai (Shitz), S., Wyner, A.D.: Information theoretic considerations for cellular mobile radio, Proc. of the Int'l Comsphere '91 Symp., Herzlia, Israel, Dec. 91, pp. 10.1.1-10.1.4; also to appear in IEEE Trans. on Veh. Technology.
2. Matsumoto, T., Higashi, A.: Performance analysis of RS-coded M-ary FSK for frequency-hopping mobile radios, IEEE Trans. on Veh. Tech., Vol. 41, No. 3, pp. 266-270, Aug. 1992.
3. Proakis, J.G.: Digital Communications, Ch. 7. New York: McGraw-Hill, 2nd edit., 1990.
4. Geist, J.M., Cain, J.B.: Viterbi decoder performance in Gaussian and periodic erasure bursts, IEEE Trans. on Commun., Vol. COM-28, No. 8, pp. 1417-1422, Aug. 1980.
5. Divsalar, D., Simon, M.K.: The design of trellis coded MPSK for fading channels: performance criteria, IEEE Trans. on Commun., Vol. 36, pp. 1004-1012, Sept. 1988.
6. Kadishevitz, Y.: Convolutional codes for channels with memory, M.Sc. dissertation, Dept. of Elect. Eng., Technion, Israel, Dec. 1991.
7. Trumpis, B.D.: Convolutional coding for M-ary channels. Ph.D. dissertation, UCLA, 1975.
8. Ryan, W.E., Wilson, S.G.: Two classes of convolutional codes over $GF(q)$ for q-ary orthogonal signaling", IEEE Trans. on Commun., Vol. 39, No. 1, pp. 30-40, Jan. 1991.
9. Lapidoth, A.: The performance of convolutional codes on the block erasure channel using various finite interleaving techniques", submitted to IEEE Trans. on Inform. Theory; presented at Int'l Conf. on Inform. Theory, San Antonio, Jan. 1993.
10. Piret, P.: Convolutional Codes – An Algebraic Approach, Ch. 11. Cambridge: MIT Press, 1988.
11. Kovalev, S.I.: Non-binary low rate convolutional codes with almost optimum weight spectrum, presented at the 1st French-Soviet Workshop, Algebraic Coding, Paris, France, pp. 82-86, July 1991.
12. Rodgers, W.E., Lackey, R.B.: Burst error correcting convolutional codes with short constraint lengths, IEEE Trans. on Inform. Theory, Vol. IT-26, No. 3, pp. 354-359, May 1980.

13. Hagenauer, J., Lutz, E.: Forward error correction coding for fading compensation in mobile satellite channels, IEEE Journal on Select. Areas in Commun., Vol. SAC-5, No. 2, pp. 215-225, Feb. 1987.
14. Viterbi, A.J., Omma, J.K.: Principles of Digital Communication and Coding. New York: McGraw-Hill, 1979.
15. Kofman, Y., Zehavi, E., Shamai (Shitz), S.: Convolutional codes for noncoherent detection, Proc. of ISSSE'92 Conf., Paris, pp, 121-123, Sept., 1992.
16. Massey, J.L., Sain, M.K.: Inversion of linear sequential circuits, IEEE Trans. on Computers, Vol. C-18, No. 4, pp. 330-337, Apr. 1968.
17. Massey, J.L., Costello, D.J., Justesen, J.: Polynomial weights and code constructions, IEEE Trans. on Inform. Theory, Vol. IT-19, No. 1, pp. 101-110, Jan. 1973.
18. Justesen, J.: New convolutional code constructions and a class of asymptotically good time-varying codes, IEEE Trans. on Inform. Theory, Vol. IT-19, No. 2, pp. 220-225, March 1973.
19. Kaplan, G., Shamai (Shitz), S.: Achievable performance over the correlated Rician channel, accepted for publication in IEEE Trans. on Commun.
20. MacWilliams, F.J., Sloane, N.J.A.: The Theory of Error-Correcting Codes. North-Holland, 1977.

Table 1. Generators of convolutional codes (octal notation) with the optimum D_{eq} ($E_s/N_o = 6$dB).

(a) $r = \frac{1}{2}$

v	$J = 1$	$J = 2$	$J = 3$
4	(36,13)	(33,25)	several, e.g. (35,23)
5	several, e.g. (73,53)	(73,37)	(77,73)
6	(175,133)	(165,127)	(171,133)

(b) $r = \frac{1}{3}$

v	$J = 1$	$J = 2$
4	several, e.g. (37,27,15)	(37,33,25)
5	(77,67,57)	(77,73,57)
6	(173,135,57)	(171,163,133)

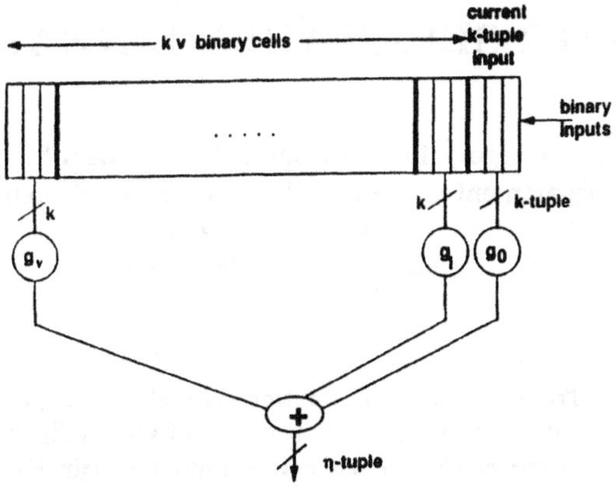

Fig. 1: A convolutional encoder for rate k/η codes.

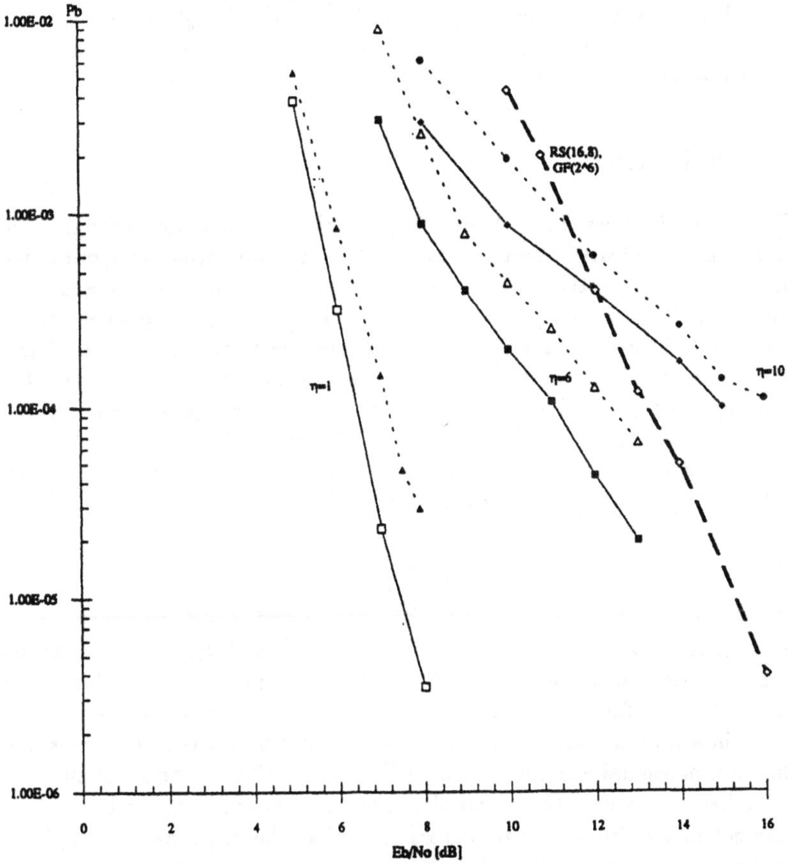

Fig. 2: Bit error probability vs. average E_b/N_o for a Rayleigh channel with $\eta = (1, 6, 10)$ symbols per fading block, for the (133, 171) rate $\frac{1}{2}$ code with/without CSI. Also shown is a lower bound on the bit error probability for the (16, 8) RS code over $GF(2^6)$.

MODULO-2 SEPARABLE LINEAR CODES

Gregory Poltyrev and Jakov Snyders *
Department of Electrical Engineering - Systems,
Tel Aviv University,
Ramat Aviv 69978, Israel

Abstract. Transmission of information through a multiple access channel is addressed for the situation that out of a fixed family of N potential users at most some m, $m < N$, are active simultaneously. We assign codes to the N users in a way that allows correction of any t or less errors and separate the $s \leq m$ messages, provided that the subset of active users is known to the receiver. The construction of codes is tightly connected with a construction of maximum distance codes. A decoding procedure is presented.

1 Introduction

This paper addresses the problem of information transmission through a multiple access channel (MAC) designed for a family of N potential users, out of which at most $m < N$ are active simultaneously (are transmitting their messages). We assume that the active users are completely synchronized, in the sense that both the signals and blocks of the signals of different users coincide in time. Furthermore, we assume that the set of active users is known to the receiver, but is unknown to the users. Such mode of transmission through a linear MAC was considered earlier in [1], where it was called "T out of N active users".

Assuming for a while no restrictions on the size of the alphabet of input signals, let the output signal y to the linear MAC be given by

$$y = \sum_{i=1}^{m} a_i x_i, \qquad (1)$$

where $x_1, x_2, \cdots, x_m$ are the m inputs to the MAC, and $a_1, a_2, \cdots, a_m$ are some real coefficients that define the MAC. Separable coding for the MAC in this case can be realized as follows. We associate with each of the N users a vector $\mathbf{v}_i$; $i = 1, \cdots, N$ in a way that any set $\{\mathbf{v}_{i_1}, \mathbf{v}_{i_2}, \cdots, \mathbf{v}_{i_m}\}$ of m different vectors is a basis of the m-dimensional Euclidean space $\mathcal{R}^m$. Clearly, this can be accomplished for any number N of users. (For example, in the case $m = 2$, $\{\mathbf{v}_i, i = 1, \cdots, N\}$ can be any set of two dimensional vectors such that the angle between $\mathbf{v}_i$ and $\mathbf{v}_j$ is neither equal to 0 nor to π for all pairs $i \neq j$). If the ith user is active, it

* This work was supported by the Ministry of Science and Technology and by the Basic Research Foundation administrated by the Israel Academy of Sciences and Humanities.

transmits $\mathbf{x}_i = b_i\mathbf{v}_i$, where b_i is an information symbol. The receiver expands the received vector $\mathbf{y} = (y^1, y^2, \cdots, y^m)$ on the basis $\{\mathbf{v}_{i_1}, \mathbf{v}_{i_2}, \cdots, \mathbf{v}_{i_m}\}$, where $\{i_1, i_2, \cdots, i_m\}$ is the index set of active users, $y^j = \mathbf{a}\mathbf{x}_j, \mathbf{a} = (a_{i1}, a_{i_2}, \cdots, a_{i_m}))$ and $\mathbf{x}_j = (x^j_{i_1}, x^j_{i_2}, \cdots x^j_{i_m})^t$, where the superscript t stands for transposition (recall that the set of active users is known to the receiver). Consequently, the receiver obtains the vectors $\mathbf{z}_{i_j} = a_{i_j} b_{i_j} \mathbf{v}_{i_j}$. If the receiver knows also the channel transmission vector $\mathbf{a}$, it can recover the information symbols $b_{i_j}, j = 1, \cdots, m$ without errors. Otherwise, the receiver can nevertheless obtain without errors the information symbols provided they are binary, for example, $b_i \in \{0,1\}$. It should be pointed out that this method of transmission achieves the transmission rate that would be obtainable also by time sharing between the m active users. However, time sharing can not be realized in our case, because the set of active users is known to none of the users.

The problem becomes complicated for MAC with binary *inputs*, i.e., $x_i \in \{0,1\}$. It can be shown that the maximum number N of binary vectors of length m, such that any m of them are linearly independent, is given by $N = m + 1$. Our main idea for increasing the number of users is the following. We choose binary $(n, k), n = mk$, codes $C_i; i = 1, 2, \cdots, N$, one for each user, such that for any set of m users the direct sum of the corresponding codes is equal to the whole binary space $GF(2)^n$, i.e., $C_{i_1} \oplus C_{i_2} \oplus \cdots \oplus C_{i_m} = GF(2)^n$. It is clear that if G_i is a generator matrix of the code C_i, then for any set of different indices $\{i_1, i_2, \cdots, i_m\}, 1 \le i_j \le N$, the rows of the matrix $G = (G^t_{i_1}, G^t_{i_2}, \cdots, G^t_{i_m})^t$ are lineary independent and form a basis for the n dimensional Euclidean space $\mathcal{R}^n$. Consequently, the receiver can expand the received sequence $\mathbf{y}$ of length n on the basis consisting of the rows of G and obtain, without errors, the information bits of all active users. For the case of the binary adder channel, i.e., $\mathbf{a} = \mathbf{1}$, the receiver can operate as follows. For every received symbol y, it checks the parity and then changes y to 0, if y is even, or to 1, if y is odd. Thereafter, the receiver expands the binary sequence thus obtained on the basis G over $GF(2)$.

For a noisy channel, error protection of the information transmitted through an MAC can be achieved by partitioning not all the binary space, but some subspace of it. If this subspace is a binary code with minimum distance d then any combination of up to $t \le \lfloor (d-1)/2 \rfloor$ errors can be corrected.

Thus the following question arises: what is the maximum number of users for fixed m and k. We provide some answer to this question in Section 2, where we show that, loosely speaking, the maximum number of users is approximately 2^k. We present also a constructive method for building the codes C_i and show its tight relationship to the problem of constructing maximum distance (MD) codes. A simple example of construction is provided in Section 3. In Section 4 we describe an efficient decoding algorithm.

2 Modulo-2 separable codes

Definition 1. For some integers $m \ge 2$ and $k \ge 1$, let C be a binary linear (n, mk) code. A collection $\{C_i\}, C_i \subset C$, of N binary linear (n, k) codes is called (m, C)-separable if $C_{i_1} \oplus C_{i_2} \oplus \cdots \oplus C_{i_m} = C$ for any subset of m indices $\{i_1, i_2, \cdots, i_m\} \subset \{1, 2, \cdots, N\}$. In case that C is the (mk, mk) code we shall call this collection (m, k)-separable.

Since the subset of active users is assumed to be known to the receiver, assignment of codes belonging to a collection of (m, C)-separable codes enables the separation of the messages of any m or less active users. In addition, if the minimum distance of C is $2t+1$ then any t or less errors can be corrected.

Lemma 2. *Let C be an (n, mk) code. There exists an (m, C)-separable collection of N codes if and only if there exists an (m, k)-separable collection of N codes.*

Proof. In any linear (n, mk) code there is a set of mk positions which can be interpreted as information positions. Without loss of generality we can assume that the first mk positions are the information positions. Let $\{C_i : i = 1, 2, \cdots, N\}$ be some (m, C)-separable collection of N codes. Let $\{C'_i : i = 1, 2, \cdots, N\}$ be the collection of (mk, k) codes such that C'_i coincides with C_i on the first mk positions, $i = 1, 2, \cdots, N$. Clearly, the collection $\{C'_i : i = 1, 2, \cdots, N\}$ is (m, k)-separable. Now let $\{C'_i : i = 1, 2, \cdots, N\}$ be some (m, k)-separable collection of N codes. Let G'_i be a generator matrix of $C'_i; i = 1, 2, \cdots, N$. If for each $i = 1, 2, \cdots, N$ we choose the generator matrix G_i of C_i such that the matrices G_i and G'_i coincide on the first mk positions, and the rows of G_i are codewords of C, then we obtain an (m, C)-separable collection of N codes.

An (n, k, d) code C over $GF(q)$, where $k = log_q|C|$, is called a maximum distance (MD) code if its minimum distance d satisfies $d = n - k + 1$ (C is not assumed to be linear). The following two lemmas present a connection between MD codes and (m, k)-separable collections of codes.

Lemma 3. *If there exists an (m, k)-separable collection of N codes then there exists an $(N, N - m, m + 1)$ MD code over $GF(2^k)$.*

Proof. Assume that there is an (m, k)-separable collection $\{C_i\}$ of N codes, each code with parameters (mk, k). Let G_i be a generator matrix of $C_i; i = 1, 2, \cdots, N$. Consider a binary linear code C_{Nk} with the following $mk \times Nk$ check matrix $H = (G_1^t, G_2^t, \cdots G_N^t)$. By Definition 1, any set of columns of H belonging to not more than m different submatrices G_i^t are linearly independent. Let us partition all the binary codewords of C_{Nk} into N segments of equal length k. Now we can obtain a q-ary, $q = 2^k$, code C_N by considering each such segment as the binary representation of some element from $GF(2^k)$. Although the q-ary code C_N is not necessarily linear, its minimum Hamming distance is clearly $m + 1$, and the number of codewords of C_N is equal to $2^{Nk-mk} = q^{N-m}$. Consequently, C_N is an $(N, Nm, m + 1)$ MD code over $GF(2^k)$.

Lemma 4. *The maximum number $N(m)$ of (m, k)-separable codes satisfies the following conditions*

$$N(m) = 2^k + m - 1, \qquad m = 2, 3; \tag{2}$$

$$2^k + m - 1 \geq N(m) \geq \max\{2^k + 1; m + 1\}, \quad m > 3. \tag{3}$$

Proof. We start with the proof of the upper bound on $N(m)$ (the left inequality in (3)). We shall show first that from the existence of an (m,k)-separable collection of N codes follows the existence of an $(m-1,k)$-separable collection of $N-1$ codes, $m>2$. Let $\{C_i\}$ be an (m,k)-separable collection of N codes. We can represent the generator matrices G_i of the codes C_i as follows $G_i=(G_{1i},G_{2i},,G_{mi})$ where each G_{ji} is an $k\times k$ matrix. Without loss of generality we can assume that G_{m1} is the identity matrix, and $G_{j1}; j=1,\cdots,m-1$ are zero matrices. The collection of codes $\{C_i'\}$ with generator matrices $\{G_i'=(G_{1i},G_{2i},\cdots,G_{(m-1)i})\}; i=2,\cdots,N$, is clearly an $(m-1,k)$-separable collection of $N-1$ codes. It follows from Definition 1 that codes $C_i, i=1,\cdots,N$, belonging to any (m,k)-separable collection are intersected only on the zero codeword. Hence $N(m)(2^k-1)\leq 2^{mk}-1$, and in particular $N(2)\leq 2^k+1$. This and the previously proved inequality $N(m)\leq N(m-1)+1$ imply the left inequality in (3).

We proceed now to prove (2) and the right inequality in (3). Let H be an $m\times N$ check matrix of a Reed-Solomon code over $GF(2^k)$. Let us represent elements of $GF(2^k)$ by binary $k\times k$ matrices. Then any column of H becomes an $mk\times k$ binary matrix. Denote such matrix, which corresponds to the ith column of H, by $G_i^t; i=1,2,\cdots,N$. Let C_i be the linear span of the rows of G_i, i.e., C_i is a binary linear (mk,k) code. Since any m columns of the matrix H are linearly independent, the collection of codes $C_i; i=1,2,\cdots,N$ thus obtained is (m,k)-separable. Equation (2) and the right inequality in (3) now follow by the existence of doubly-extended Reed-Solomon codes with any number m, $2\leq m\leq 2^k+1$, of check symbols, by the existence of triply-extended Reed-Solomon codes with $m=3$ check symbols, and by the existence of repetition codes of any length over $GF(2^k)$.

Remark. Lemma 3 asserts that an $(N,N-m,m+1)$ MD code over $GF(2^k)$, which is not necessarily linear, can be constructed with the aid of every (m,k)-separable collection of N codes. Furthermore, it follows by Lemma 4 that the upper bound on the maximum length of such codes coincides with the known upper bound [2] on the maximum length of *linear* maximum distance separable (MDS) codes over $GF(2^k)$.

3 An example

We wish to allow simultaneous transmission of up to $m=3$ users. For simplicity of the example, we provide each user with a two dimensional code. Then by (2), $N(3)=6$. We start by assigning codes C_1,C_2 and C_3 to three of the users. When selecting these codes, we may take into account the required error correction capability. Suppose we would like to correct up to $t=3$ channel crossovers. Then we need to employ a code with minimum distance $d=7$. Let C_1,C_2 and C_3 be generated, respectively, by the following matrices

$$G_1=\begin{pmatrix}100000\ 11011100010\\ 010000\ 01101110001\end{pmatrix}, \tag{4}$$

$$G_2 = \begin{pmatrix} 001000\ 10110111000 \\ 000100\ 01011011100 \end{pmatrix} \tag{5}$$

and

$$G_3 = \begin{pmatrix} 000010\ 00101101110 \\ 000001\ 00010110111 \end{pmatrix}. \tag{6}$$

The linear span $\mathcal{C}$ of $\mathcal{C}_1, \mathcal{C}_2$ and $\mathcal{C}_3$ is a (17,6,7) code (a shortened version of the (23,12,7) Golay code). In order to construct the codes $\mathcal{C}_4, \mathcal{C}_5$ and $\mathcal{C}_6$ for the other $M(3) - 3 = 3$ users, consider the following check matrix of a triply-extended RS code

$$H = \begin{pmatrix} 1 & & & 1 & 1 & 1 \\ & 1 & & 1 & \alpha & \alpha^2 \\ & & 1 & 1 & \alpha^2 & \alpha \end{pmatrix}, \tag{7}$$

where α is a root of $x^2 + x + 1$. By selecting $\{1, \alpha\}$ to be the standard basis of $GF(2)^2$ over $GF(2)$, the binary version of (7) is the following

$$H = \begin{pmatrix} 10 & & & 10 & 10 & 10 \\ 01 & & & 01 & 01 & 01 \\ & 10 & & 10 & 01 & 11 \\ & 01 & & 01 & 11 & 10 \\ & & 10 & 10 & 11 & 01 \\ & & 01 & 01 & 10 & 11 \end{pmatrix}. \tag{8}$$

According to (4), (5), (6) and the fourth pair of columns of H of (8), the generator matrix for $\mathcal{C}_4$ is given by

$$G_4 = \begin{pmatrix} 101010\ 01000110100 \\ 010101\ 00100011110 \end{pmatrix}$$

Similarly, the fifth and sixth pairs of columns of H of (8) prescribe, respectively, the following generator matrices for $\mathcal{C}_5$ and $\mathcal{C}_6$

$$G_5 = \begin{pmatrix} 100111\ 10111100111 \\ 011110\ 10101111011 \end{pmatrix}$$

and

$$G_6 = \begin{pmatrix} 101101\ 00100110101 \\ 011011\ 11100010000 \end{pmatrix}.$$

Since the minimum distance of $\mathcal{C}$ is 7, the desired error correction capability is secured for any subset of up to three of the six users. Note that we could have employed for the construction the whole (23,12,7) Golay code to form, for example, a collection of $2^4 + 1 = 15$ four dimensional codes that allows any set of up to three out of the 15 users to transmit simultaneously.

4 Decoding $(m, \mathcal{C})$-separable codes

We address now the decoding of an $(m, \mathcal{C})$-separable collection of codes constructed in Section 2.

Lemma 5. *Let C be a binary linear (n,k) code and let $C = C_1 \oplus C_2$ where C_1 and C_2 are binary linear (n,k_1) and (n,k_2) codes, respectively, such that $k = k_1 + k_2$. If $\mathbf{x} = \mathbf{x}_1 \oplus \mathbf{x}_2$ where $\mathbf{x} \in C, \mathbf{x}_1 \in C_1$ and $\mathbf{x}_2 \in C_2$, then there is an $n \times n$ binary matrix A such that $\mathbf{x}_1 = \mathbf{x}A$ (mod-2).*

Proof. Consider an $(n, n-k)$ code C^* such that $C_2 \subset C^*$ and $C_1 \oplus C^* = GF(2)^n$. Let H be an $n \times k_1$ check matrix of C^*. Pick $\mathbf{b}_i \in C_1; i = 1, 2, \cdots, k_1$ such that $\mathbf{b}_i H^t = \mathbf{e}_i$, where $\mathbf{e}_i$ is the unit vector with the one at the ith position. Such set of vectors always exists because different codewords of C_1 belong to different cosets of C^*, and the number of cosets is equal to the number of codewords of C_1. Denote by B the matrix whose ith row is $\mathbf{b}_i; i = 1, 2, \cdots k_1$. Since $\{\mathbf{b}_i; i = 1, 2, \cdots, k_1\}$ is a basis for C_1, any $\mathbf{x}_1 \in C_1$ can be represented as follows $\mathbf{x}_1 = \alpha_1 \mathbf{b}_1 \oplus \alpha_2 \mathbf{b}_2 \oplus \cdots \oplus \alpha_k \mathbf{b}_{k_1}; \alpha_i = 0, 1; i = 1, 2, \cdots, k_1$. We have $\mathbf{x}H^t B = \mathbf{x}_1 H^t B = (\alpha_1, \alpha_2, \cdots, \alpha_{k_1})B = \mathbf{x}_1$. For completing the proof, set $A = H^t B$. @

We proceed now to describe a decoding algorithm for $(m, \mathcal{C})$-separable collection of codes. Let $\mathbf{j}_m = \{j_1, j_2, \cdots, j_m\}$ be some set of m different indices; $j_i = 1, 2, \cdots, N; i = 1, 2, \cdots, m$. Denote $\mathbf{j}_{m-1,i} = \mathbf{j}_m \backslash \{j_i\}$. By applying Lemma 5 with $C_1 = C_{j_i}$ and $C_2 = C_{j_1} \oplus \cdots \oplus C_{j_{i-1}} \oplus C_{(j_{i+1}} \oplus C_{j_m}$, we obtain a matrix $A(\mathbf{j}_{m-1,i})$ such that $\mathbf{x}A(\mathbf{j}_{m-1,i}) \in C_{j_i}$ for $\mathbf{x} \in C$. We have to construct $\binom{N}{m}$ such matrices. Now the decoding procedure can be described as follows.

Decoding Algorithm:

Step 1 (*Correction of channel errors*). Decode the output sequence $\mathbf{y}$ of the channel into a codeword $\mathbf{x} \in C$.

Step 2 (*Determination of the transmitted codewords*). Assuming the subset $\mathbf{j}_s, s \leq m$, of indices of active users is known, select $\mathbf{j}_m$ such that $\mathbf{j}_s \subset \mathbf{j}_m$ and $\mathbf{x}'_{j_i} = \mathbf{x}A(\mathbf{j}_{m-1,i}) \in C'_{j_i}$ for all $j_i \in \mathbf{j}_s$.

References

[1] P. Mathys, "A class of codes for a T active users out of N multiple access communication system," *IEEE Trans. Information Theory*, vol.IT 36, no.6, pp. 1206-1219, Nov. 1990.

[2] E.J. MacWilliams, and N.J.A. Sloane, *The Theory of Error Correcting Codes*, Amsterdam: North-Holland, 1977.

Estimation of the size of the list when decoding over an arbitrarily varying channel

V. Blinovsky and M. Pinsker

IPPI, Moscow

In paper [1] the list-of-L decoding algorithm for a deterministic code was introduced: the context was transmission over an arbitrarily varying channel (AVC). In that work necessary and sufficient conditions for the list capacity C_L to coincide with the random code capacity C under the average probability of error criterion were obtained.

In this paper we obtain an estimation on the volume L of the list for which $C_L = C$. The list-of-L decoding algorithm consists in the first step of an algorithm taken from [2].

Let us introduce some definitions. Let $\mathcal{X}, \mathcal{Y}, \mathcal{S}$ be finite sets, $\mathcal{X}, \mathcal{Y}$ input and output alphabets, and $\mathcal{S}$ the state alphabet of the AVC.

An AVC is defined by the family of conditional distributions $w(y \mid x, s)$ on $\mathcal{Y}$. The probability of the outer vector $\bar{y} = (y(1), \ldots, y(n)) \in \mathcal{Y}^n$ when $\bar{x} = (x(1), \ldots, x(n)) \in \mathcal{X}^n$ is transmitted over AVC and $\bar{s} = (s(1), \ldots, s(n)) \in \mathcal{S}^n$ is the state of the AVC is defined by the transition function

$$w^n(\bar{y} \mid \bar{x}, \bar{s}) = \prod_{i=1}^{n} w(y(i) \mid x(i), s(i)).$$

Call list-of-L decoded code the set $K^n = \{(\bar{x}_i, A_i), i \in 1, \ldots, M\}$ of pairs $(\bar{x}_i, A_i)$, where $\bar{x}_i \in \mathcal{X}^n, A_i \in \mathcal{Y}^n$ and $\bigcap_{i \in J} A_i = \emptyset$ when $\mid J \mid \geq L + 1$; here M is the number of messages. Message i is encoded into vector $\bar{x}_i$ and then transmitted over AVC. When AVC is in state $\bar{s} \in \mathcal{S}^n$ outer sequence $\bar{y} \in \mathcal{Y}^n$ has probability $w^n(\bar{y} \mid \bar{x}, \bar{s})$. Outer sequence $\bar{y} \in \mathcal{Y}^n$ is decoded into a list of no more than L messages $\{i\} \subset \{1, \ldots, M\}$ for which $\bar{y} \in A_i$. Define the average error probability for deterministic list-of-L decoded code K^n when state $\bar{s} \in \mathcal{S}^n$ has occured by

$$\bar{P}_L(\bar{S}) = M^{-1} \sum_{i=1}^{M} \sum_{\bar{y} \in \mathcal{Y}^n, \bar{y} \notin A_i} w^n(\bar{y} \mid \bar{x}, \bar{s}) \tag{1}$$

Define the capacity C_L of AVC in the list-of-L decoding case and the average error criterion by the following formula

$$C_L \triangleq \sup_{\bar{e}_L(R)=0} R, \tag{2}$$

where $\bar{e}_L(R) \triangleq \min_{K^n : \log M \geq Rn} \max_{\bar{s} \in \mathcal{S}^n} \bar{P}_L(\bar{s})$.

The random code capacity is defined by the formula

$$C = max_{p \in P} \min_{q \in Q} \sum_{x \in \mathcal{X}, y \in \mathcal{Y}} \frac{w_q(y \mid x)}{\sum_{x' \in \mathcal{X}} w_q(y \mid x')p(x')}, \tag{3}$$

here P, Q are sets of distributions p, q on $\mathcal{X}, \mathcal{S}$ respectively;

$$w_q(y \mid x) = \sum_{s \in \mathcal{S}} w(y \mid x, s)q(s).$$

The main result of this paper is the proof of the following

Theorem 1 *Let* $C > 0$, *then for* $L > log \mid \mathcal{S} \mid /C$,$C_L = C$.

Before proving the theorem we introduce some definitions. Let $\mathcal{X}_1, \ldots, \mathcal{X}_m$ be finite sets and $X_1, \ldots, X_m$ random variables with distribution $p(x_1, \ldots, x_m) = p_{X_1,\ldots,X_m}(x_1, \ldots, x_m)$, $x_i \in \mathcal{X}_i; i \in \{1, \ldots, m\}$.Denote

$$I(X_1, \ldots, X_m) = \sum_{x_i \in \mathcal{X}_i; i \in \{1,\ldots,m\}} p(x_1, \ldots, x_m) \log \frac{p(x_1, \ldots, x_m)}{p(x_1) \ldots p(x_m)}.$$

Introduce also the following distributions and functions :$p_{X_1|X_2} = p_{X_1X_2}/p_{X_2}-$ conditional probability;

$$H(X_1, X_2, \ldots, X_m) = - \sum_{x_i \in \mathcal{X}_i; i \in \{1,\ldots,m\}} p(x_1, \ldots, x_m) \log p(x_1, \ldots, x_m)$$

is the entropy function ;

$$H(X_1 \mid X_2) = H(X_1, X_2) - H(X_2);$$
$$I(X_1; X_2 \mid X_3) = H(X_1 \mid X_3) + H(X_2 \mid X_3) - H(X_1, X_2 \mid X_3).$$

We define the type of the sequence $\bar{x} = (x(1), \ldots, x(n)) \in \mathcal{X}^n$. The type of the sequence $\bar{x} \in \mathcal{X}^n$ is the distribution $p_{\bar{x}}$ on $\mathcal{X}$, which is defined by the following relation $p_{\bar{x}}(x) = n_x/n$, where n_x is the number of elements in the sequence $\bar{x}$ equal to x. Similarly one can define the joint type $p_{\bar{x}\bar{y}\bar{s}}$ of the sequence $(\bar{x}, \bar{y}, \bar{s})$ which is defined on space product $\mathcal{X} \otimes \mathcal{Y} \otimes \mathcal{S}$ by the formula $p_{\bar{x}\bar{y}\bar{s}}(x, y, s) = n_{xys}/n$, where n_{xys} is number of triples $(x(i), y(i), s(i)), i \in \{1, \ldots, n\}$ which are equal to (x, y, s). We associate with types random variables $X; (X, Y, S)$ with distributions $p_X = p_{\bar{x}}; p_{XYS} = p_{\bar{x}\bar{y}\bar{s}}$ and use the following definitions

$$\tau_X = \{\bar{x} \in \mathcal{X}^n : p_{\bar{x}} = p_X\},$$
$$\tau_{XY} = \{(\bar{x}, \bar{y}) \in \mathcal{X}^n \otimes \mathcal{Y}^n : p_{\bar{x}\bar{y}} = p_{XY}\},$$
$$\tau_{XYS} = \{(\bar{x}, \bar{y}, \bar{s}) \in \mathcal{X}^n \otimes \mathcal{Y}^n \otimes \mathcal{S}^n : p_{\bar{x}\bar{y}\bar{s}} = p_{XYS}\},$$
$$\tau_{Y|X}(\bar{x}) = \{\bar{y} : (\bar{x}, \bar{y}) \in \tau_{XY}\},$$
$$\tau_{Y|XS}(\bar{x}, \bar{s}) = \{\bar{y} : (\bar{x}, \bar{y}, \bar{s}) \in \tau_{XYS}\},$$

$D(P_{XYS} \,\|\, p_{XS}w)$, where $p_{XS}w(x, y, s) = P_{XS}w(y \mid x, s)$ is the relative entropy of the distributions $p_{XYS}, p_{XS}w$.

The following relations are true ([2])

$$(n+1)^{-|\mathcal{X}|}2^{nH(X)} \leq| \tau_X |\leq 2^{nH(X)}, \tag{4}$$

$$(n+1)^{-|\mathcal{X}||\mathcal{Y}|}2^{nH(Y|X)} \leq| \tau_{Y|X} |\leq 2^{nH(Y|X)}, \tag{5}$$

$$\sum_{\bar{y}\in\tau_{Y|XS}(\bar{x},\bar{s})} w^n(\bar{y} \mid \bar{x}, \bar{s}) \leq 2^{-nD(p_{XYS}||p_{XS}w)}. \tag{6}$$

The following lemma is true (see [1]; this lemma is the direct generalisation of a result by R.Ahlswede, which is proved in [3] in the case $L = 1$).

Lemma 1 *Capacity C_L is either $C_L = C$ or else $C_L = 0$.*

Hence to prove the theorem it is enough to prove the following lemma

Lemma 2 *Let $C > 0, L > \log \mid \mathcal{S} \mid /C$, then there exists $\nu > 0$, such that for all $R \in (0,\nu), \bar{e}_L(R) = 0$.*

Indeed if for all $R \in (0,\nu)$ equality $\bar{e}_L(R) = 0$ holds, then it is possible to transmit over AVC with error probability tending to zero when $n \to \infty$ and the transmission rate differs from zero. In this case due to lemma 1 maximum of the transmission rate is equal to C.

Proof of lemma 2. We consider the set of code vectors $K^n_{\bar{x}} = \{x_1,\ldots,x_M\}$ which have the same type coincide with type of fixed vector $\bar{x}$ which maximize right part of (3) : $p_X(x) = p_{\bar{x}}(x)$. Define $A^n_{\bar{x}}$ - ensemble, generating by codes, defined above with uniform distribution on them. Decoding algorithm which we use for this codes consists in choosing decoding areas $\{A_i\} \subset \mathcal{Y}^n$ and is as follows. Compose list of vectors $\Gamma = \{\bar{x}_{i_1},\ldots,\bar{x}_{i_N}\} \subset K^n_{\bar{x}}$ such that for every $\bar{x}_i \in \Gamma$ there exists at least one vector $\bar{s}_i \in \mathcal{S}^n$ with the following property

$$D(p_{\bar{x}_i\bar{s}_i\bar{y}} \mid\mid p_{\bar{x}_i}p_{\bar{s}_i}w) \leq \delta_1; \tag{7}$$

in this case set $\bar{y} \in A_i$. If $N > L$ we propose an error.

It is easy to prove that for transmitted vector $\bar{x}_i$ and state of channel $\bar{s}$ probability of relation (7) tends to 1 in $n \to \infty$ (see in [1] for a proof).

Result of lemma 2 follows from statement 1.

Statement 1 *For $L > \log \mid \mathcal{S} \mid /C$ and $\nu > 0$ small enough there exist sequence list-of-L decoding codes $K^n_{\bar{x}}$ with rate $R \in (0,\nu)$ such that*

$$\lim_{n\to\infty} \max_{\bar{s}\in\mathcal{S}^n} \bar{p}_L(\bar{s}) = 0. \tag{8}$$

Next we introduce two useful lemmas and a statement.

Lemma 3 *Let $(X.Y),(X,Y')$ pairs of random variables with distributions*

$$p_{XY}(x,y) = \sum_{s\in\mathcal{S}} p_{X,S,Y}(x,s,y),$$

$$p_{XY'}(x,y) = \sum_{s\in\mathcal{S}} p_X(x)p_S(s)w(y \mid x,s),$$

then

$$I(X;Y) = I(X;Y') + o(1) \tag{9}$$

where $o(1) \to 0$ uniformly on $\bar{x}_i \in K_{\bar{x}}^n$ if $\delta_1 \to 0$.

Corollary 1

$$I(X_i;Y) \geq C - o(1), n \to \infty, \delta_1 \to 0. \tag{10}$$

Lemma 4 *Let $A_{\bar{x}}^n$ be an ensemble of sets of code vectors $K_{\bar{x}}^n$. The probability that for all sets of $L+1$ different vectors $\{\bar{x}_{i_1}, \ldots, \bar{x}_{i_{L+1}}\} \subset K_{\bar{x}}^n$ the inequality*

$$I(X_{i_1}, \ldots, X_{i_{L+1}}) \leq 3L\delta_2 \tag{11}$$

holds tends to 1 when $n \to \infty$.

We omit proofs of lemmas 3 and 4 (result of lemma 4 can be found in [1]) .

Statement 2 *For arbitrarily $\gamma, \delta_3 > 0$, n large enough with probability not less, than $1 - 2^{-\gamma n}$ for all $\bar{z}, \bar{s} \in S^n$ following relation is true*

$$| \{i : \bar{x}_i \in \tau_{Z|S}(\bar{s})\} | \leq 2^{n(|R - I(Z;S)|^+ + \delta_3)},$$

where

$$| a |^+ = \begin{cases} a \text{ ,if } a \geq 0 \\ 0 \text{ ,if } a < 0. \end{cases}$$

the proof of this statement is an easy corollary of the Chernoff estimation (see [2] formula (A 8)).

Let us now estimate error probability of the list- of $-L$ decoding. Note that from statement 2 follows that with probability not less that $1 - 2^{-\gamma n}$ for codes $K_{\bar{x}}^n$ and for all $\bar{s} \in S^n$ following inequality is true

$$2^{-nR} | \{i : \bar{x}_i \in K_{\bar{x}}^n, \bar{x}_i \in \tau_{Z|S}(\bar{s})\} | \leq 2^{-\epsilon n},$$

$R \geq 2\epsilon > 0$ if $I(Z;S) \geq \epsilon$. So we suppose that

$$I(X_i;S) \leq \epsilon. \tag{12}$$

Next

$$D(p_{X_i S_i Y} \,||\, p_{X_i S_i} w) = D(p_{X_i S_i Y} \,||\, p_{X_i} p_{S_i} w) - I(X_i;S_i)$$

and from (6) we obtain

$$\sum_{p_{X_i S_i Y}} \sum_{\bar{y} \in \tau_{Y|X_i S_i}(\bar{x}_i, \bar{s})} w^n(\bar{y} \mid \bar{x}_i, \bar{s}_i) \leq m 2^{-nD(p_{X_i S_i Y} || p_{X_i S_i} w)} =$$

$$m 2^{-n(D(p_{X_i S_i Y} || p_{X_i} p_{S_i} w) - I(X_i;S_i))} \leq m 2^{-n(\delta_1 - \epsilon)}. \tag{13}$$

Because the number m of types $p_{X_i S_i Y}$ is a polynomial function of n, the RHS expression in (13) tends to 0 exponentially in n when $\delta_1 > 1$.

Hence when estimating error probability it is enough to consider only such $\bar{x}_i$ for which relation (12) and (7) are true. In other case error probability does not exceed $2^{-\nu n}, \nu > 0$ .Let

$$e_{X,Z_1,\ldots,Z_L,S,Y}(i,\bar{s}) = \sum_{\bar{y}:(\bar{x}_i,\bar{x}_{j_1},\ldots,\bar{x}_{j_L},\bar{s},\bar{y})\in\tau_{XZ_1\ldots Z_LSY}} w^n(\bar{y} \mid \bar{x}_i,\bar{s}).$$

Then the following estimation for error probability is true

$$\bar{p}_L(\bar{s}) \le 2^{-\nu n} + M^{-1}\sum_{i=1}^{M}\sum_{p_{XZ_1\ldots Z_LSY}} e_{X,Z_1,\ldots,S,Y}(i,\bar{s}). \tag{14}$$

Here the inner sum includs summation over distributions of $X, Z_1 \ldots Z_L SY$ and is a polynomial in n. We have

$$e_{X,Z_1,\ldots,S,Y}(i,\bar{s}) = \sum_{j_1,\ldots,j_L:(\bar{x}_i,\bar{x}_{j_1},\ldots,\bar{x}_{j_L},\bar{s})} \sum_{y\in\tau_{Y|XZ_1\ldots Z_LS}(\bar{x},\bar{x}_{j_1},\ldots,\bar{x}_{j_L},\bar{s})} w^n(\bar{y} \mid \bar{x}_i,\bar{s}).$$

Because $w^n(\bar{y} \mid \bar{x}_i,\bar{s})$ is constant when $y \in \tau_{Y|XS}(\bar{x}_i,\bar{s})$ and this constant does not exeed $\mid \tau_{Y|XS}(\bar{x}_i,\bar{s}) \mid^{-1}$, the inner sum has the upper bound value $\mid \tau_{Y|XZ_1\ldots Z_LS}(\bar{x}_i,\bar{x}_{j_1},\ldots,\bar{x}_{j_L},\bar{s}) \mid \mid \tau_{Y|XS}(\bar{x}_i,\bar{s}) \mid^{-1}$ which for n large enough is less than $2^{-n(I((Y;(Z_1,\ldots Z_L)|X,S)-\kappa)}, \kappa > o$.

Hence

$$M^{-1}\sum_{i=1}^{M}\sum_{p_{XZ_1\ldots Z_LSY}} r_{X,Z_1,\ldots,Z_L,S,Y}(i,\bar{s}) \le$$
$$M^{-1}\sum_{i=1}^{M}\sum_{p_{XZ_1\ldots Z_LSY}} 2^{n[RL-I(Y;(Z_1,\ldots,Z_L)|X,S)+\kappa]}.$$

Next we obtain

$$I(Y;(Z_1,\ldots,Z_L) \mid X,S) = I(Y;(X,S,Z_1,\ldots,Z_L)) - I(Y;(X,S) \ge$$
$$I(Y;(X,Z_1,\ldots,Z_L)) - I(Y;S) + I(Y;S \mid X) \ge I(Y;X) +$$
$$\sum_{i=1}^{L} I(Y,Z_i) - 3L\delta_2 - I(Y;X) - H(S) \ge LC - \log \mid S \mid -\delta_4. \tag{15}$$

Here in second inequality we use (11). In last inequality we use estimation (10). Also $I(Y;S \mid X) \le H(S) \le \log \mid S \mid$. Because $\delta_4, R > 0$ are arbitrarily, we obtain from (15) that when

$$L > \log \mid S \mid /C, \bar{p}_L(\bar{s}) < e^{-\nu n} + e^{-\mu n} \to 0, n \to \infty$$

and hence list–of–L decoding with the algorithm introduced above and size of the list greater than $\log \mid S \mid /C$ when $R > 0$ is small enough , leads to the arbitrarily small probability of error when $n \to \infty$. Consequently according to lemma 1 $C_L = C$.

References

1. V.Blinovsky,P.Narayan,M.Pinsker *'The Capacity Of The AV Channel In List Decoding Case'*,to appear .
2. I.Csiszar,P.Narayan *'The Capacity Of The Arbitrarily Varying Channel Revisited, Positivity, Constrains'*,IEEE Trans.Inf.Th., v.34,N2,1988 pp. 181-193.
3. R.Ahlswede*Elimination Of Correlation In Random Codes For Arbitrary Varying Channels'*,Z.Wahrscheinlichkeitstheorie,44,1978, pp.159-175 .

A Lower Bound on Binary Codes with Covering Radius One

Iiro Honkala
Department of Mathematics
University of Turku
20500 Turku 50, Finland
e-mail: honkala@sara.utu.fi

Abstract. We give a lower bound on $K(n, R)$, the minimum cardinality of a binary code of length n and covering radius R, when $n \equiv 5 \pmod 6$. In particular, our bound implies that $K(17, 1) \geq 7399$.

1 Introduction

If $C \subseteq \mathbf{F}_2^n$, then we say that the covering radius of C is the smallest integer R such that the Hamming balls of radius R centered at the codewords of C cover the whole space $\mathbf{F}_2^n$. We denote by $K(n, R)$ the smallest possible cardinality of a binary code of length n and covering radius R. Many lower bounds on $K(n, R)$ have been presented, see e.g., the references of [5], and [1] and [7]. In this paper we deal exclusively with the case $R = 1$, and show how a recent lower bound by van Wee can be improved further. Indeed, in Section 2 we prove the following theorem, where $A(n, d)$ denotes the maximum number of codewords in a binary code of length n and minimum distance d, and $V(n, r)$ denotes the cardinality of a Hamming ball of radius r in $\mathbf{F}_2^n$.

Theorem: For every n with $n \equiv 5 \pmod 6$ we have

$$K(n, 1) \geq \frac{(V(n, 2) + 5)2^{n+1} - 9A(n, 3)(n + 1)}{(2V(n, 2) - 3)(n + 1)}.$$

In particular, for $n = 17$ we get (as we know that $A(17, 3) \leq 6552$ [3, p. 674]) the lower bound $K(17, 1) \geq 7399$ compared to 7391 in [5], 7378 in [6], and 7377 in [4].

2 The Proof

Assume that $C \subseteq \mathbf{F}_2^n$. As usual, we denote $B_r(x) = \{y \in \mathbf{F}_2^n \mid d(y, x) \leq r\}$, the Hamming ball of radius r, and denote its cardinality by $V(n, r)$. As in [4], we say that the r-excess on $V \subseteq \mathbf{F}_2^n$ by a code C is

$$E_C^r(V) = \sum_{c \in C} | B_r(c) \cap V | - | \bigcup_{c \in C} B_r(c) \cap V | .$$

In this paper we always have $r = 1$, and we omit the superscript r.

Assume now that $C \subseteq \mathbf{F}_2^n$ has covering radius 1 and that $n \equiv 5 \pmod 6$, and denote $Z = \{x \in \mathbf{F}_2^n \mid\, \mid B_1(x) \cap C \mid \geq 2\}$, i.e., Z is the set of those points that are covered by more than one codeword of C. It has been shown in [5] that for every $x \in \mathbf{F}_2^n$

$$E_C(B_2(x)) \equiv 2 \pmod 3 \tag{1}$$

and in particular

$$E_C(B_2(x)) \geq 2. \tag{2}$$

It is furthermore shown that for every x such that

$$\mid B_2(x) \cap Z \mid \geq 2 \tag{3}$$

we have

$$E_C(B_2(x)) \geq 5. \tag{4}$$

As in [5] let $C' = \{c'_1, c'_2, \ldots, c'_{M'}\}$ be a maximal 1-error-correcting subcode of C, i.e., the spheres $B_1(c'), c' \in C'$ are disjoint, and every sphere $B_1(c), c \in C \backslash C'$ intersects at least one of the spheres $B_1(c'), c' \in C'$, in two points. Denote

$$B_1(C') = \bigcup_{c' \in C'} B_1(c').$$

The definition of C' now implies that every point $x \in \mathbf{F}_2^n \backslash B_1(C')$ satisfies (3) and hence (4).

We now try to estimate

$$\sum_{x \in B_1(C')} E_C(B_2(x)).$$

Define

$$f_i = \mid B_2(c'_i) \cap (C \backslash C') \mid, \text{ for } i = 1, 2, \ldots, M'$$

and further

$h_i =$ the number of indices $j = 1, 2, \ldots, M'$ for which $f_j = 2i$ or $2i - 1$.

Clearly

$$\sum_{i=0}^{\infty} h_i = \mid C' \mid \tag{5}$$

and

$$\sum_{i=0}^{\infty} 2ih_i \geq \mid C \mid - \mid C' \mid \tag{6}$$

because $\mid B_2(c) \cap C' \mid \geq 1$ for every $c \in C \backslash C'$.

Consider now a fixed j and suppose

$$B_2(c'_j) \cap (C \backslash C') = \{c_1, \ldots, c_{f_j}\}.$$

Then for any $x \in B_1(c'_j)$ we have

$$E_C(B_2(x)) \geq \sum_{k=1}^{f_j} E_{\{c'_j, c_k\}}(B_2(x)) \geq 2f_j \tag{7}$$

since $d(c'_j, c_k) \leq 2$ implies that there are two points $y_1, y_2 \in B_1(c'_j) \cap B_1(c_k)$ and $d(y_1, x) \leq 2$ and $d(y_2, x) \leq 2$. In particular, (3) holds if $f_j \geq 1$.

We now claim that if $f_j = 2i$ or $2i - 1$, and $x \in B_1(c'_j)$, then

$$E_C(B_2(x)) \geq 3i + 2.$$

Indeed, for i = 0, 1, 2, 3 this follows from (1), (2), (4) and (7). When $i \geq 4$, our claim immediately follows from (7) because $2(2i - 1) \geq 3i + 2$.

Therefore

$$\sum_{x \in B_1(C')} E_C(B_2(x)) \geq \sum_{i=0}^{\infty} (n+1)(3i+2)h_i$$

$$\geq 3(n+1)(\mid C \mid - \mid C' \mid)/2 + 2(n+1) \mid C' \mid.$$

We now again proceed as in [5] to obtain (see [5] for details)

$$5(2^n - \mid C' \mid (n+1)) + 3(n+1)(\mid C \mid - \mid C' \mid)/2 + 2(n+1) \mid C' \mid$$

$$\leq \sum_{x \in \mathbf{F}_2^n \backslash B_1(C')} E_C(B_2(x)) + \sum_{x \in B_1(C')} E_C(B_2(x))$$

$$= V(n,2)(\mid C \mid (1+n) - 2^n)$$

and therefore

$$(V(n,2) - 3/2)(n+1) \mid C \mid \geq (V(n,2) + 5)2^n - 9A(n,3)(n+1)/2$$

as claimed.

Acknowledgment: The author would like to thank Gerhard van Wee for sending a preprint of [5].

References:

[1] W. Chen and D. Li, "New lower bounds for binary covering codes," IEEE Trans. Inform. Theory, submitted.

[2] I. S. Honkala, "Modified bounds for covering codes," IEEE Trans. Inform. Theory 37 (1991) 351-365.

[3] F. J. MacWilliams and N. J. A. Sloane, The Theory of Error-Correcting Codes. Amsterdam: North-Holland, 1977.

[4] G. J. M. van Wee, "Improved sphere bounds on the covering radius of codes," IEEE Trans. Inform. Theory 34 (1988) 237-245.
[5] G. J. M. van Wee, "Some new lower bounds for binary and ternary covering codes," IEEE Trans. Inform. Theory, submitted.
[6] Z. Zhang, "Linear inequalities for covering codes: Part I - Pair covering inequalities," IEEE Trans. Inform. Theory 37 (1991) 573-582.
[7] Z. Zhang and C. Lo, "Linear inequalities for covering codes: Part II - Triple covering inequalities," IEEE Trans. Inform. Theory 38 (1992) 1648-1662.

On Some Mixed Covering Codes of Small Length

E.Kolev, I.Landgev
Institute of Mathematics,
Bulgarian Academy of Sciences,
8 G.Bonchev str.
Sofia 1113, Bulgaria
e-mail sectmoi@bgearn.bitnet

Abstract

We find some new lower bounds for the cardinality of mixed covering codes having length $n = 6, 7$, or 8 and covering radius up to 3. Some exact values for these numbers are also computed.

1.PRELIMINARIES

Let $\mathcal{F}_{t,b} = \{(x_1, \ldots, x_t, x_{t+1}, \ldots, x_{t+b}) \,|x_i \in \mathbf{F}_3, i = 1, \ldots, t,\ x_j \in \mathbf{F}_2,\ j = t+1, \ldots, t+b\}$, where $\mathbf{F}_2$ (resp. $\mathbf{F}_3$) are the fields with two (resp. three) elements, and t, b are nonnegative integers. The set $\mathcal{F}_{t,b}$ is an abelian group under coordinatewise addition. Its elements are called *words*. The *Hamming distance* between two words from $\mathcal{F}_{t,b}$, say $\mathbf{x} = (x_1, x_2, \ldots, x_{t+b})$ and $\mathbf{y} = (y_1, y_2, \ldots, y_{t+b})$, is defined as the number of positions in which these words differ

$$d(\mathbf{x}, \mathbf{y}) = \left|\{i | x_i \neq y_i, i \in \{1, 2, \ldots, t+b\}\}\right|. \tag{1.1}$$

Every nonempty subset C of $\mathcal{F}_{t,b}$ is called a (*mixed*) *code* and its elements are referred to as codewords. The number $n = t + b$ is the *length* of the code.

Let C be a code in $\mathcal{F}_{t,b}$. The *covering radius* $R(C)$ of C is defined as the minimal nonnegative integer R such that for every $\mathbf{x} \in \mathcal{F}_{t,b}$ there exists a codeword $\mathbf{c}$ from C with $d(\mathbf{x}, \mathbf{c}) \leq R$. By $K(t, b, R)$ we denote the minimal cardinality of a code in $\mathcal{F}_{t,b}$ with covering radius R.

The general problem of determining the numbers $K(t, b, R)$ has been extensively studied in the last few years [1]-[4] . However, the exact values for these numbers are known for relatively small n only (excluding those lengths where binary and ternary perfect codes do exist). In this paper we describe techniques for finding lower bounds on $K(t, b, R)$ and determine the exact covering radius for some mixed codes of length 6,7, and 8.

The sphere $\mathcal{B}_R(\mathbf{x})$ with center $\mathbf{x} \in \mathcal{F}_{t,b}$ and radius R is defined by

$$\mathcal{B}_R(\mathbf{x}) = \{\mathbf{y} | \mathbf{y} \in \mathcal{F}_{t,b}; d(\mathbf{x},\mathbf{y}) \leq R\}. \tag{1.2}$$

The word $\mathbf{x}$ is said to *R-cover* $\mathbf{y}$ iff $\mathbf{y}$ belongs to $\mathcal{B}_R(\mathbf{x})$ and does not belong to $\mathcal{B}_{R'}(\mathbf{x})$ for some $R' < R$. We say that $\mathbf{x}$ *covers* $\mathbf{y}$ iff $\mathbf{x}$ 1-covers $\mathbf{y}$ or $\mathbf{x} = \mathbf{y}$. For a set $\mathcal{A} = \{\mathbf{x}_1, \ldots, \mathbf{x}_a\} \subset \mathcal{F}_{t,b}$ let

$$\mathcal{C}_R^{(k)}(\mathcal{A}) = \bigcup_{i_1,\ldots,i_k} \bigcap_{j=1}^{k} \mathcal{B}_R(\mathbf{x}_{i_j}), \tag{1.3}$$

$$\mathcal{C}_R(\mathcal{A}) = \bigcup_{k \geq 1} \mathcal{C}_R^{(k)}(\mathcal{A}), \tag{1.4}$$

$$\mathcal{N}_R(\mathcal{A}) = \mathcal{C}_R^{(0)}(\mathcal{A}) = \mathcal{F}_{t,b} \setminus \mathcal{C}_R(\mathcal{A}). \tag{1.5}$$

If the lower index R is omitted it is considered to be equal to one.

Let

$$\varphi : \left\{ \begin{array}{ccc} \mathcal{F}_{t,b} & \rightarrow & \mathcal{F}_{t,b} \\ x & \rightarrow & x^{\varphi} \end{array} \right. \tag{1.6}$$

be a bijective mapping from $\mathcal{F}_{t,b}$ onto $\mathcal{F}_{t,b}$. It is said to be *distance preserving* iff for every two vectors $\mathbf{x}, \mathbf{y} \in \mathcal{F}_{t,b}$, $d(\mathbf{x},\mathbf{y}) = d(\mathbf{x}^{\varphi}, \mathbf{y}^{\varphi})$. The following mappings are distance preserving:

(i) $\varphi_\sigma : \mathbf{x} = (x_1, x_2, \ldots, x_n) \rightarrow \mathbf{x}^{\varphi_\sigma} = (x_{\sigma(1)}, x_{\sigma(2)}, \ldots, x_{\sigma(n)})$, where $\sigma \in S_t \times S_b$ (S_n is the symmetric group of degree n);

(ii) $\psi_{\mathbf{a}} : \mathbf{x} \rightarrow \mathbf{x}^{\psi_{\mathbf{a}}} = \mathbf{x} + \mathbf{a}$, where $\mathbf{a} \in \mathcal{F}_{t,b}$;

(iii) $\mu_{i,b} : \mathbf{x} = (x_1, x_2, \ldots, x_n) \rightarrow \mathbf{x}^{\mu_{i,b}} = (x_1, x_2, \ldots, bx_i, \ldots, x_n)$, where $0 \neq b \in \mathbf{F}_3$, and $i \in \{1, 2, \ldots, t\}$.

Let $G = < \varphi_\sigma, \psi_a, \mu_{i,b} | \sigma \in S_t \times S_b, \mathbf{a} \in \mathcal{F}_{t,b}, i \in \{1, 2, \ldots, t\}, 0 \neq b \in \mathbf{F}_3 >$. Given a code $\mathcal{C}$ and a $\varphi \in G$ we denote $\mathcal{C}^{\varphi} = \{\mathbf{x}^{\varphi} | \mathbf{x} \in \mathcal{C}\}$. Two codes $\mathcal{C}_1$ and $\mathcal{C}_2$ are said to be equivalent if there exists a $\varphi \in G$ with $\mathcal{C}_1^{\varphi} = \mathcal{C}_2$. We define $Aut\,\mathcal{C} = \{\varphi \in G | \mathcal{C}^{\varphi} = \mathcal{C}\}$.

Proposition 1.1. For every $\mathcal{A} \subset \mathcal{F}_{t,b}$ and for every two integers $k \geq 0$, $R \geq 1$ we have $Aut\,\mathcal{A} < Aut\,\mathcal{C}_R^{(k)}\mathcal{A}$ and $Aut\,\mathcal{C}_R(\mathcal{A}) = Aut\,\mathcal{N}_R(\mathcal{A})$.$\diamond$

The mutually disjoint sets $\mathcal{S}_j \subset \mathcal{F}_{t,b}$, $j = 1, 2, \ldots, m$, are said to define a *regular partition on $\mathcal{F}_{t,b}$* if

(i) $\mathcal{F}_{t,b} = \cup_{j=1}^{m} \mathcal{S}_j$, and

(ii) for every $i, j \in \{1, 2, \ldots, m\}$, and every integer $R \geq 1$ the number of words from $\mathcal{S}_j$ at a distance R from any $\mathbf{x} \in \mathcal{S}_i$ is a constant, not depending on the choice of $\mathbf{x}$. We denote this constant by r_{ij}^R.

Proposition 1.2. Let $\mathcal{C} \subset \mathcal{F}_{t,b}$ be a code with $R(\mathcal{C}) = R$, and let $\{\mathcal{S}_j | j = 1, 2, \ldots, m\}$ be a regular partition on $\mathcal{F}_{t,b}$. Then for every $i \in \{1, 2, \ldots, m\}$

$$(1.7) \qquad \sum_{j=1}^{m}(|\mathcal{C} \cap \mathcal{S}_j| \sum_{k=0}^{R} r_{ij}^{R}) \geq |\mathcal{S}_i|. \diamond$$

In the case of $m = 1$ this gives the well-known *sphere covering bound.*

Given $\mathcal{F}_{t,b}$, choose $t' \leq t$ ternary and $b' \leq b$ binary coordinates, say those with numbers $1, \ldots, t'$, and $t+1, \ldots, t+b'$. Let $\mathbf{v} = (v_1, v_2, \ldots, v_{t'}, v_{t'+1}, \ldots, v_{t'+b'}) \in \mathcal{F}_{t',b'}$, and

(1.8)

$$\mathcal{S}_{\mathbf{v}} = \{(x_1, x_2, \ldots, x_n) \in \mathcal{F}_{t,b} | x_i = v_i, i = 1, \ldots, t'; x_{t+j} = v_{t'+j}, j = 1, \ldots, b'\}.$$

$\{\mathcal{S}_{\mathbf{v}} | \mathbf{v} \in \mathcal{F}_{t',b'}\}$ defines a regular partition on $\mathcal{F}_{t,b}$. In order to save space we denote the number of words from $\mathcal{F}_{t-t',b-b'}$ contained in a sphere of radius i by σ_i, by definition $\sigma_0 = 1$.

Let $\mathcal{U} \subset \mathcal{F}_{t',b'}$ so that $\{\mathcal{U}, \mathcal{F}_{t'b'} \setminus \mathcal{U}\}$ is a regular partition on $\mathcal{F}_{t',b'}$. The constants from (ii) for this partition will be denoted by ρ_{ij}^{R}, $i, j \in \{1, 2\}$ (the set $\mathcal{U}$ is considered as "first").

Proposition 1.3. Let $\mathcal{C} \subset \mathcal{F}_{t,b}$ be a code of covering radius R, and let $\{\mathcal{S}_{\mathbf{v}} | \mathbf{v} \in \mathcal{F}_{t',b'}\}$ be the regular partition defined in (1.8). Then

(1.9)

$$\sum_{i=0}^{\min\{R,t'+b'\}} (\sum_{\mathbf{v} \in \mathcal{F}_{t',b'} \setminus \mathcal{U}} \sigma_{R-i} \rho_{21}^{i} |\mathcal{C} \cap \mathcal{S}_{\mathbf{v}}| + \sum_{\mathbf{u} \in \mathcal{U}} \sigma_{R-i} \rho_{11}^{i} |\mathcal{C} \cap \mathcal{S}_{\mathbf{u}}|) \geq 3^{t-t'} 2^{b-b'} |\mathcal{U}|.$$

Proof. Sum up both sides of (1.7) for all $\mathbf{u} \in \mathcal{U}$. $\diamond$

Corollary 1.4 Let $b' = 2, t' = 0$ and let $\mathcal{C} \subset \mathcal{F}_{t,b}$ be a code with covering radius $R = 1$. Then for every $\mathbf{u} \in \mathcal{F}_{t',b'}$

$$(1.10) \qquad |\mathcal{C} \cap \mathcal{S}_{\mathbf{v}}| \geq \frac{3^t 2^{b-2} - |\mathcal{C}|}{2t + b - 3}.$$

Proof. Let $\mathbf{u} = (00), \mathbf{u}' = (11), \mathbf{v} = (10), \mathbf{v}' = (01)$, and put $\mathcal{U} = \{\mathbf{u}, \mathbf{u}'\}$. From (1.10) one gets

$$\begin{aligned} \sigma_1(|\mathcal{C} \cap \mathcal{S}_{\mathbf{u}}| + |\mathcal{C} \cap \mathcal{S}_{\mathbf{u}'}|) + \sigma_0(|\mathcal{C} \cap \mathcal{S}_{\mathbf{v}}| + |\mathcal{C} \cap \mathcal{S}_{\mathbf{v}'}|) &\geq 3^t 2^{b-1}, \\ |\mathcal{C} \cap \mathcal{S}_{\mathbf{u}}| + |\mathcal{C} \cap \mathcal{S}_{\mathbf{u}'}| &\geq \frac{3^t 2^{b-1} - 2|\mathcal{C}|}{\sigma_1 - 2}. \end{aligned}$$

Now

$$\begin{aligned}|\mathcal{C}| &= |\mathcal{C}\cap(\cup_{\mathbf{w}}\mathcal{S}_{\mathbf{w}})| = |\cup_{\mathbf{w}}(\mathcal{C}\cap\mathcal{S}_{\mathbf{w}})| \\ &\geq |\mathcal{C}\cap\mathcal{S}_{\mathbf{u}}| + |\mathcal{C}\cap\mathcal{S}_{\mathbf{u}'}| + |\mathcal{C}\cap\mathcal{S}_{\mathbf{v}}| + |\mathcal{C}\cap\mathcal{S}_{\mathbf{v}'}| \\ &\geq |\mathcal{C}\cap\mathcal{S}_{\mathbf{u}}| + \frac{3^t2^{b-1} - 2|\mathcal{C}|}{\sigma_1 - 2} - |\mathcal{C}\cap\mathcal{S}_{\mathbf{u}}| + 3^t2^{b-2} - \sigma_1|\mathcal{C}\cap\mathcal{S}_{\mathbf{u}}|,\end{aligned}$$

whence the assertion follows.◇

Let $\mathcal{T} = \{\mathcal{T}_1, \ldots, \mathcal{T}_p\}$ be some partition on the set of coordinate positions with the property that each $\mathcal{T}_j$ contain either binary or ternary position numbers but not both. Given a word $\mathbf{u} \in \mathcal{F}_{t,b}$ denote by $\mathbf{u}|_{\mathcal{T}_j}$ the word obtained from $\mathbf{u}$ by taking the coordinates from $\mathcal{T}_j$ only. Let $\mathbf{a} = (a_1, a_2, \ldots, a_p)$, where the a_i's are nonnegative integers, $a_i \leq |\mathcal{T}_i|$. Then $\{\mathcal{S}_{\mathbf{a}}\}$, where

$$\text{(1.12)} \qquad \mathcal{S}_{\mathbf{a}} = \{\mathbf{u} \in \mathcal{F}_{t,b} | wt(\mathbf{u}|_{\mathcal{T}_j}) = a_j, j = 1, 2, \ldots, p\}$$

and $\mathbf{a}$ runs all admissible p-tuples, is aregular partition on $\mathcal{F}_{t,b}$.

More generally, let H be a subgroup of G. The orbits of H on $\mathcal{F}_{t,b}$ define a regular partition. The partitions defined by (1.8) and (1.12) are of this kind.

2. THE CASE R = 1

In this section we prove lower bounds for mixed codes of length 6 and covering radius 1. Let $\mathcal{C} \subset \mathcal{F}_{t,b}$, and let $\mathcal{S}_{\mathbf{v}}$ be as in (1.8). Denote by $\mathcal{C}_{\mathbf{v}}$ the set of words from $\mathcal{F}_{t-t',b-b'}$ obtained from $\mathcal{C} \cap \mathcal{S}_{\mathbf{v}}$ by deleting the fixed coordinates. Let $c_{\mathbf{v}} = |\mathcal{C}_{\mathbf{v}}|$. We start with a simple observation.

Proposition 2.1. Let $\mathcal{C} \subset \mathcal{F}_{t,b}$ be a code of covering radius 1, $b \geq 2$, and let $b' = 2, t' = 0$. Then $\mathcal{C}_{\mathbf{u}+(01)} \cup \mathcal{C}_{\mathbf{u}+(10)} \supset \mathcal{N}(\mathcal{C}_{\mathbf{u}})$. Further, if $\mathcal{C}_{\mathbf{u}\cup(01)} \cup \mathcal{C}_{\mathbf{u}+(10)} = \mathcal{N}(\mathcal{C}_{\mathbf{u}}) \cup \mathcal{B}$ then $\mathcal{C}_{\mathbf{u}+(11)} \cup \mathcal{C}_{\mathbf{u}} \supset \cup_{k<2}\mathcal{C}^{(k)}(\mathcal{N}(\mathcal{C}_{\mathbf{u}}) \bigcup \mathcal{B})$.◇

The case t = 1, b = 5

It is known that $13 \leq K(1,5,1) \leq 16$ [1],[3]. Let us assume that there exists a code $\mathcal{C} \subset \mathcal{F}_{1,5}$ of covering radius 1 with $|\mathcal{C}| = 15$. Now Corollary (1.4) yields $c_{\mathbf{u}} \geq 3$ for every $\mathbf{u} \in \mathcal{F}_{0,2}$.

Proposition 2.2 Let $\mathcal{A} \subset \mathcal{F}_{1,3}, |\mathcal{A}| = 3$. Then $|\mathcal{C}(\mathcal{A})| \leq 18, |\mathcal{C}(\mathcal{A})| \neq 17$.

(i) If $|\mathcal{C}(\mathcal{A})| = 18$ then up to equivalence $\mathcal{A} = \{(0100), (1010), (2001)\}$.

(ii) If $|\mathcal{C}(\mathcal{A})| = 16$ then up to equivalence $\mathcal{A} = \{(0000), (1111), 2011)\}$, or $\{(0011), (0101), (1110)\}$.

(iii) If $|\mathcal{C}(\mathcal{A})| = 15$ then up to equivalence $\mathcal{A} = \{(0000), (1000), 2110)\}$, $\{(0000), (1000), (0111)\}$, or $\{(0000), (1000), (2111)\}$.◇

Without loss of generality we can assume that $c_{00} = 3$. Now Propsition 2.1 yields that (ii) and (iii) of Proposition 2.2 are impossible. If $\mathcal{C}_{00}$ is equivalent to the $\mathcal{A}$ from (i) then the same proposition gives $c_{01} + c_{10} = 8$, or 9.

First, we consider the case $c_{01} + c_{10} = 9$. With no loss of generality we have $c_{11} = 3$, $c_{01} = 4$, $c_{10} = 5$. Note that $\mathcal{A}$ from Proposition 2.2(i) has an automorphism group of order 3. It is generated by $\tau = \varphi_\sigma \psi_{\mathbf{a}}$, where $\sigma = (234) \in S_3$, and $\mathbf{a} = (1000)$. It is readily seen that there are only two (up to equivalence) choices for $\mathcal{C}_{11}$, so that $|\mathcal{N}(\mathcal{A}) \cup \mathcal{N}(\mathcal{C}_{11})| \le 9$:

(a) $\mathcal{C}_{11} = \mathcal{A}^{\psi_{\mathbf{a}}}$;

(b) $\mathcal{C}_{11} = \mathcal{A}$.

(a) In this case we know exactly the words from $\mathcal{C}_{01} \cup \mathcal{C}_{10}$. Then a straightforward check leads to a contradiction with Proposition 2.1.

(b) Now three of the words from are $\mathcal{C}_{01} \cup \mathcal{C}_{10}$ unknown. In order to cover the words $(*01000)$ and $(*10000)$ two of the words in $\mathcal{C}_{01} \cup \mathcal{C}_{10}$ must be of the shape $(*000)$. They can be taken as (0000), and (1000). There are two possibilities for the third word in $\mathcal{C}_{01} \cup \mathcal{C}_{10}$. In both cases we cannot obtain a code of covering radius 1 by any partitioning of the words in $\mathcal{C}_{01} \cup \mathcal{C}_{10}$.

Now consider the case $c_{01} + c_{10} = 8$. Obviously, $c_{01} = c_{10} = 4$. Denote by $\rho^{p,q}_{\alpha,\beta}$, $\alpha, \beta \in \mathbf{F}_2$, $1 \le p < q \le 5$, the number of words in $\mathcal{C}$, containing α in the p-th and β in the q-th binary positions. We have proved so far that for every choice of p, q, $1 \le p < q \le 5$, one can find a binary pair (α, β) with $\rho^{p,q}_{\alpha,\beta} = 3$, and $\rho^{p,q}_{\alpha+i,\beta+j} = 4$ for $(i,j) \in \{(01),(10),(11)\}$. With no loss of generality the binary parts of the codewords of $\mathcal{C}$ are

$$\begin{array}{ccccc} 0&0&1&0&0\\ 0&0&0&1&0\\ 0&0&0&0&1\\ 0&1&*&*&*\\ 0&1&*&*&*\\ 0&1&*&*&*\\ 0&1&*&*&* \end{array} \qquad \begin{array}{ccccc} 1&0&*&*&*\\ 1&0&*&*&*\\ 1&0&*&*&*\\ 1&0&*&*&*\\ 1&1&*&*&*\\ 1&1&*&*&*\\ 1&1&*&*&*\\ 1&1&*&*&* \end{array}$$

Assume that $\rho^{1,3}_{0,1} = 3$. Consequently, $\rho^{2,3}_{0,1} = 3$, and it is straightforward that the words in S_{001} cannot be covered by $\mathcal{C}$. Hence $\rho^{1,3}_{0,1} = 4$, and $\rho^{1,3}_{0,0} = 3$. In the same way one can prove that $\rho^{1,i}_{0,0} = 3$, $i = 4, 5$, whence $\rho^{i,j}_{0,0} = 3$ for every i, j, $1 \le i, j \le 5$. Furthermore, every one of the last three columns contains exactly three zeros in the words $(11 * **)$. Therefore, the binary parts of the words in $\mathcal{C}$ are (up to equivalence)

$$\begin{array}{ccccc} 0&0&1&0&0\\ 0&0&0&1&0\\ 0&0&0&0&1\\ 0&1&0&1&1\\ 0&1&1&0&1\\ 0&1&1&1&0\\ 0&1&1&1&1 \end{array} \qquad \begin{array}{ccccc} 1&0&0&1&1\\ 1&0&1&0&1\\ 1&0&1&1&0\\ 1&0&1&1&1\\ 1&1&1&0&0\\ 1&1&0&1&0\\ 1&1&0&0&1\\ 1&1&0&0&0 \end{array}$$

Now the words $(*11111)$ from $\mathcal{F}_{1,5}$ cannot be covered by $\mathcal{C}$. Thus we have proved the following result.

Proposition 2.3. $K(1,5,1) = 16.\ \diamond$

The case $\mathbf{t = 2, b = 4}$

In this case $17 \leq K(2,4,1) \leq 20$ [1],[3]. We suppose that $\mathcal{C}$ is a code with $|\mathcal{C}| = 19$. As in the previous case corollary 1.4 gives $c_{\mathbf{u}} \geq 4$ for every $\mathbf{u} \in \mathcal{F}_{0,2}$.

Proposition 2.4. Let $\mathcal{A} \subset \mathcal{F}_{2,2}$, $|\mathcal{A}| = 4$. Then $|\mathcal{C}(\mathcal{A})| \leq 28$, $|\mathcal{C}(\mathcal{A})| \neq 27$.

(i) If $|\mathcal{C}(\mathcal{A})| = 28$ then up to equivalence $\mathcal{A} =${(0000),(1101),(2110),(0211)}.

(ii) If $|\mathcal{C}(\mathcal{A})| = 26$ then up to equivalence $\mathcal{A} =${(0000),(1100),(2210),(1011)}, {(0000),(0011),(1110),(2101)}, {(0000),(0011),(1110),(2201)}, {(0000),(0101),(1110),(2011)}, {(0000),(0101),(1110),(2211)} , or {(0000),(0101),(1210),(2011)}.

(iii) If $|\mathcal{C}(\mathcal{A})| = 25$ then up to equivalence $\mathcal{A} =${(0000),(1000),(2110),(2201)}, or {(0000),(1000),(2110),(0211)}.$\diamond$

The cases (ii) and (iii) can be rejected as above (using Proposition 2.1). In the same way one can prove that (i) is impossible whenever $c_{01} + c_{10} = 8, 9$, or 11. In the case $c_{01} + c_{10} = 11$ one has to use the obvious fact that $Aut\,\mathcal{A} > \{id, \mu_{1,2}\varphi_\sigma, \mu_{2,2}\psi_{\mathbf{a}}\psi_{\mathbf{b}}\varphi_\sigma, \mu_{2,2}\psi_{\mathbf{a}}\psi_{\mathbf{b}}\mu_{1,2}\}$, $\mathbf{a} = (0200)$, $\mathbf{b} = (0011)$, $\sigma = (34)$.

Only the case $c_{01} + c_{10} = 10$ remains to be considered. Obviously, $c_{01} = c_{10} = 5$, and $\mathcal{C}_{00} = \mathcal{A}$, where $\mathcal{A}$ is taken from (i). We can assume, that this holds true for any choise of the binary coordinate positions. The binary parts of the codewords of $\mathcal{C}$ are

$$
\begin{array}{cccc}
 & & & \\
0 & 0 & 0 & 0 \\
0 & 0 & 0 & 1 \\
0 & 0 & 1 & 0 \\
0 & 0 & 1 & 1 \\
0 & 1 & * & * \\
0 & 1 & * & * \\
0 & 1 & * & * \\
0 & 1 & * & * \\
0 & 1 & * & *
\end{array}
\qquad\qquad
\begin{array}{cccc}
1 & 0 & * & * \\
1 & 0 & * & * \\
1 & 0 & * & * \\
1 & 0 & * & * \\
1 & 0 & * & * \\
1 & 1 & * & * \\
1 & 1 & * & * \\
1 & 1 & * & * \\
1 & 1 & * & * \\
1 & 1 & * & *
\end{array}
$$

With no loss of generality one can suppose that $\rho_{0,0}^{i,j} = 4$ for every i, j, $1 \le i < j \le 4$ (because we can invert zeros and ones in a coordinate position if necessary). The third and the fourth column are easily seen to contain two zeros and three ones in the words $(01**)$. The same holds true for the words $(10**)$. The word (0100) must occur among the words of the shape $(01**)$, otherwise the elements of $\mathcal{F}_{2,4}$ cannot be covered by $\mathcal{C}$. Similarly, (1000) is among the words $(10**)$. Then (up to equivalence) the binary parts of the words in $\mathcal{C}$ become

$$
\begin{array}{cccc}
 & & & \\
0 & 0 & 0 & 0 \\
0 & 0 & 0 & 1 \\
0 & 0 & 1 & 0 \\
0 & 0 & 1 & 1 \\
0 & 1 & 0 & 0 \\
0 & 1 & 0 & 1 \\
0 & 1 & 1 & 0 \\
0 & 1 & 1 & 1 \\
0 & 1 & 1 & 1
\end{array}
\qquad\qquad
\begin{array}{cccc}
1 & 0 & 0 & 0 \\
1 & 0 & 0 & 1 \\
1 & 0 & 1 & 0 \\
1 & 0 & 1 & 1 \\
1 & 0 & 1 & 1 \\
1 & 1 & 0 & 0 \\
1 & 1 & 0 & 1 \\
1 & 1 & 0 & 1 \\
1 & 1 & 1 & 0 \\
1 & 1 & 1 & 0
\end{array}
$$

Now $\mathcal{C}$ cannot cover all words of the shape $(**1111)$. Thus, we have proved the following result.

Proposition 2.5. $K(2,4,1) = 20$. ◇

The case $\mathbf{t = 4, b = 2}$

It is known that $30 \le K(4,2,1) \le 36$ [1],[3]. We shall prove that $K(4,2,1) \ge 34$. Suppose that $\mathcal{C} \subset \mathcal{F}_{4,2}$ is a code of covering radius 1 with $|\mathcal{C}| = 33$. Now we choose $t' = 1$, and $b' = 2$.

Proposition 2.6. (i) If $\mathbf{u}, \mathbf{v} \in \mathcal{F}_{1,2}$ with $d(\mathbf{u},\mathbf{v}) = 3$ then $c_{\mathbf{u}} + c_{\mathbf{v}} \ge 4$.
(ii) For every $\mathbf{u} \in \mathcal{F}_{1,2}$ $c_{\mathbf{u}} \ge 2$.

Proof. (i) Put $\mathbf{u} = (000)$, $\mathbf{v} = (111)$. The words in $\mathcal{S}_{\mathbf{u}} \cup \mathcal{S}_{\mathbf{v}}$ has to be covered by $\mathcal{C}$, whence

$$7(c_{\mathbf{u}} + c_{\mathbf{v}}) + \sum_{d(\mathbf{u},\mathbf{w})=1} c_{\mathbf{w}} + \sum_{d(\mathbf{v},\mathbf{w})=1} c_{\mathbf{w}} \geq 2.27.$$

But

$$c_{\mathbf{u}} + c_{\mathbf{v}} + \sum_{d(\mathbf{u},\mathbf{w})=1} c_{\mathbf{w}} + \sum_{d(\mathbf{v},\mathbf{w})=1} c_{\mathbf{w}} \leq |\mathcal{C}|.$$

Thus $c_{\mathbf{u}} + c_{\mathbf{v}} \geq 4$.

(ii) This statement is proved similarly using (i). ◇

It follows from Proposition 2.6 that there exists a $\mathbf{u} \in \mathcal{F}_{1,2}$ such that $c_{\mathbf{u}} = 2$. Without loss of generality $\mathbf{u} = (000)$. Suppose $c_{111} = 2$, or $c_{211} = 2$. Then

$$|\mathcal{C}| \geq c_{000} + c_{111} + \sum_{d(000,\mathbf{w})=1} c_{\mathbf{w}} + \sum_{d(111,\mathbf{w})=1} c_{\mathbf{w}} + c_{210} + c_{201} \geq 34,$$

a contradiction. Therefore, if $d(\mathbf{u},\mathbf{v}) = 3$ then $c_{\mathbf{u}} + c_{\mathbf{v}} \geq 5$. Now we get

$$|\mathcal{C}| \geq c_{000} + c_{111} + \sum_{d(000,\mathbf{w})=1} c_{\mathbf{w}} + (c_{210} + c_{101}) + (c_{201} + c_{110}) + c_{211} + c_{011} \geq 33.$$

Hence, $c_{111} = c_{211} = 3$, $c_{011} = 2$, $c_{110} + c_{201} = c_{101} + c_{201} = 5$, and for every $\mathbf{u} \in \mathcal{F}_{1,2}$ such that $c_{\mathbf{u}} = 2$, $\sum_{d(\mathbf{u},\mathbf{w})=1} c_{\mathbf{w}} = 13$. Two cases have to be considered:

(a) $c_{110} = 2$, $c_{201} = 3$, $c_{101} = 2$, $c_{210} = 3$;
(b) $c_{110} = 2$, $c_{201} = 3$, $c_{101} = 3$, $c_{210} = 2$.

(a) In this case

$$\sum_{d(110,\mathbf{w})=1} c_{\mathbf{w}} = \sum_{d(101,\mathbf{w})=1} c_{\mathbf{w}} = \sum_{d(011,\mathbf{w})=1} c_{\mathbf{w}} = 13.$$

Therefore, $c_{100}+c_{010} = c_{100}+c_{001} = c_{010}+c_{001} = 7$, whence $2(c_{100}+c_{010}+c_{001}) = 21$, a contradiction.

(b) Once again

$$\sum_{d(110,\mathbf{w})=1} c_{\mathbf{w}} = \sum_{d(210,\mathbf{w})=1} c_{\mathbf{w}} = \sum_{d(011,\mathbf{w})=1} c_{\mathbf{w}} = 13.$$

Therefore, $c_{100} + c_{010} = c_{200} + c_{010} = 8, c_{010} + c_{001} = 7$, whence $c_{100} + 3c_{010} + c_{001} + c_{200} = 23$, and $c_{010} = 5$. Furthermore, $c_{001} = 2$, and $\sum_{d(001,\mathbf{w})=1} c_{\mathbf{w}} = 13$, a contradiction, because $c_{000} + c_{101} + c_{201} + c_{011} = 10$. Hence, we have proved the following proposition.

Proposition 2.7 $K(4,2,1) \geq 34$. ◇

3. THE CASE R > 1

In this section we describe some results for the numbers $K(t, b, R)$ for higher values of R.

Proposition 3.1. For every integer $R \geq 1$ $K(1, 2R+1, R) = 6$.

Proof. It is well known that $K(1,3,1) = 6$. Suppose that $K(1, 2R'+1, R') \geq 6$ for every $R' < R$. Let $\{\mathcal{S}_{\mathbf{u}} | \mathbf{u} \in \mathcal{F}_{0,2}\}$ be the regular partition from (1.8), and let $\mathcal{C} \subset \mathcal{F}_{1,2R+1}$ be a code of covering radius R. By the induction hypothesis $|\mathcal{C} \cap \mathcal{S}_{\mathbf{u}}| \geq 1$ for every $\mathbf{u} \in \mathcal{F}_{0,2}$. Suppose $|\mathcal{C}| = 5$. (If $\mathcal{C}$ contains less than five words it can be extended by adding an arbitrary element from $\mathcal{F}_{1,2R+1}$.) Any two binary coordinate positions contain every word from $\mathcal{F}_{0,2}$ at least once. So without loss of generality their binary parts are

$$\begin{array}{cccccc} 0 & 0 & 0 & 0 & \ldots & 0 \\ 0 & 0 & * & * & \ldots & * \\ 0 & 1 & \alpha & * & \ldots & * \\ 1 & 0 & \beta & * & \ldots & * \\ 1 & 1 & \gamma & * & \ldots & * \end{array}$$

It is easily seen that $\alpha = \beta \neq \gamma$, whence $2R+1 \leq 4$. This proves that $K(1, 2R+1, R) \geq 6$.

The code

$$\mathcal{C} = \{(i\underbrace{000\ldots00}_{2R+1}), (i\underbrace{111\ldots11}_{2R+1}) \mid i \in \mathbf{F}_3\} \tag{3.1}$$

has the desired parameters and is of covering radius R, therefore $K(1, 2R+1, R) = 6$.◇

Proposition 3.2. For evrey integer $R \geq 1$ $K(2, 2R-1, R) = 4$.

Proof. One can prove easily by induction that $K(2, 2R-1, R \geq 4$. The code

$$\mathcal{C} = \{(00\underbrace{000\ldots00}_{2R-1}), (00\underbrace{111\ldots11}_{2R-1}), (11\underbrace{000\ldots00}_{2R-1}), (22\underbrace{111\ldots11}_{2R-1})\} \tag{3.2}$$

proves $K(2, 2R-1, R) = 4$. ◇

Proposition 3.3. For every integer $R \geq 1$ $K(2, 2R, R) = 6$.

Proof. The proof of $K(2, 2R, R) \geq 6$ repeats the steps from Proposition 3.1. Examining the binary parts of the codewords we obtain $2R \leq 4$. As $K(2,2,1) = 6$, we have only to prove that $K(2,4,2) \geq 6$. Suppose there exists a code $\mathcal{C} \subset \mathcal{F}_{2,4}$ with $|\mathcal{C}| = 5$ and $R(\mathcal{C}) = 2$. Now up to equivalence the codewords are

$$\begin{array}{cccccc} * & * & 0 & 0 & 0 & 0 \\ * & * & 0 & 0 & 1 & 1 \\ * & * & 0 & 1 & 0 & 1 \\ * & * & 1 & 0 & 0 & 1 \\ * & * & 1 & 1 & 1 & 0 \end{array}$$

Now it can be easily checked that the words from $\mathcal{F}_{2,4}$ of the shape $(**1111)$ cannot be covered by spheres of radius 2.

Let C contain the following words:

$$\begin{array}{cccc} 0 & 0 & 0\ldots0 & 0\ldots0 \\ 0 & 0 & 1\ldots1 & 1\ldots1 \\ 1 & 2 & 0\ldots0 & 1\ldots1 \\ 2 & 1 & 0\ldots0 & 1\ldots1 \\ 1 & 1 & 1\ldots1 & 0\ldots0 \\ 2 & 2 & \underbrace{1\ldots1}_{p} & \underbrace{0\ldots0}_{q} \end{array}$$

where p, q are odd, $p+q=2R$. We are going to prove that $R(C)=R$.

Fix the regular partition $\{\mathcal{S}_{\mathbf{u}} | \mathbf{u} \in \mathcal{F}_{2,0}\}$. The words from $\mathcal{S}_{00}$ are obviously covered. Let $\mathbf{x} \in \mathcal{F}_{2,2R}$ be an uncovered word from one of $\mathcal{S}_{01}$, $\mathcal{S}_{02}$, $\mathcal{S}_{10}$, $\mathcal{S}_{20}$. Because of the symetry of the code we may suppose that

$$\mathbf{x} = (01\,\underbrace{11\ldots1}_{a}\,\underbrace{00\ldots0}_{p-a}\,\underbrace{11\ldots1}_{b}\,\underbrace{00\ldots0}_{p-b}) \tag{3.2}$$

If $a+b \neq R$ then $\mathbf{x}$ is covered by $C \cap \mathcal{S}_{00}$, hence $a+b=R$. The words from $C \cap \mathcal{S}_{21}$ and $C \cap \mathcal{S}_{11}$ do not cover $\mathbf{x}$ therefore

$$\begin{array}{rcl} a+q-b & \geq & R \\ p-a+b & \geq & R \end{array} \tag{3.3}$$

From $b=R-a$ we obtain

$$\begin{array}{rcl} 2a+q-R & \geq & R \\ p-2a+R & \geq & R \end{array} \tag{3.4}$$

i.e. $2a \geq 2R-q=p$ and $2a \leq p$. Thus $2a=p$, which is impossible.

Let once again $\mathbf{x}$ be an uncovered by C which is conttained in one of $\mathcal{S}_{11}$, $\mathcal{S}_{12}$, $\mathcal{S}_{21}$, $\mathcal{S}_{22}$. Now we can assume that up to equivalence

$$\mathbf{x} = (11\,\underbrace{11\ldots1}_{a}\,\underbrace{00\ldots0}_{p-a}\,\underbrace{11\ldots1}_{b}\,\underbrace{00\ldots0}_{p-b}) \tag{3.5}$$

As above, the fact that $\mathbf{x}$ remains uncovered by $C \cap \mathcal{S}_{11}$ and by $C \cap \mathcal{S}_{12}$ yields

$$\begin{array}{rcl} p-a+b & \geq & R+1 \\ a+q-b & \geq & R \end{array} \tag{3.6}$$

whence $p+q \geq 2R+1$, a contradiction. ◇

Proposition 3.4. For every integer $R \geq 1$ $K(3, 2R-2, R) = 5$.

Proof. One can easily prove by induction that $K(3,2R-2,R) \geq 5$. Now we are going to prove that the code $\mathcal{C}$, containing the words

$$(3.7) \qquad \begin{array}{cccccc} 0 & 0 & 0 & 0\ldots0 & 0\ldots0 \\ 1 & 1 & 1 & 1\ldots1 & 1\ldots1 \\ 1 & 1 & 1 & 0\ldots0 & 1\ldots1 \\ 2 & 2 & 2 & 1\ldots1 & 1\ldots1 \\ 2 & 2 & 2 & \underbrace{1\ldots1}_{p} & \underbrace{0\ldots0}_{q} \end{array}$$

where p, q are odd, $p+q = 2R-2$, has a covering radius R.

Suppose there exists a vector $\mathbf{x} \in \mathcal{F}_{3,2R-2}$, uncovered by $\mathcal{C}$. Let

$$(3.8) \qquad \mathbf{x} = (*\,*\,*\,\underbrace{11\ldots1}_{a}\,\underbrace{00\ldots0}_{p-a}\,\underbrace{11\ldots1}_{b}\,\underbrace{00\ldots0}_{p-b})$$

and denote by d_i, $i = 0,1,2$, the distance of the ternary part of $\mathbf{x}$ to the vector (iii). Obviously, $\sum_{i=0}^{2} d_i = 6$ for every $\mathbf{x} \in \mathcal{F}_{3,2R-2}$. As $\mathbf{x}$ is uncovered from each of the five words of the code we can write

$$(3.9) \qquad \begin{array}{lcl} a+b & \geq & R+1-d_0 \\ p+q-a-b & \geq & R+1-d_1 \\ a+q-b & \geq & R+1-d_1 \\ p+q-a-b & \geq & R+1-d_2 \\ p-a+b & \geq & R+1-d_2 \end{array}$$

whence

$$(3.10) \qquad 3(p+q)-(a+b) \geq 5R+5+d_0-2\sum_{i=0}^{2} d_i = 5R-7+d_0.$$

Now $a+b+d_0 \leq R+1$ with equality iff we have "=" in all inequalities of (3.9). On the other hand, $a+b+d_0 \geq R+1$ for otherwise $\mathbf{x}$ would be covered by $(000\underbrace{0\ldots0}_{p}\underbrace{0\ldots0}_{q})$. Therefore, $a+b+d_0 = R+1$. From the last two inequalities of (3.9) we get $p+q-a-b = p-a+b$, i.e. $q = 2b$, a contradiction. ◇

At last we are going to discuss briefly the case $t = 5, b = 0, R = 2$. It is known that $5 \leq K(5,0,2) \leq 8$ [1][3]. Below we prove the following result.

Proposition 3.5. $K(5,0,2) = 8$.

Proof. Suppose that $\mathcal{C} \subset \mathcal{F}_{5,0}$ is a code of covering radius 2 with $|\mathcal{C}| = 7$. Fix the regular partition $\{\mathcal{S}_{\mathbf{u}} | \mathbf{u} \in \mathcal{F}_{2,0}\}$.

Without loss of generality we can put $c_{00} = 0$. It follows from Proposition 1.3 that $c_{01}+c_{10}+c_{02}+c_{20} \geq 4$. But four spheres of radius one in $\mathcal{F}_{3,0}$ leave at least four words uncovered. Hence, for every $\mathbf{u} \in \mathcal{F}_{3,0}$ with $c_{\mathbf{u}} = 0$ we have

$$(3.10) \qquad \sum_{d(\mathbf{u},\mathbf{v})=1} c_{\mathbf{v}} \geq 5.$$

There exist (up to equivalence) two possibilities:
(1) $c_{00} = 0,\ c_{11} = 0,\ c_{22} = 0$;
(2) $c_{00} = 0,\ c_{12} = 0,\ c_{22} = 0$.
In the first case (3.10) yields

$$\begin{array}{ccccccccc} c_{01} & + & c_{02} & + & c_{10} & + & c_{20} & \geq & 5 \\ c_{01} & + & c_{10} & + & c_{12} & + & c_{21} & \geq & 5 \\ c_{02} & + & c_{20} & + & c_{12} & + & c_{21} & \geq & 5 \end{array} \tag{3.11}$$

whence $2\sum_{i\neq j} c_{ij} \geq 15$, a contradiction to $|C| = 7$.
In the second case $c_{11} = c_{21} = 1$, and

$$\begin{array}{ccccccccc} c_{01} & + & c_{02} & + & c_{10} & + & c_{20} & \geq & 5 \\ c_{10} & + & c_{02} & + & c_{11} & + & c_{22} & \geq & 5 \\ c_{02} & + & c_{20} & + & c_{12} & + & c_{21} & \geq & 5 \end{array} \tag{3.12}$$

which has the following solutions

$$\begin{array}{llll} c_{02} = 4, & c_{01} = 1, & c_{10} = 0, & c_{20} = 0; \\ c_{02} = 3, & c_{01} = 0, & c_{10} = 1, & c_{20} = 1. \end{array}$$

The first solution contradicts (3.10) for $\mathbf{u} = (10)$. Similarly to the previous section, let $\rho^{p,q}_{\alpha\beta}$, $\alpha, \beta \in \mathbf{F}_3$, $1 \leq p < q \leq 5$ denote the number of words in C containing α in the p-th, and β in the q-th position. Up to this moment we have proved that for every pair $(p, q), 1 \leq p < q \leq 5$, of coordinate positions
- there exists a pair $(\gamma, \delta), \gamma, \delta \in \mathbf{F}_3$ with $\rho^{p,q}_{\gamma,\delta} = 3$;
- $\rho^{p,q}_{\alpha,\beta} = 0$ for every pair $(\alpha, \beta), \alpha, \beta \in \mathbf{F}_3$ with $d((\alpha, \beta), (\gamma, \delta)) = 1$;
- $\rho^{p,q}_{\alpha,\beta} = 1$ for every pair $(\alpha, \beta), \alpha, \beta \in \mathbf{F}_3$ with $d((\alpha, \beta), (\gamma, \delta)) = 2$.

Now we can assume that up to equivalence C contains the words

$$\begin{array}{ccccc} 0 & 0 & \alpha_1 & * & * \\ 0 & 0 & \alpha_2 & * & * \\ 0 & 0 & \alpha_3 & * & * \end{array} \tag{3.13}$$

Now it is easy to prove that for any choice of $(\alpha_1, \alpha_2, \alpha_3)$ we get a contradiction to the restrictions on the numbers $\rho^{p,q}_{\alpha,\beta}$. For example, if $\alpha_1 \neq \alpha_2 \neq \alpha_3 \neq \alpha_1$ we get a contradiction for $p = 1$, $q = 3$. $\diamond$

Remark. Similarly one can prove that $K(4, 1, 2) = 6$.

Acknowledgements. This research was partially supported by the Bulgarian NSF Contract I-35/1993.

References

[1] H.O.HÄMÄLÄINEN, S.RANKINEN, Upper bounds for football pool problems and mixed covering codes, *J. Combin. Theory Ser.A* **56** (1991), 84-95.

[2] I.HONKALA, Lower bounds for binary covering codes, *IEEE Trans. Inform. theory***34** (1988),326-329.

[3] J.H. VAN LINT, JR., G.J.M. VAN WEE, Generalized bounds on binary/ternary mixed packing and covering codes, *J. Combin. Theory Ser.A* **57**(1991), 130-134.

[4] G.J.M. VAN WEE, Improved sphere bounds on the covering radius of codes, *J. Combin. Theory Ser.A* (1991), 117-129.

The Length Function: A Revised Table

Antoine Lobstein
Centre National de la Recherche Scientifique
Télécom Paris, Dpt INF
46 rue Barrault, 75634 Paris Cedex 13
France

and

Vera Pless *
Mathematics Department
University of Illinois at Chicago
Chicago, IL 60680
USA

Abstract. We give a table with the most current available information for the shortest length of a binary code with codimension m and covering radius r for $2 \le m \le 24$ and $2 \le r \le 12$.

An important parameter of a binary linear code is its covering radius, and it is of much interest to determine parameters of codes which cover the space best. The length function, $l(m,r)$, introduced in [1] is an efficient way to record such data. The first table of $l(m,r)$ for $1 \le r \le 12$ and $1 \le m \le 12$ was given in [1]. An improved table for $1 \le r \le 12$ and $1 \le m \le 24$ was given in [2], see Table 1. Using many new results we update this table to Table 2.

For $1 \le r \le m$, $l(m,r)$ is the smallest length of a binary code C of codimension m and covering radius r.

If C has codimension m, covering radius r and length $l(m,r)$, C is called a *short* code.

We give some properties, described in [1,2], of the length function. The first three and seventh inequalities are easy to establish.

1) $m \le l(m,r)$ with equality if and only if $r = m$ (the zero code).
2) $l(m,r) \le 2^m - 1$ with equality if and only if $r = 1$ (the Hamming codes).
3) $m < l(m,r) < 2^m - 1$ for $1 < r < m$.
4) $l(m,r+1) \le l(m,r)$.
5) $l(m,r) < l(m+1,r)$.
6) $l(m+1,r+1) \le l(m,r) + 1$.

* This work was supported in part by NSA Grant MDA 904-91-H-0003.

7) (The sphere-covering bound)

$$\min \{ n : \sum_{0 \le i \le r} \binom{n}{i} \ge 2^m \} \le l(m,r).$$

8) $l(m,r) = m+1$ for $\lceil \frac{m}{2} \rceil \le r < m$.
9) $l(2s+1,s) = 2s+5$ for $1 \le s$.
10) $l(2s,s-1) = 2s+6$ for $4 \le s$.
11) $l(2s+1,s-1) = 2s+7$ for $6 \le s$.

The following bound was conjectured in [1] and proven in [9].
12) $2^m + 1 \le l(2m-1,2)$ for $3 \le m$.

Equalities 8 through 11 give the parameters for all the short codes above the jagged line in Table 1. These values are clearly the same in Table 2. The fact that $l(11,3) = 23$ is due to the existence of the perfect Golay [23,12] code. It was first shown in [5] that $l(6,2) = 13$.

Most of the values in the table do not represent short codes but give intervals for $l(m,r)$. Inequality 7 always gives a lower bound for such an interval although most of the lower bounds in Table 2 are larger than that given by this inequality. An upper bound always represents a construction; some of these are ADS sums [5] of other codes in the table. Naturally one has to be careful to take ADS sums only of codes which are known to be normal. Inequality 4 says that the values in a row of a table of $l(m,r)$ are non-increasing. Inequality 5 says that values are strictly increasing going down columns of such a table. This gives many lower bounds. Similarly inequality 6 gives bounds in a table of $l(m,r)$. Equality 12 improves the sphere-covering bound for $l(m,2)$ with m odd.

Many improvements in Tables 1 and 2 are given by showing that an $[l(m), l(m) - m]$ code with covering radius r cannot exist. See [1, 2, 9, 10] and the references listed therein for the various arguments used. In particular, in [12] it is shown that $l(7,2)$ cannot be 18 so that by Table 1, $l(7,2) = 19$.

Table of lower and upper bounds for $l(m,r)$, $2 \le r \le 12$, $2 \le m \le 24$, from: *On the length of codes with a given covering radius*, by R.A. Brualdi and V.S. Pless - with two corrections: $l(16,5) \le 31$ (instead of 30) and $l(23,7) \le 46$ (instead of 45).

m/r	2	3	4	5	6	7	8	9	10	11	12
2	2										
3	4	3									
4	5	5	4								
5	9	6	6	5							
6	13	7	7	7	6						
7	18-19	11	8	8	8	7					
8	24-26	14	9	9	9	9	8				
9	33-41	16-18	13	10	10	10	10	9			
10	46-58	19-22	16	11	11	11	11	11	10		
11	65-90	23	17-20	15	12	12	12	12	12	11	
12	91-119	30-38	19-24	18	13	13	13	13	13	13	12
13	128-182	37-55	23-25	19	17	14	14	14	14	14	14
14	181-246	47-71	27-29	20-26	20	15	15	15	15	15	15
15	256-374	59-88	31-37	23-27	21	19	16	16	16	16	16
16	362-495	74-120	36-51	26-31	22-25	22	17	17	17	17	17
17	512-750	93-149	43-66	30-35	24-29	23	21	18	18	18	18
18	724-1006	117-181	51-81	34-41	27-33	24-29	24	19	19	19	19
19	1024-1518	147-244	61-98	38-48	29-36	25-31	25	23	20	20	20
20	1448-2015	185-308	72-115	43-63	33-40	27-31	26-31	26	21	21	21
21	2048-2976	233-372	85-147	49-80	36-44	30-38	27-32	27	25	22	22
22	2896-4062	294-500	101-179	56-112	40-45	33-42	29-33	28-31	28	23	23
23	4096-6048	370-621	120-208	65-127	45-60	35-46	30-37	29-34	29	27	24
24	5793-8127	466-749	143-237	74-156	50-75	39-47	34-41	30-35	30-35	30	25

Table 1.

Explanation of Table 2, modified table of lower and upper bounds for $l(m,r)$, $2 \le r \le 12$, $2 \le m \le 24$.
All upper bounds can be achieved by a normal code. All DS and ADS sums have been computed.
Key to the Table (we mention the earliest reference): unmarked = same as in Table 1.

lower bounds: a = [12] b = [11] c = [6] d = [7] e = [13] f = [14] g = [8] h = [9] i = [10].
upper bounds: q = ADS of $[15,11]r = 1$, with $[39,30]r = 2$ or $[59,39]r = 5$ codes r = [5] s = [4] t = [3].

m/r	2	3	4	5	6	7	8	9	10	11	12
2	2										
3	4	3									
4	5	5	4								
5	9	6	6	5							
6	13	7	7	7	6						
7	a 19	11	8	8	8	7					
8	h 25-26	14	9	9	9	9	8				
9	h 34-39 t	h 17-18	13	10	10	10	10	9			
10	d 47-53 t	h 21-22	16	11	11	11	11	11	10		
11	65-79 t	23	17-20	15	12	12	12	12	12	11	
12	i 92-107 t	c 31-38	19-24	18	13	13	13	13	13	13	12
13	h 129-159 t	e 38-53 q	23-25	19	17	14	14	14	14	14	14
14	h 182-215 t	47-63 t	27-29	c 21-26	20	15	15	15	15	15	15
15	h 257-319 t	d 60-75 t	i 32-37	23-27	21	19	16	16	16	16	16
16	h 363-431 t	i 75-95 t	e 37-49 s	i 27-31	22-25	22	17	17	17	17	17
17	h 513-639 t	93-126 t	d 44-62 t	30-35	24-29	23	21	18	18	18	18
18	h 725-863 t	117-153 t	d 53-77 t	34-41	27-33	24-27 r	24	19	19	19	19
19	h 1025-1279 t	i 148-205 t	i 62-84 t	i 39-48	f 30-36	25-30 r	25	23	20	20	20
20	h 1449-1727 t	i 187-255 t	i 73-93 t	e 44-59 s	33-40	g 28-31	26-29 r	26	21	21	21
21	h 2049-2559 t	i 235-308 t	i 86-125 t	i 51-75 t	e 37-44	30-38	27-32	27	25	22	22
22	h 2897-3455 t	i 295-383 t	i 103-150 t	e 57-88 t	d 41-45	33-42	29-33	28-31	28	23	23
23	h 4097-5119 t	i 371-511 t	i 122-174 t	65-98 t	i 46-60	i 37-46	b 31-37	29-34	29	27	24
24	h 5794-6911 t	i 467-618 t	i 144-190 t	i 76-107 t	e 51-73 q	d 40-47	34-41	30-35	30-33 r	30	25

Table 2.

References

1. R.A. Brualdi, V.S. Pless, R.M. Wilson: Short codes with a given covering radius. IEEE IT-35, 99-109 (1989)
2. R.A. Brualdi, V.S. Pless: On the length of codes with a given covering radius. In: Coding Theory and Design Theory. Part I: Coding Theory. Springer-Verlag 1990, pp. 9-15
3. A.A. Davydov, A.Y. Drozhzhina-Labinskaya: Constructions, families and tables of binary linear covering codes. IEEE IT, to appear
4. R. Dougherty, H. Janwa: Covering radius computations for binary cyclic codes. Mathematics of Computation 57, 415-434 (1991)
5. R.L. Graham, N.J.A. Sloane: On the covering radius of codes. IEEE IT-31, 385-401 (1985)
6. X.D. Hou: Covering radius and error correcting codes. Ph. D. thesis, University of Illinois, Chicago 1990, 77 p.
7. X.D. Hou: New lower bounds for covering codes. IEEE IT-36, 895-899 (1990)
8. D. Li, W. Chen: New lower bounds for binary covering codes. IEEE IT, to appear
9. R. Struik: On the structure of linear codes with covering radius two and three. IEEE IT, submitted
10. R. Struik: An improvement of the van Wee bound for linear covering codes. IEEE IT, submitted
11. G.J.M. van Wee: Improved sphere bounds on the covering radius of codes. IEEE IT-34, 237-245 (1988)
12. O. Ytrehus: Binary [18,11]2 codes do not exist - nor do [64,53]2 codes. IEEE IT-37, 349-351 (1991)
13. Z. Zhang, C. Lo: Lower bounds on $t[n,k]$ from linear inequalities. IEEE IT-38, 194-197 (1992)
14. Z. Zhang, C. Lo: Linear inequalities for covering codes: Part II -- triple covering inequalities. IEEE IT-38, 1648-1662 (1992)

On the covering radius of convolutional codes

Irina. E. Bocharova and Boris. D. Kudryashov

St.-Petersburg Academy of Airspace Instrumentation,
Bolshaia Morskaia str.,67, St.-Petersburg,190000, Russia,
e-mail:liap@sovam.com

Abstract. We consider a problem of calculating covering capabilities for convolutional codes. An upper bound on covering radius for convolutional code is obtained by random coding arguments. The estimates on covering radius for some codes with small constraint length are presented.

1 Introduction

The covering radius is the smallest integer such that spheres of this radius centered on codewords cover the space. A survey of results and detailed tables for covering radii of block codes are presented in [1].

Convolutional codes with the Viterbi encoding algorithm [2] may also be used as covering codes. In this paper we study covering capabilities for convolutional codes.

Our paper is organized as follows. In part 2 we introduce some definitions and notations concerning covering convolutional codes. In part 3 an asymptotical behavior of these codes is discussed. We derive a nonconstructive upper bound on the covering radius of convolutional codes analogous to the asymptotic bound for linear block codes [3]. The estimates for covering radius of some convolutional codes are presented in part 4. These estimates found by computer search coincide with the results obtained in [4] which was unknown to the authors when they prepared present work. In part 4 we also investigate block codes obtained from the short convolutional codes. These codes have the same parameters as the best covering codes from [1] but some of them have smaller encoding complexity than codes from [1] due to the fact that they are derived from convolutional codes of small constraint length.

2 Definitions

We consider binary convolutional codes. Our terminology follows [2]. Any convolutional code may be determined by the semi-infinite generating matrix which may be represented in different forms. The first form is

$$G_L = \left(\begin{array}{cccccccccccccc}
G_\nu^{(0)} & 0 & \dots & 0 & 0 & 0 \dots 0 & 0 & \dots & 0 & \dots & 0 & \dots \\
G_{\nu-1}^{(1)} & G_\nu^{(1)} & \dots & 0 & 0 & 0 \dots 0 & 0 & \dots & 0 & \dots & 0 & \dots \\
\dots & \dots & \dots & \dots & \dots & \dots\dots\dots & \dots & \dots & \dots & \dots & \dots & \dots \\
G_0^{(\nu)} & G_1^{(\nu)} & \dots & G_\nu^{(\nu)} & 0 & 0 \dots 0 & 0 & \dots & 0 & \dots & 0 & \dots \\
0 & G_0^{(\nu+1)} & \dots & G_{\nu-1}^{(\nu+1)} & G_\nu^{(\nu+1)} & 0 \dots 0 & 0 & \dots & 0 & \dots & 0 & \dots \\
\dots & \dots & \dots & \dots & \dots & \dots\dots\dots & \dots & \dots & \dots & \dots & \dots & \dots \\
0 & 0 & \dots & 0 & 0 & 0 \dots 0 & 0 & \dots & G_0^{(i)} & \dots & G_\nu^{(i)} & \dots \\
\dots & \dots & \dots & \dots & \dots & \dots\dots\dots & \dots & \dots & \dots & \dots & \dots & \dots
\end{array} \right), \tag{1}$$

where $G_i^{(j)}$ are matrices of size $k_0 \times n_0$ with elements from $GF(2)$, 0 is zero matrix of the same size.

The other form of the generating matrix is the following

$$G_U = \left(\begin{array}{cccccccccccccc}
G_0^{(\nu)} & G_1^{(\nu)} & \dots & G_\nu^{(\nu)} & 0 & 0 \dots 0 & 0 & \dots & 0 & \dots & 0 & \dots \\
0 & G_0^{(\nu+1)} & \dots & G_{\nu-1}^{(\nu+1)} & G_\nu^{(\nu+1)} & 0 \dots 0 & 0 & \dots & 0 & \dots & 0 & \dots \\
\dots & \dots & \dots & \dots & \dots & \dots\dots\dots & \dots & \dots & \dots & \dots & \dots & \dots \\
0 & 0 & \dots & 0 & 0 & 0 \dots 0 & 0 & \dots & G_0^{(i)} & \dots & G_\nu^{(i)} & \dots \\
\dots & \dots & \dots & \dots & \dots & \dots\dots\dots & \dots & \dots & \dots & \dots & \dots & \dots
\end{array} \right). \tag{2}$$

Matrix G_L corresponds to the nonzero initial encoder state (the first $k_0\nu$ symbols are in the encoder register at the beginning of the encoding) and matrix G_U corresponds to the initial zero encoder state. Two sequences of block codes obtained from the convolutional code by truncating correspond to the above two representations of this code. The block codes are determined by the matrices

$$G_L^{(l)} = \left(\begin{array}{ccccccccc}
G_\nu^{(0)} & 0 & \dots & 0 & 0 & 0 \dots 0 & 0 & \dots & 0 \\
G_{\nu-1}^{(1)} & G_\nu^{(1)} & \dots & 0 & 0 & 0 \dots 0 & 0 & \dots & 0 \\
\dots & \dots & \dots & \dots & \dots & \dots\dots\dots & \dots & & \dots \\
G_0^{(\nu)} & G_1^{(\nu)} & \dots & G_\nu^{(\nu)} & 0 & 0 \dots 0 & 0 & \dots & 0 \\
0 & G_0^{(\nu+1)} & \dots & G_{\nu-1}^{(\nu+1)} & G_\nu^{(\nu+1)} & 0 \dots 0 & 0 & \dots & 0 \\
\dots & \dots & \dots & \dots & \dots & \dots\dots\dots & \dots & & \dots \\
0 & 0 & \dots & 0 & 0 & 0 \dots 0 & 0 & \dots & G_0^{(l+\nu-1)}
\end{array} \right), \tag{3}$$

$$G_U^{(l)} = \left(\begin{array}{ccccccccc}
G_0^{(\nu)} & G_1^{(\nu)} & \dots & G_\nu^{(\nu)} & 0 & 0 \dots 0 & 0 & \dots & 0 \\
0 & G_0^{(\nu+1)} & \dots & G_{\nu-1}^{(\nu+1)} & G_\nu^{(\nu+1)} & 0 \dots 0 & 0 & \dots & 0 \\
\dots & \dots & \dots & \dots & \dots & \dots\dots\dots & \dots & & \dots \\
0 & 0 & \dots & 0 & 0 & 0 \dots 0 & 0 & \dots & G_0^{(l+\nu-1)}
\end{array} \right). \tag{4}$$

These matrices are of size $k_0(\nu+l) \times n_0 l$ and $k_0 l \times n_0 l$ respectively. Let denote as $A_L^{(l)}$ and $A_U^{(l)}$ the code sequences determined by the matrices $G_L^{(l)}$ and $G_U^{(l)}$ and as $R_L^{(l)}$ and $R_U^{(l)}$ the corresponding covering radii. Consider a time-invariant convolutional code for which $G_t^{(l)}$ do not depend on $t, i = 0, \ldots, \nu$. It follows from the representation of the code $A_U^{(l)}$ in the form (4) that

$$R_U^{(m)} + R_U^{(l)} - \nu n_0 \leq R_U^{(m+l)} \leq R_U^{(m)} + R_U^{(l)}. \tag{5}$$

By inequality (5) it is easy to show that there exists the

$$\lim_{l \to \infty} R_U^{(l)}/(l n_0) = \rho, \tag{6}$$

which we call the normalized covering radius of the convolutional code. Moreover it follows from (5) that

$$\rho \leq R_U^{(l)}/(l n_0), l = 1, 2, \ldots. \tag{7}$$

Consider the sequence $R_L^{(l)}, l = 1, 2, \ldots$. Since $A_U^{(l)} \subset A_L^{(l)}, l = 1, 2, \ldots$. then $R_U^{(l)} \geq R_L^{(l)}, l = 1, 2, \ldots$. The codewords of $A_L^{(l))}$ differ from the codewords of $A_U^{(l))}$ at most in the νn_0 first symbols. Hence

$$R_U^{(l)} - \nu n_0 \leq R_L^{(l)} \leq R_U^{(l)}. \tag{8}$$

¿From (6) and (8) we have

$$\lim_{l \to \infty} R_L^{(l)}/(l n_0) = \rho,$$

It follows from (4) that

$$R_L^{(m)} + R_L^{(l)} \leq R_L^{(m+l)} \leq R_L^{(m)} + R_L^{(l)} + \nu n_0. \tag{9}$$

and

$$\rho \geq R_L^{(l)}/(l n_0), l = 1, 2, \ldots. \tag{10}$$

Formulas (7) and (10) determine the computational procedure for calculating the covering radius of the convolutional code. One can obtain the covering radius ρ calculating covering radii for the code sequences $A_L^{(l)}$ and $A_U^{(l)}$ respectively.

Note that the constraint length characterizes the decoding complexity for the convolutional code. The size of the decoder memory and the number of computations per code symbol is proportional to $2^{\nu k_0}$ and they do not depend on l. To guarantee the decoding complexity $2^{\nu k_0}$ one should use the Viterbi algorithm [2].

3 Asymptotic performance for convolutional codes

It is known [3] that there exists (n,k)-code with the covering radius R such that

$$k/n \le 1 - \mathcal{H}(R/n) + 2(\log_2 n)/n, \tag{11}$$

where $\mathcal{H}(x) = -x\log_2 x - (1-x)\log_2(1-x)$. The decoding complexity for this code is proportional to $\exp_2\{\min\{k, n-k\}\}$ [5]. In this part we present an asymptotic bound on the covering radius for convolutional codes. This bound asymptotically coincides with bound (11) for block codes but convolutional codes are preferable in the sense of the decoding complexity.

Theorem 1 *For ν sufficiently large there exists the convolutional code with parameters n_0 , k_0 , ν such that*

$$k_0/n_0 \le 1 - \mathcal{H}(\rho - o_1(\nu)) + o_2(\nu), \tag{12}$$

where $o_i(\nu) = C_i\nu e^{-\delta_i\nu}$, and C_i, $\delta_i \ge 0$, $i = 1, 2, \ldots$ denote constants on ν.

Proof. Consider code $A_{lk_0} = A_L^{(l)}$ determined by the matrix $G_L^{(l)}$. The elements of matrices $G_j^{(i)}$ are assumed to be independent equally distributed binary random numbers. Basically we follow the terminology and the approach of [3].

Let $S_R(a)$ be a sphere with radius R centered at point a. Note that for some codewords $a \in A_{lk_0}$ some code symbols are equal to zero with the probability one. These code symbols correspond to the zero subblocks of rows from matrix $G_L^{(l)}$ and their linear combinations. Introduce function

$$\varphi(a) = \begin{cases} 0 \text{ if there are some zero subblocks in } \mathbf{a}, \\ 1 \text{ otherwise.} \end{cases}$$

Let $\{g_i\}$, $i = 1, \ldots, lk_0$ be the random set of codewords $g_i = m_i G_L^{(l)}$, where m_i are random sequences of length $k_0 l$ equally distributed on $\{0,1\}^{k_0 l}$.

Consider the following sequence of linear codes embedded in the code A_{lk_0}

$$A_0 = \{(0, \ldots, 0)\}, A_j = A_{j-1} \cup (A_{j-1} + g_j),$$

where $A + g = \{a + g, a \in A\}$. Denote $S_R(A) = \cup_{a \in A} S_R(a)$.

Let Q_j be the probability of the event $x \in S_R^c(A_j)$, where x denotes a random sequence of length $n_0 l$ equally distributed over $\{0,1\}^{n_0 l}$. The inequality $Q_j < 2^{-n_0 l}$ means that there exists the code with covering radius at most R. We have

$$Q_j = \Pr(x \in S_R^c(A_j)) = \Pr[(x \in S_R^c(A_{j-1})) \cap (x \in S_R^c(A_{j-1} + g_j))]$$

$$\le \Pr(x \in S_R^c(A_{j-1})) \Pr(x \in S_R^c(A_{j-1} + g_j))^{\varphi(g_j)}$$

The inequality is evident for $\varphi(g_j) = 0$ and it results from the independent events $x \in S_R^c(A_{j-1})$ and $x \in S_R^c(A_{j-1} + g_j)$ for $\varphi(g_j) = 1$ since in this case

g_j represents a random sequence of $n_0 l$ independent symbols equally distributed on $\{0,1\}$. Continue partitioning we get

$$Q_j = \prod_{u=(u_1,\dots,u_j)} [\Pr(x \in S_R^c(\sum_{i=1}^{j} u_i g_i))]^{\prod_{i=1}^{j}(\varphi(g))^{u_i}}.$$

where u_i ,$i = 1,\dots,j$ denote coefficients of the expansion $a = \sum_{i=1}^{j} u_i g_i$. Here we put $\varphi^0(\cdot) = 1$ for any $\varphi(\cdot)$.

The probabilities $\Pr(x \in S_R^c(A_j))$ are equal for all vectors a and may be determined as

$$\Pr(x \in S_R^c(A_j)) = 1 - 2^{-n_0 l} \sum_{w=0}^{R} \binom{l n_0}{w} \triangleq P.$$

Hence we obtain

$$Q_{lk_0} \le \exp_P \{\sum_u \prod_{i=1}^{lk_0} \varphi(g_i)^{u_i}\} = \exp_P \{\prod_{i=1}^{lk_0} \sum_{u_i} \varphi(g_i)^{u_i}\} =$$

$$\exp_P \{\prod_{i=1}^{lk_0} (1 + \varphi(g_i))\}. \tag{13}$$

Let show that there exists such set $\{m_1,\dots,m_{lk_0}\}$ for that $\prod_{i=1}^{lk_0}(1+\varphi(g_i))$ is so large that $Q_{lk_0} \le 2^{-ln_0}$. Averaging the power over the set of vectors m_i we obtain

$$E[\prod_{i=1}^{lk_0}(1+\varphi(g_i))] = (1 + E[\varphi(g)])^{lk_0}.$$

The value $E[\varphi(g)]$ represents the fraction of vectors g for those $\varphi(g) = 1$. Consider the matrix $G_L^{(l)}$. It is easy to see that vectors g can contain zero trivial subblocks if and only if the corresponding vector m contains series of νk_0 zero symbols. So to estimate $E[\varphi(g)]$ it is necessary to count up the number N_{lk_0} of sequences m containing not more than νk_0 zeros one after the other. It is easy to obtain the following estimate

$$N_{lk_0} \ge (2^{k_0} - \frac{2^{k_0}-1}{2^{\nu k_0}}(\frac{\nu+1}{\nu})^{\nu})^{l}.$$

After simplification we have

$$E[\varphi(g)] \ge (1 - e2^{-\nu k_0})^{l}.$$

Hence there exists a set $\{m_1,\dots,m_{lk_0}\}$ such that

$$\prod_{i=1}^{lk_0}(1+\varphi(g_i)) \ge (1 + (1 - e2^{-\nu k_0})^{l})^{lk_0}. \tag{14}$$

Combining (13) and (14) with estimates for $\sum \binom{n}{m}$ ([5], p.302) we get that for this set $\{m_1, \ldots, m_{lk_0}\}$

$$Q_{lk_0} \leq \left(1 - 2^{-ln_0[1-\mathcal{H}(R/(ln_0))]} / \sqrt{2ln_0}\right)^{[1+(1-e2^{-\nu k_0})^l]^{lk_0}}. \tag{15}$$

Let find the minimal code rate k_0/n_0 which permit the inequality $Q_{lk_0} \leq 2^{-ln_0}$. Using (15) obtain the condition

$$k_0/n_0 \geq (1 - \mathcal{H}(R/(ln_0)) + (\log(ln_0))/(ln_0)) / \log(1 + (1 - e2^{-\nu k_0})^l). \tag{16}$$

Using (7) and (8) we have

$$\rho \leq (R + \nu n_0)/(ln_0) \tag{17}$$

Let $l = e^{\delta\nu}$, $\delta \in (0, k_0 \ln 2)$. Substituting (17) into (16) after simple transformation we get the establishment of the Theorem 1.

4 Examples

Examples for the good covering convolutional codes with rate 1/2 are given in Table 1.

Table 1. Covering radii for some convolutional codes.

ν	$G_0, G_1, \ldots, G_\nu$	ρ
1	[11],[10]	1/4
2	[11],[01],[11]	$1/6 \leq \rho \leq 7/38$
3	[11],[11],[01],[11]	$1/6 \leq \rho \leq 7/34$
4	[11],[01],[01],[10],[11]	$6/38 \leq \rho \leq 1/5$
5	[11],[10],[01],[11],[01],[11]	$6/38 \leq \rho \leq 1/5$
6	[11],[01],[11],[11],[00],[10],[11]	$6/38 \leq \rho \leq 1/5$

Note, that the code with $\nu = 2$ was considered in [4]. It has been established that for this code $\rho = 1/6$.

Some good covering block codes can be obtained by truncating of convolutional codes and deleting the coordinates corresponding to the zero or to the identical columns of the check matrix. Examples are presented in Table 2. Codes of rate less or equal to 1/2 are obtained using generating matrix (3), other codes are obtained using generating matrix (4).

Codes in table 2 are optimal in the sense of having the smallest covering radii of any linear code of that length and dimension. It is easy to see that the encoding complexity for convolutional codes (2^ν) is significantly less than Wolf's estimate [5] for the encoding complexity for corresponding block codes ($\min\{2^k, 2^{(n-k)}\}$).

Table 2. Block codes obtained by truncating of convolutional codes.

n	k	R	ν	$G_0, G_1, \ldots, G_\nu$	n	k	R	ν	$G_0, G_1, \ldots, G_\nu$
8	3	3	2	[01],[11],[11]	17	8	4	2	[11],[01],[11]
8	4	2	1	[01],[11]	17	10	3	3	[11],[11],[01],[11]
9	4	3	1	[01],[11]	18	9	4	2	[01],[11],[11]
10	4	3	2	[01],[11],[11]	18	10	3	2	[11],[01],[11]
10	5	2	2	[01],[11],[11]	19	9	4	2	[11],[01],[11]
11	5	3	1	[01],[11]	19	11	3	3	[11],[11],[01],[11]
12	6	3	1	[01],[11]	20	11	3	2	[11],[01],[11]
12	7	2	2	[11],[01],[11]	20	12	3	4	[11],[11],[10],[00],[11]
13	6	3	2	[11],[01],[11]	21	13	3	4	[11],[01],[10],[00],[11]
14	5	4	4	[10],[00],[11],[11],[11]	22	10	5	3	[11],[11],[01],[11]
14	7	3	2	[01],[11],[11]	22	13	3	4	[11],[11],[00],[10],[11]
14	8	2	2	[11],[01],[11]	22	14	3	4	[11],[01],[00],[10],[11]
15	9	2	3	[11],[11],[01],[11]	23	11	5	2	[11],[01],[11]
16	7	4	4	[11],[10],[01],[00],[11]	26	14	4	2	[11],[01],[11]
16	9	3	2	[11],[01],[11]	28	16	4	4	[11],[11],[10],[00],[11]

References

[1] R.L. Graham, N.J. Sloane, "On the covering radius of codes," *IEEE Trans. Inform. Theory*, vol. IT-31, pp.385-401, 1985.

[2] G.D. Forney, "Convolutional codes, II Maximum likelihood decoding," *Inform. and Control*, vol.25, pp.222-266, 1974.

[3] G.D. Cohen, "A nonconstructive upper bound on covering radius'" *IEEE Trans. Inform. Theory*, vol. IT-29, pp. 352-353, 1983.

[4] A.R. Calderbank and P.C. Fishbum, A. Rabinovich, "Covering properties of convolutional codes and associated lattices", submitted to *IEEE Trans. Inform. Theory*, May 1992.

[5] J.K. Wolf, "Efficient maximum likelihood decoding of linear block codes using a trellis,"*IEEE Trans. Inform. Theory*, vol. IT-24, pp.76-80, 1978.

[5] F.J. MakWilliams and N.J.A. Sloane, *The theory of error-correcting codes*. New York: North-Holland, 1977.

Efficient multi-signature schemes for cooperating entities

Olivier Delos [1] and Jean-Jacques Quisquater [2]

[1] Dept of Computer Sc. (INFO)
University of Louvain
Place Sainte-Barbe, 2
B-1348 Louvain-la-Neuve Belgium
delos@info.ucl.ac.be

[2] Dept of Elect. Eng. (DICE)
University of Louvain
Place du Levant, 3
B-1348 Louvain-la-Neuve Belgium
jjq@dice.ucl.ac.be

Abstract

Sharing signature power may be required in many occasions. Moreover a multisigning operation may be required to be performed simultaneously (in some sense) by all the involved cosigning parties.

We describe a complete cooperation-based signature scheme achieving such requirements. It is based on mental games and in particular on the Guillou-Quisquater zero-knowledge scheme. In our scheme, the cosigners interact with an intermediate entity, the *combiner*, to produce a multisignature. Only one interactive exchange is required. The scheme is practical and secure. A cheating combiner can only prevent the operation from happening but it will be immediately detected. No impersonation, substitution, or coalition attacks are possible.

1 Introduction

In this paper we focus on the concept of cooperation-based signature schemes, that is, signature schemes in which several signers interact with a verifier using zero-knowledge tools [Sh85, GMRa89, BGKW88]. The idea of shared generation of signatures already appears in [DF92] but is limited to RSA schemes.

In many occasions the power to sign may be desired to be shared (*e.g.* multi-signing cheques is a policy of many companies). Furthermore it may be required that the multisigning operation is performed *simultaneously* by all the involved cosigning parties. For example in a legal contract, the lawyer's role may be double: that of a cosigner and a verifier. Such requirements are achieved with our cooperation-based signature scheme.

Our scheme is based on mental games [SRA81] and uses the Guillou-Quisquater zero-knowledge scheme [GQ89a]. The original contribution of this paper is to describe and prove a complete scheme.

An interesting feature of our scheme is the role of our intermediate entity, the *combiner*, which processes information received from the signers. It is no need to trust the combiner. In our model no interaction between the cosigners is necessary, but the cosigners must interact with the combiner. Only one interactive exchange is requested, so the scheme is very practical. [1]

Clearly this setting is more efficient than having each signer creating and sending his own signature. There is no increase in bandwidth overhead, less calculations for the verifier and a smaller key directory to manage. A cheating combiner can only prevent the operation from happening but it will be immediately detected.

We first describe the Guillou-Quisquater proof and signature scheme [GQ89b] and then present our scheme. The scheme is easily extended to a setting involving several verifiers, and to a setting with each cosigner having several identities each of them signing a different message or authenticating themselves to the verifier(s). A real life application of a many signers–many verifiers signature scheme is again the notarial contract in which each cosigner may choose his own lawyer, so many verifiers are involved in the operation.

2 Cooperation-based scheme including a GQ "appendix" signature scheme

2.1 The GQ zero-knowledge proof

Multiple-input zero-knowledge proofs were introduced in the context of cooperation [GQ89b, dWQ90, GUQ91]. We describe a "multiple-input" variant of the Guillou-Quisquater proof [GQ89a]. This scheme uses a public composite modulus $n = p \cdot q$, where p and q are appropriate secret primes. It is a zero-knowledge proof of knowledge of vth residues, where v is an appropriate public exponent. Its security is based on the RSA assumption if the exponent v is coprime to $\phi(n)$ [Bu93].

Each user has an identity I : the shadowed identity J is the identity I with some added redundancy [GQ88]. This considerably enhances the intractability of an identity fraud. The authority center computes the secrets as follows: a redundancy function Red is applied to the original identity I (256 bits) such that the left and right parts of the result (512 bits) match a particular pattern. The Red identity J is then used to extract a secret D using the signing function S of the authority [GQ89a]. For a 512 bits long modulus and an identity I, the authority thus computes :

$$I \text{ (256 bits)}$$
$$\text{— } J = Red(I) \quad \text{(512 bits)}$$
$$\text{— } D = S(Red(I)), \text{ where } D^v \cdot J \equiv 1 \pmod{n}$$

To prove knowledge of vth residues of J_i in Z_n, for $i = 1, 2, \ldots, k$, we use the following protocol where P and V denote respectively the prover and the verifier :

---

[1] The verifier may play the role of the combiner.

GQ multiple-input zero-knowledge proof:

Input: $(J_1, J_2, \ldots, J_k; v, n)$.

$P \rightarrow V$: $T = r^v \mod n$, where $r \in_R Z_n$.[2]

$V \rightarrow P$: $e_1, e_2, \ldots, e_k$, where $e_i \in_R Z_v$ for $i = 1, 2, \ldots, k$.

$P \rightarrow V$: $t = r \cdot D_1^{e_1} \cdot D_2^{e_2} \cdots D_k^{e_k} \mod n$, where $D_i^v \cdot J_i \equiv 1 \pmod{n}$.

Verification: $T \stackrel{?}{\equiv} t^v \cdot J_1^{e_1} \cdot J_2^{e_2} \cdots J_k^{e_k} \pmod{n}$ (and $T \neq 0$).

2.2 Combining single input GQ proofs

Let P_1 and P_2 be two independent provers with

- secret key D_1 and D_2, respectively, such that

$$D_1^{v_1} \cdot J_1 \equiv 1 \pmod{n} \quad and \quad D_2^{v_2} \cdot J_2 \equiv 1 \pmod{n}.$$

- public key $(J_1; v_1, n)$ and $(J_2; v_2, n)$, respectively.

GQ single-input zero-knowledge proofs:

Input: $(J_1; v_1, n)$.

$P_1 \rightarrow V : T_1 = r_1^{v_1} \mod n$,
$V \rightarrow P_1 : e_1 \in Z_{v_1}$,
$P_1 \rightarrow V : t_1 = r_1 \cdot D_1^{e_1} \mod n$.

Verification: $T = t^{v_1} \cdot J_1^{e_1} \mod n$.

Input: $(J_2; v_2, n)$.

$P_2 \rightarrow V : T_2 = r_2^{v_2} \mod n$,
$V \rightarrow P_2 : e_2 \in Z_{v_2}$,
$P_2 \rightarrow V : t_2 = r_2 \cdot D_2^{e_2} \mod n$.

Verification: $T = t^{v_2} \cdot J_2^{e_2} \mod n$.

The corresponding multiple-input proof is :

GQ multiple-input proof:

Input: $(J_1^{v_2} \mod n, J_2^{v_1} \mod n; v_1 \cdot v_2, n)$.

$(P_1, P_2) \rightarrow V$: $T = T_1^{v_2} \cdot T_2^{v_1} \mod n$.

$V \rightarrow (P_1, P_2)$: $(e_1, e_2) \in (Z_{v_1} \times Z_{v_2})$,

$(P_1, P_2) \rightarrow V$: $t = t_1 \cdot t_2 \mod n$.

[2] $a \in_R A$ means that the element a is selected randomly from the set A with uniform distribution.

Verification: $T \stackrel{?}{=} t^{v_1 v_2} \cdot (J_1^{v_2})^{e_1} \cdot (J_2^{v_1})^{e_2} \mod n$

Clearly,

$$\begin{aligned} t^{v_1 v_2} \cdot (J_1^{v_2})^{e_1} \cdot (J_2^{v_1})^{e_2} &\equiv (t_1 \cdot t_2)^{v_1 v_2} \cdot (J_1^{v_2})^{e_1} \cdot (J_2^{v_1})^{e_2} \\ &\equiv (r_1 r_2)^{v_1 v_2} \cdot D_1^{v_1 v_2 e_1} \cdot D_2^{v_1 v_2 e_2} \cdot (J_1^{v_2})^{e_1} \cdot (J_2^{v_1})^{e_2} \\ &\equiv (r_1 r_2)^{v_1 v_2} (D_1^{v_1} \cdot J_1)^{e_1 v_2} \cdot (D_2^{v_2} \cdot J_2)^{e_2 v_1} \\ &\equiv (r_1 r_2)^{v_1 v_2} \equiv T_1^{v_2} \cdot T_2^{v_1} \equiv T \pmod n \end{aligned}$$

The probability of error per round is bounded by : $1/\min\{v_1, v_2\}$.

2.3 The GQ signature scheme

The Guillou-Quiquater signature scheme [GQ89b] is based on the GQ zero-knowledge proof. It uses an appropriate hash function. [3] The message M and a test value T are hashed to obtain an element $d \in Z_v$ which corresponds to the question of the GQ proof.

The prime factors v_i of the exponent v must be sufficiently large to prevent a legitimate signer to collide two different inputs to the same hashing value, *i.e.* two messages with the same signature. We shall require the least prime factor v_1 of v to be at least super-polynomial in $|n|$, *e.g.* $v_1 \geq n^c$ ($c > 0$ a small constant). The GQ signature scheme can be described as follows :

GQ signature scheme:

Input: $(I^*; v, n)$.

Step 1 The signer selects $r \in_R Z_n$, computes $T = r^v \mod n$, $d = h(T, M)$ and $t = r \cdot D^d \mod n$.

Then he sends the verifier the signature of the message M, sgn= $(M \mid I^* \mid T \mid t)$ which is the concatenation of M together with I^*, T and t. [4]

Step 2 The verifier obtains J from I^* ($J = Red(I^*)$) and then computes $t^v \cdot J^d \mod n$, where $d = h(T, M)$. He checks that $d = h(t^v \cdot J^d \mod n, M)$ (and that $\gcd(t, n) = 1$).

---

[3] See discussion at the end of this section.

[4] In this paper, $a \mid b$ means that a is concatenated with b.

Schematically,

Signer		Verifier
$I^* =$ identity		Computes : $J = Red(I^*)$
$M =$ message	$\xrightarrow{\text{sgn}}$	$d = h(T, M)$
		$t^v \cdot J^d \mod n$
Computes : $r \in_R Z_n$		
$T = r^v \mod n$		
$d = h(T, M)$		Checks :
$t = r \cdot D^d \mod n$		$d \stackrel{?}{=} h(t^v \cdot J^d \mod n, M)$

We must append to the signer's identity I a certain flag which restricts the use of I to either only authentication schemes, or only signature schemes, or "open multi-signature" schemes. Otherwise any zero-knowledge authentication scheme can be transformed by the verifier in a signature scheme without the knowledge of the prover. The prover is not aware of this misuse. The verifier by simply asking a question $d = h(T, M)$ for the message M, can construct a signature for M, as illustrated below.

A misuse of a Zero-knowledge authentication

$P \rightarrow V : T$

$V \rightarrow P : d = h(T, M)$

$P \rightarrow V : t$

Signature : $M \,|\, I \,|\, T \,|\, t$

Solution to prevent the abuse: $I \;\Rightarrow\; I^* \equiv I +$ flag

We assume that the hashing function h is suitable [Da88]. That is, h must be:

- a one-way function in both arguments,
- collision resistant (often called collision free), and
- uniformly distributed.

Guillou and Quisquater proposed as hash function in [GQ89b]:

$$d = h(T, M) = J^M \cdot T^{v^k} \mod n$$

where $M \leq v^k$. The set of the messages is supposed to be large enough. Finally, as pointed out earlier, v must be large enough. For example, at least 128 bits (to face the attacks related to the birthdays' paradox [FO89]) if n is 512 bits long.

2.4 Extending the GQ scheme to a cooperation based multi-signature scheme

Let us consider two tamper-resistant devices [5] (Smart Cards), each one storing its unique secret number D_1 and D_2, respectively, related to its own identity, I_1^* and I_2^*, by the following equations :

$$D_1^{v_1} \cdot J_1 \equiv 1 \pmod{n} \qquad D_2^{v_2} \cdot J_2 \equiv 1 \pmod{n}$$

where $J_i = Red(I_i^*)$ for $i = 1, 2$ and v_1, v_2 are different primes.

During the cooperation-based multisigning "operation" the intermediate combiner (typically a Personal Computer) simulates an entity having identity $(J_1^{v_2} \bmod n, J_2^{v_1} \bmod n)$. [6]

Both cosigning entities will interact with the combiner to prove to the verifier their knowledge of their secret number D_1 and D_2, respectively. It is no need to trust the combiner whose the only wicked activity would be to prevent the operation from happening. But this would be immediately detected.

2.4.1 Initialisation

During a prior but non-cryptographic protocol, the cosigning entities agree on a certain message M to sign, which is concatenated with the identities I_1^* and I_2^* (*i.e.* the classical identity concatenated with a flag **and** with the corresponding public exponent) of the cosigners. This message to be signed by the two cosigners is denoted by $M|I_1^*|I_2^*$. The initialisation step of the scheme is achieved as follows :

The entities wishing to take part in a cooperation-based multisignature scheme send their identities to the Center. Giving this information, the Center computes the (global) exponent $v = v_1 \cdot v_2$ of the simulated entity and computes the unique D_i secret numbers of each signing entity, as $D_1 = 1/J_1^{1/v_1} \bmod n$ and $D_2 = 1/J_2^{1/v_2} \bmod n$ respectively. The Center gives each D_i to the corresponding entity.

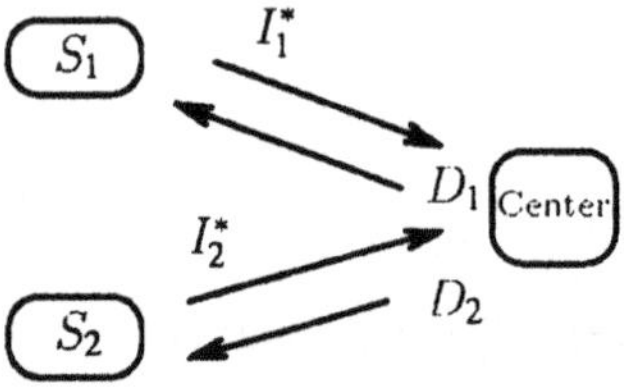

$v = v_1 \cdot v_2$, $J_1 = Red(I_1^*)$, $J_2 = Red(I_2^*)$,
$D_1^{v_1} \cdot J_1 \equiv 1 \pmod{n}$, $D_2^{v_2} \cdot J_2 \equiv 1 \pmod{n}$

The secret numbers D_1 and D_2 may be used for any signing or multi-signing operation (depending on the flag) and are not restricted to a particular instance.

2.4.2 The scheme

After the signers agree on the message to be signed, they interact with the combiner to perform the multisignature.

---

[5] This is necessary to protect the secret keys, but also to prevent coalition frauds between signers [DQ87, GUQ91].

[6] If $v_1 = v_2$, it is not necessary to raise J_1, J_2 to any exponent.

Each signing entity proposes an initial test and sends it to the combiner. Then the combiner computes the global test as the product of all these initial tests and sends it to each cosigner. Each signer then hashes the global test with the message to be signed in an integer global challenge. Each signer extracts its question from this global challenge. He then replies with an individual answer computed using its question and its secret. The combiner collects all these answers and combines them in a global answer. Finally the combiner sends to the verifier the signature of the message which is the message concatenated with all the identities and a signature appendix consisting of the global test and the global answer.

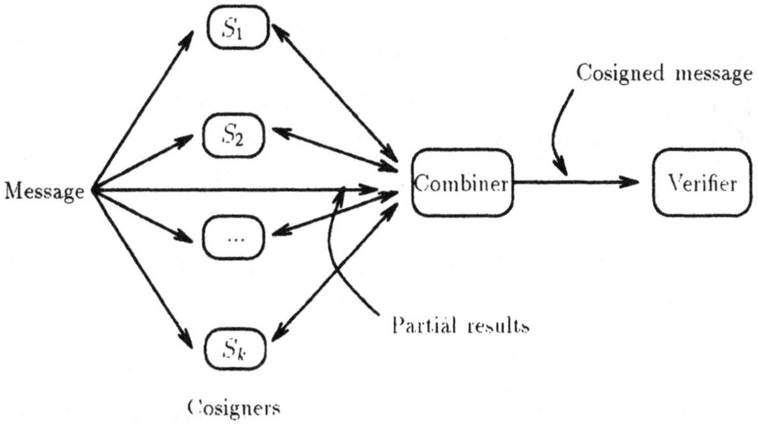

The verification is performed in the usual way : The verifier computes the public exponents; the global final check is computed as the product mod n of the vth power of the global answer and the public inputs raised to the appropriate "questions". Finally the global final challenge is obtained as the hashing of the global final check and the message to be signed. The signature is accepted if and only if initial and final global challenges are equal.

2.4.3 The complete protocol

The complete cooperation based multi-signature protocol consists of the following steps :

Cooperation based multi-signature protocol :

Inputs: $x = (J_1^{v_2} \bmod n, J_2^{v_1} \bmod n;\ v_1 \cdot v_2\,;\ n)$, $M|I_1^*|I_2^*$ = the message to be signed.

The verifier checks that v_1 and v_2 are primes (Rabin [R80]), and that the identities I_1^* and $I_2^* \in Z_n^*$.[7] If any one of these checks fails then the verifier halts and rejects.

[7] Z_n^* is the set of "invertible" elements of Z_n. Burmester and Desmedt have shown that if this is not checked then the proof is unsound [BD89].

Step 1 The cosigners S_1 and S_2 select respectively r_1 and r_2 randomly in Z_n and compute their test T_1 and T_2 respectively as $T_i = r_i^v \mod n$, where $v = v_1 \cdot v_2$ and $i = 1, 2$. They send then T_1 and T_2 respectively (or at least a part of them) to the combiner (with their identity I_i^*).

Step 2 The combiner computes the common test T as follows:

$$T = \prod_{i=1}^{k} T_i^{\nu_i} \mod n$$
$$= T_1^{v_2} \cdot T_2^{v_1} \mod n$$

where $\nu_i = v/v_i$ and sends it to each cosigner. [8]

Step 3 Each cosigner hashes the common test T with the message $M|I_1^*|I_2^*$ in the integer d, the global "question". With this d being such that $d = d_1 \mid d_2$ where $|d_1| = |v_1|$ and $|d_2| = |v_2|$. The cosigner S_1 replies with $t_1 = r_1 D_1^{d_1} \mod n$ and S_2 with $t_2 = r_2 D_2^{d_2} \mod n$.

$$
\boxed{S_i} \quad
\begin{aligned}
d &= h(T, M|I_1^*|I_2^*) \\
d &= d_1|d_2 \text{ where } |d_i| = |v_i| \\
t_i &= r_i D_i^{d_i} \mod n
\end{aligned}
\quad \xrightarrow{\ t_i\ } \quad
\boxed{\text{Combiner}} \quad t = t_1 \cdot t_2 \mod n
$$

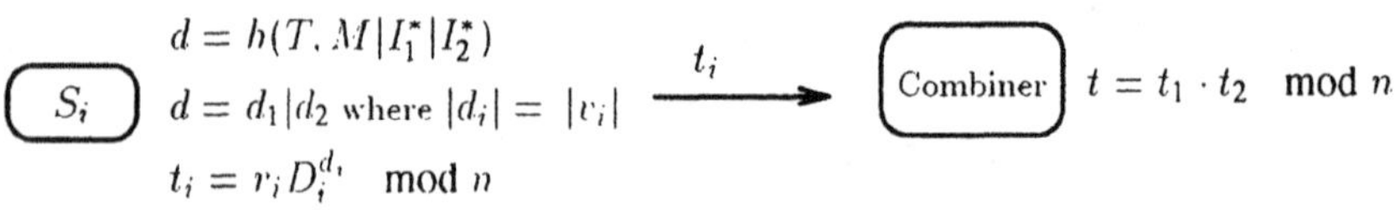

Step 4 The combiner computes the global answer t as the product in Z_n of the individual answers t_1 and t_2.

Step 5 The combiner sends the signed message $M|I_1^*|I_2^*|T|t$ to the verifier.

Verification: The verifier computes each shadowed identity $J_1 = Red(I_1^*)$, $J_2 = Red(I_2^*)$ and the global public exponent $v = v_1 \cdot v_2$. He performs the global question d and the individual question d_1 and d_2 as the signers did. He then checks that: $d = h(t^v \cdot (J_1^{v_2})^{d_1} \cdot (J_2^{v_1})^{d_2} \mod n, M|I_1^*|I_2^*)$ and that $gcd(t, n) = 1$.

This cooperation-based signature scheme may easily be extended to any number of cooperating signers in a set of k.

2.4.4 Security

Note that it is a redundancy in I_1^* and I_2^* which assures a part of the security of the scheme. It prevents fake signatures as in the original scheme.

The security of this system relies on the security of the zero-knowledge proof on which it is based (see Section 2.1 and 2.2). We assume that the hash function used is appropriate.

To prevent the combiner to recycle (or replay) old cosigned messages, we could need a time synchronization. The cosigners and the combiner have to agree with a common stamp related to a certain date and hour to include in the message M. The

---

[8] Observe that the order of the factors of T is not important.

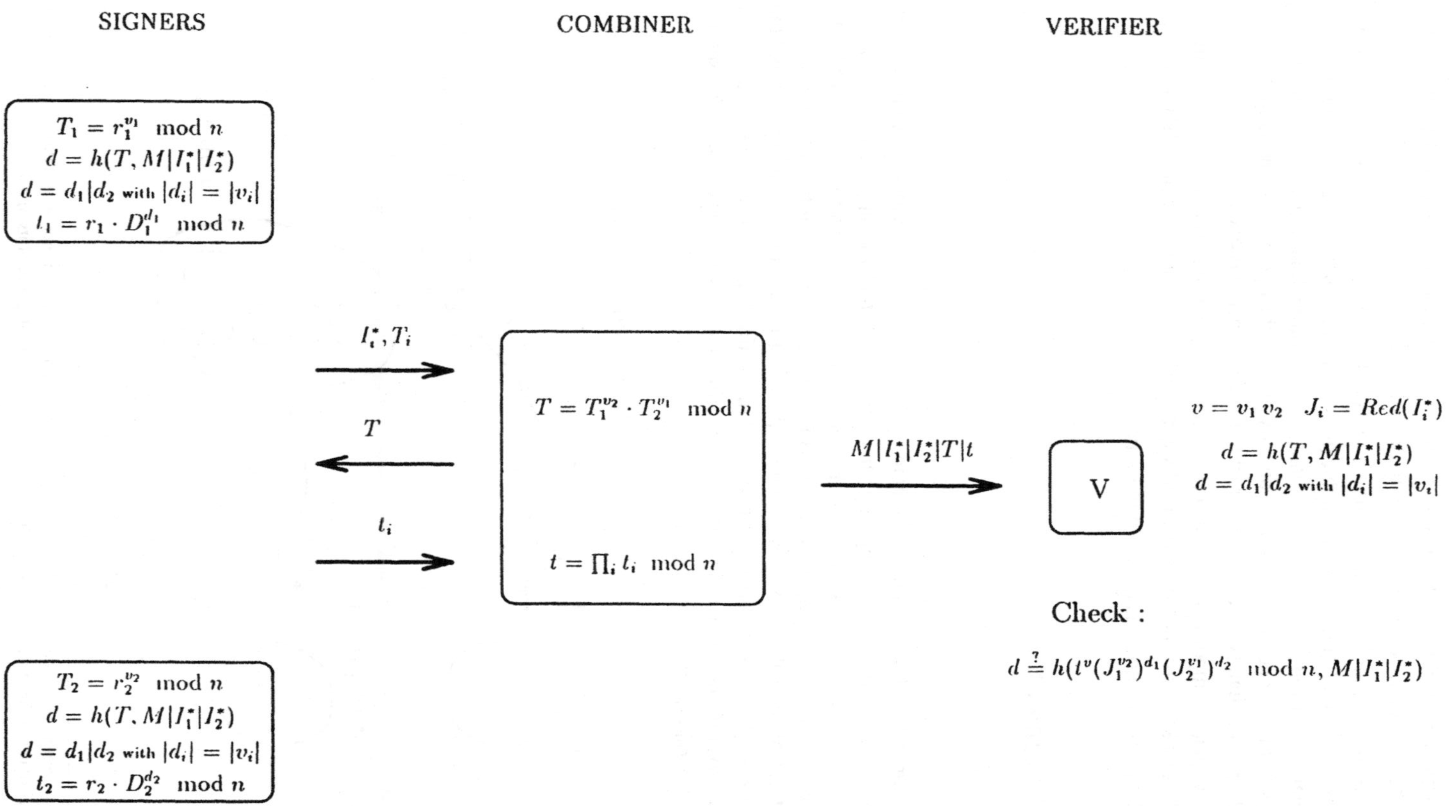

Figure 1 : The complete protocol.

cosigners may also decide to enclose in the r_i a redundancy related to the time which could be verified (opened) by a judge in case of conflict.

A likelier and more elegant solution is to include in the message M all the cosigners' identities. A prior but non-cryptographic protocol insures each cosigner to know all the cosigners' identities.[9] This prevents, thanks to the hashing function, to add some signers in the scheme. A cheating combiner can only prevent the operation from happening but it will be immediately detected. No trust in the combiner is needed.

Note that an important consequence of "flagged identities" is to prevent any "in the middle attack" performed by a crooked signer against an honest verifier [BBDGQ91].

The global protocol is depicted in Figure 1.

3 Extensions

3.1

The cooperation-based signature scheme presented in the previous section may be extended to a setting involving several verifiers. A real-life application of such a scheme could be the following: some account opening of great importance, huge loans, big contracts involving many banks and companies, *etc*, may require the simultaneous signatures of several individuals. From the provers to the main verifier, throughout the network, some extra "verifiers" may come "on the line" (gateways, intermediaries, ...) in order to stop "non authentic" messages. For facility the serial aspect of such a setting may be parallelized as depicted in Figure 2.

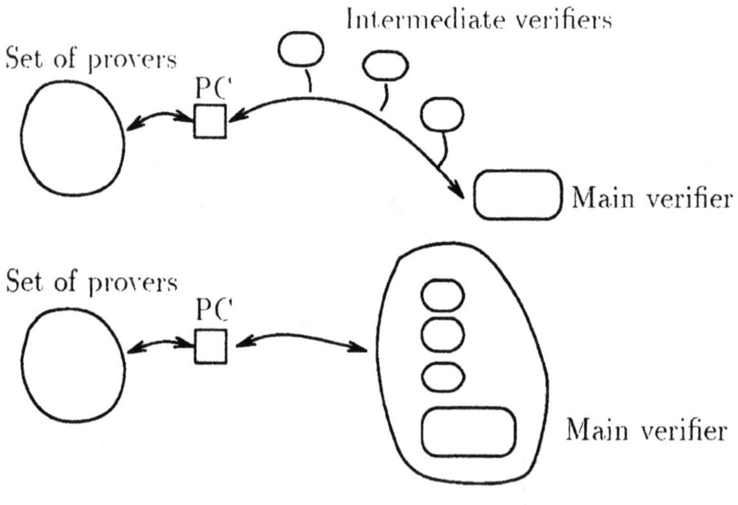

Figure 2: Application

[9] We remind the reader that each identity signed by the authority consists not only of the classical identity of the cosigner in the system but also of the "open multisignature" flag and of its public exponent as well.

3.2 Inside a cooperation based protocol

We may insert our cooperation based signature scheme in a more generalized protocol. In this setting many provers (including groups of cosigners working as entities) which may be provided with several identities are negociating authentication transactions with many verifiers.

4 Conclusion

In this paper we focused on the concept of cooperation-based signature schemes, that is, signature schemes in which several signers interact with a combiner using zero-knowledge tools. We described complete schemes which allow quasi-simultaneous multisigning operations and are very efficient. These are based on zero-knowledge proofs and are secure.

We have real cooperation-based signature scheme if the identity of each cosigner is included in the message to be signed. The size of the identities must be large enough. The goal of cooperation is then reached.

5 Acknowledgements

The authors would like to thank Mike Burmester and Adi Shamir for discussions and comments about this work.

References

[BBDGQ91] S. Bengio, G. Brassard, Y. Desmedt, C. Goutier and J.-J. Quisquater. Secure Implementation of Identification Systems. *Journal of Cryptology* (1991) 4, pp. 175–183.

[BGKW88] M. Ben-Or, S. Goldwasser, J. Killian and A. Wigderson. Multi-prover interactive proofs: How to remove intractability assumptions. In *Proceedings of the twentieth annual ACM Symp. Theory of Computing, STOC*, pp. 113–131, May 2–4,1988.

[BD89] M. V. D. Burmester and Y. G. Desmedt. Remarks on Soundness of Proofs. *Electronic letters*, 26th October 1989, Vol. 25, N° 22, pp. 1509–1510.

[Bu93] M. V. D. Burmester. *To appear.*

[Da88] I. B. Damgård. Collision-Free Hash Functions and Public-Key Signature schemes. *Advances in cryptology, Proceedings of EUROCRYPT '87, Lecture Notes in Computer Science*, N° 304, pp. 203–216, Springer-Verlag, 1988.

[DF92] Y. Desmedt and Y. Frankel. Shared Generation of Authenticators and Signatures. *Advances in cryptology, Proceedings of CRYPTO '91, Lecture Notes in Computer Science*, N° 576, pp. 457–469, Springer-Verlag, 1992.

[DQ87] Y. Desmedt and J.-J. Quisquater. Public-key systems based on the difficulty of tampering. *Advances in cryptology. Proceedings of CRYPTO '86, Lecture Notes in Computer Science*, N° 263, pp. 186–194, Springer-Verlag, 1987.

[dWQ90] D. de Waleffe and J.-J. Quisquater. Better login protocols for computer networks. *Proceedings of ESORICS '90*, pp. 163–172, October 1990.

[FO89] Ph. Flajolet and A. M. Odlyzko. Random Mapping Statistics. *Advances in cryptology, Proceedings of EUROCRYPT '89, Lecture Notes in Computer Science*, N° 434, pp. 329–354, Springer-Verlag.

[GMRa89] S. Goldwasser, S. Micali and C. Rackoff. The Knowledge Complexity of Interactive Proof Systems. *Siam J. Comput.*, 1989, Vol. 18, N° 1, pp. 186–208.

[GQ88] L. C. Guillou and J.-J. Quisquater. Efficient digital public-key signatures with shadow. *Advances in cryptology, Proceedings of CRYPTO '87, Lecture Notes in Computer Science*, N° 304, p. 223, Springer-Verlag, 1988.

[GQ89a] L. C. Guillou and J.-J. Quisquater. A practical zero-knowledge protocol fitted to security microprocessor minimizing both transmission and memory. In C. G. Günther, editor, *Advances in Cryptology, Proceedings of EUROCRYPT '88, Lecture Notes in Computer Science*, N° 330, pp. 123–128, Springer-Verlag, May 1988. Davos, Switzerland.

[GQ89b] L. C. Guillou and J.-J. Quisquater. A "paradoxical" identity-based signature scheme resulting from zero-knowledge. In *Advances in cryptology, Proceedings of CRYPTO '88*, N° 403, pp. 216–231, Springer-Verlag, 1989.

[GUQ91] L. C. Guillou, M. Ugon and J.-J. Quisquater. The Smart Card: A standardized Security Device Dedicated to Public Cryptology. Contemporary Cryptology: The Science Information Integrity, edited by G. J. Simmons, IEEE Press, 1991.

[R80] M.O. Rabin. Probabilistic algorithms for testing primality. *J. Number theory*, Vol. 12, pp. 128–138, 1980.

[Sh85] A. Shamir. Identity-based cryptosystems and signatures schemes. *Advances in cryptology, Proceedings of CRYPTO '84, Lecture Notes in Computer Science*, N° 196, pp. 47–53, Springer-Verlag, 1985.

[SRA81] A. Shamir, R. Rivest and L. Adleman. Mental Poker. *The Mathematical Gardner*, edited by D. A. Klarner, Wadsworth International, 1981.

Montgomery-Suitable Cryptosystems

David NACCACHE and David M'RAÏHI

Gemplus Card International, Crypto-Team
1 Place de Navarre, F-95200, Sarcelles, France.
Email : {100142.3240 and 100145.2261}@compuserve.com

Abstract. Montgomery's algorithm [8], hereafter denoted $Mn(\cdot,\cdot)$, is a process for computing $Mn(A,B) = ABN \bmod n$ where N is a constant factor depending only on n.

Usually, $AB \bmod n$ is obtained by $M_n(M_n(A,B), N^{-2} \bmod n)$ but in this article, we introduce an alternative approach consisting in pre-integrating N into cryptographic keys so that a single $Mn(\cdot,\cdot)$ will replace directly each modular multiplication.

Except the advantage of halving the number of Montgomery multiplications, our strategy skips the pre-calculation (and the storage) of the constant $N^{-2} \bmod n$ and reveals to be particularly efficient when a hardware device implementing $Mn(\cdot,\cdot)$ is the basic computational tool at one's command.

1. Introduction

Since the discovery of public-key cryptography, considerable endeavours were invested in the design of efficient modular reduction algorithms. Reducing the time and memory complexities of the operation $a \bmod n$ is a challenging problem on which relies the practical feasibility of most signature, identification and encryption schemes.

In 1987, Montgomery [8] introduced a very elegant $O(\log^2(n))$-per-constant-hardware algorithm for computing $A\,B\,2^{-l(n)} \bmod n$ where $l(n)$ denotes the length of n in bits.
As this scheme outputs $A\,B\,2^{-l(n)}$ mod n and not $A\,B$ mod n, a correction of the result is done by a re-multiplication of the output by the pre-calculated constant $K = 4^{l(n)} \bmod n$ (in other words, standard modular reduction is reconstituted by $AB \bmod n = M_n(K, M_n(A,B))$ where $M_n(A,B)$ denotes the Montgomery operator).

In itself, the algorithm $M_n(A,B)$ is very simple :

```
input  : an odd n, A = Σ_{i=0}^{l(n)-1} 2^i A[i], B

output : A B 2^{-l(n)} mod n

R ← A

for i← 0 to l(n)-1

        if A[i]=1 then R ← R+B

        if R is odd then R ← R+n

        R ← R/2

if R<n then return(R) else return(R-n)
```

A deeper explanation of the method and its variants can be found in [1], [4], [6], [8] and [14].

Replacing the modular multiplications in the 'square & multiply' exponentiation algorithm by $M_n(A,B)$s we get a new operator $E_n(A,p)$ wherein *p-1* parasite factors are accumulated with the final result (the accumulation of *p-1* factors, and not *p*, is a a consequence of the fact that for multiplying *p* numbers one must perform *p-1* multiplications) :

```
input  : an odd n, A, p>0
output : A^p 2^(-l(n)(p-1)) mod n

e ← A

first ← #True

while(p≠0)

    if p is odd then

        if first then

            z←e

            first ← #False

        else z←Mn(z,e)

    p ← ⌊p/2⌋

    e←Mn(e ,e)

return(z)
```

In the next sections, we will show how to use $E_n(A,p)$ and $M_n(A,B)$ in the implementation of a variety of cryptosystems without getting rid of the *N*s nor pre-calculating the constant *K*. This approach is new since instead of designing computational tools for the comfortable execution of protocols we modify the cryptosystems to meet the computational particularities of a modular reduction tool. Technically, this is achieved by pre-integrating the *N*s directly into the keys.

[1] shows that the hardware, memory and time complexities of $M_n(A,B)$ are equivalent to that of the operation *AB* (without modular reduction). From this remark, it appears that a transformation of cryptosystems wherein modular multiplications are replaced by M_ns is equivalent in practice to performing these schemes when multiplications are done in $\mathbf{Z}$ instead of $\mathbf{Z}_n$.

Hereafter, we will successively apply the protocol re-writing strategy to the RSA [12], Diffie-Hellman [3], Fiat-Shamir [7], Guillou-Quisquater [11], Schnorr [13], El-Gamal [5], OSS (Naccache's repaired version) [9], DSS [10] and Benaloh-de Mare's one-way accumulators [2].

In this paper, n is an RSA modulus, $N = 2^{-l(n)} \bmod n$, p and q are primes, $P = 2^{-l(p)} \bmod p$ and $Q = 2^{-l(q)} \bmod q$.

2. An RSA-like scheme

Let *m* and *c* represent a message and its ciphertext and denote by $\{e, d\}$ a couple of RSA [12] keys.
The sender encrypts m by computing $c = E_n(m,e) = m^e N^{e-1} \bmod n$ and the receiver decrypts c by calculating :

$$E_n(c,d) \equiv c^d N^{d-1} \equiv (m^e N^{e-1})^d N^{d-1} \equiv m^{ed} N^{ed-1} \equiv m \bmod n$$

The security of the modified scheme, formally equivalent to that of the original RSA, is based on the fact that the encryption operation is a successive application of :

① A public permutation of the message space (multiplication by the public factor N)
② Encryption with a standard RSA
③ And a public reverse-permutation of the ciphertext space (division by N).

The decryption consists in applying successively the inverse transformations.

3. A Diffie-Hellman-like scheme

Alice and Bob share the same prime p and a common exponentiation base a [3].

Alice computes $r_1 = E_p(a,x) \equiv a^x p^{x-1} \bmod p$

Bob computes $r_2 = E_p(a,y) \equiv a^y p^{y-1} \bmod p$

Alice and Bob swap r_1 and r_2

Alice computes $k = E_p(r_2,x) \equiv (a^y p^{y-1})^x p^{x-1} = a^{xy} p^{xy-1} \bmod p$

Bob computes $k = E_p(r_1,y) \equiv (a^x p^{x-1})^y p^{y-1} = a^{xy} p^{xy-1} \bmod p$

Again, the security of this Montgomery variant is based on the observation that the scheme is equivalent to a standard Diffie-Hellman with a new exponentiation base $\alpha = a\ P$ where both parties obtain the common key $\alpha^{xy} \bmod p$ and then perform a public permutation of the key space (division by the public constant P).

4. A Fiat-Shamir-like protocol

Redefine the Fiat-Shamir keys [7] by : $v_j\ s_j^2\ N^3 \equiv 1 \bmod n$

① The prover sends $z = M_n(r,r) \equiv r^2 N \bmod n$ to the verifier
② The verifier sends a random binary word e of size k (let $\hbar(e)$ be the Hamming weight of e).
③ The prover sends $y = M_n(s_{i_1}, M_n(s_{i_2}, \ldots M_n(s_{i_{\hbar(e)}}, r)\ldots)$ where the i_js denote the $\hbar(e)$ indices selected by vector e.

④ The verifier compares z and $M_n(v_{i_1}, M_n(v_{i_2}, \ldots M_n(v_{i_{\hbar(e)}}, M_n(y,y))\ldots)$

Note that this can be applied to the digital signature as well.

As in the original Fiat-Shamir, one can show that breaking this protocol is equivalent to the extraction of modular roots and thus to factoring. For simplicity, we will only discuss the case $k=1$ since generalisation is straightforward and only involves heavier notations :

Cheating implies the existence of an efficient algorithm $\mathsf{A}(v,n) = \{Z, y_0, y_1\}$ where :

$$y_0^2 N \bmod n \equiv Z \quad \text{and} \quad y_1^2 NvN \equiv Z \bmod n.$$

From these two equations, we get directly : $\dfrac{y_0}{y_1} = \sqrt{vN} \bmod n$.

To transform A to a root-extracting algorithm, firstly obtain $\sqrt{N} \bmod n$ from $\mathsf{A}(1, n)$ (this is needed only if $l(n)$ is odd. If not, compute directly $\sqrt{N} = 2^{-\bar{n}/2} \bmod n$) and then calculate $\sqrt{v} = \dfrac{y_0}{y_1 \sqrt{N}} \bmod n$ from $\mathsf{A}(v, n)$.

5. A Naccache-Ong-Schnorr-Shamir-like signature scheme

The relation between the secret keys and the public keys is $v_j\, s_j^2\, N^3 \equiv 1 \bmod n$.

① The prover sends $x = r + M_n(m, \frac{1}{r} \bmod n) \equiv r + \frac{mN}{r} \bmod n$ and the verifier replies with e.

② The prover sends $y = M_n(s_{i_1}, M_n(s_{i_2}, \ldots M_n(s_{i_{\hbar(e)}}, r - M_n(m, \frac{1}{r} \bmod n))\ldots)$.

③ The verifier computes $z = M_n(x,x) - M_n(v_{i_1}, M_n(v_{i_2}, \ldots M_n(v_{i_{\hbar(e)}}, M_n(y,y))\ldots)$ and compares if : $z = M_n(4,m)$.

The proof of security is similar to that of the section 4 (see [9] as well).

6. A Guillou-Quisquater-like protocol

The new relation between J, B and v [11] is : $f(ID)\, B^v\, N^{v+1} \equiv 1 \bmod n$, modular multiplications are replaced by *Ms* and exponentiations by *Es*.

① The prover picks r and sends *ID* and $T = E(r,v)$ to the verifier.
② The verifier picks $d<v$, computes $J = f(ID)$ and replies with d.
③ The prover sends $U = M_n(E_n(B,d),r)$
④ The verifier checks that $T = M_n(E_n(J,d), E_n(U,v))$

The zero-knowledge property of the scheme is preserved since the external view of the protocol can be simulated by a probabilistic Turing machine that picks d and U from a random tape, computes $T = J^d U^v N^{d+v-1}$ and outputs the triple $\{T, d, U\}$ which is indistinguishable from a communication between the prover and the verifier.

7. A Schnorr-like Protocol

Schnorr's protocol [13], presented in Crypto'89, is a DLP-based system where the relation between the public (v) and the secret (s) keys are : $v = \alpha^{-s} \bmod p$ and $\alpha^q \equiv 1 \bmod p$.

Redefine : $v\, p^{s+1} \alpha^s \equiv 1 \bmod p$ and give to the signer a new secret-key $\sigma = sQ^{-1} \bmod q$

① The prover picks r and sends $x = E_p(\alpha, r)$
② The verifier picks e and sends it to the prover.
③ The prover sends back $y = r + M_q(\sigma, e) \bmod q = r + se \bmod q$
④ The verifier tests that $x = M_p(E_p(\alpha,y), E_p(v,e))$

Security is guaranteed by a zero-knowledge simulator that picks at random y and e, computes $x = \alpha^y v^e p^{y+e-1} \bmod p$ and outputs the indistinguishable triple $\{x, e, y\}$.

8. An El-Gamal-like signature scheme

In El-Gamal's scheme [5] the public key is $p = g^s \bmod q$ (where s is the secret key), q and g are system parameters and m is signed by $\{u, v\}$ where : $u = g^k \bmod q$, k is random and v is such that $m = s\,u + k\,v \bmod (q\text{-}1)$. Upon reception, the signature is checked by comparing $p^u u^v$ to g^m (modulo q).

To make El-Gamal's scheme Montgomery-suitable, redefine $p = E_q(g,s)$ but let v remain unchanged.

For signing m, compute $u = E_q(g,k)$ and for checking the signature, make sure that :

$$M_q(E_q(p,u),E_q(u,v)) = E_q(g,m).$$

It is still unclear if this scheme is strictly equivalent to the original El-Gamal in terms of security.

9. A Data Signature Standard-like algorithm

The Digital Signature Algorithm [10] was proposed by the US National Institute of Standards and Technology to *provide an appropriate core for applications requiring a digital rather than written signature.*

DSA parameters are :

① A prime modulus p where $2^{L-1} < p < 2^L$ for $512 \leq L \leq 1024$ and $L = 64\ \alpha$ for some α.
② A prime q such that $2^{159} < q < 2^{160}$ and p-1 is a multiple of q.
③ A number g, of order q modulo p such that $g = h^{\frac{p-1}{q}} \bmod p$, where h is any integer such that $1 < h < p$-1 and $g = h^{\frac{p-1}{q}} \bmod p > 1$.
④ A number x, generated randomly and a number y defined by the relation : $y = g^x \bmod p$.
⑤ A number k generated randomly such that $0 < k < q$.

The integers p, q and g are system parameters and can be public or common to a group of users. The signer's private and public keys are respectively x and y. Parameters x and k are used for signature generation only and must be kept secret. Parameter k must be regenerated for each signature.

In order to sign a message m (hashed value of a primitive file M), the signer computes the signature $\{r, s\}$ by :

$$r = \left(g^k \bmod p\right) \bmod q \quad \text{and} \quad s = \frac{m + xr}{k} \bmod q$$

After making sure that $r \neq 0$ and $s \neq 0$, the signature $\{r, s\}$ is sent to the verifier who computes :

① $w = \frac{1}{s} \bmod q$
② $u_1 = m\,w \bmod q$
③ $u_2 = r\,w \bmod q$
④ $v = \left(g^{u_1} y^{u_2} \bmod p\right) \bmod q$
⑤ And checks if v and r match to accept or reject the signature.

Redefine $r = E_p(g, M_q(x,1)) = g^{Qx} p^{Qx-1} \bmod p$

The signer computes :

$$r = M_q(E_p(g,k),1) = (g^k p^{k-1} \bmod p)\, Q \bmod q \quad \text{and} \quad s = M_q(\tfrac{1}{k}, m + M_q(x,r)) = \frac{mQ + xrQ^2}{k} \bmod q$$

and sends $\{r, s\}$ to the verifier who controls the signature by computing $w = s^{-1} \bmod q$ and comparing $M_q(M_p(E_p(g,M_q(m,w)),E_p(y,M_q(r,w))),1)$ to r.

Note that the strategy of using M_q "upstairs" and M_p "downstairs" can also be applied to El-Gamal's algorithm upon a proper modification of the keys. Also, modular inverses can be computed by $M_q(M_q(E_q(x,q-2),1),1) = x^{-1} \bmod q$ or $M_q(E_q(x,q-2),Q) = x^{-1} \bmod q$ (if the storage of Q is believed to be a good trade-off compared to one M_q).

As in section 9, it is unclear if this scheme is equivalent to the original DSA in terms of security.

10. Using E_n as a one-way accumulator

At Eurocrypt'93, Benaloh and de Mare [2] introduced a new family of one-way functions which satisfy a quasi-commutativity property that allows them to be used as accumulators. This property allows to design protocols in which the need for a trusted central authority can be eliminated.
Typical examples of practical applications of the concept are document time-stamping and membership testing.

Accumulators should satisfy the property : $h(h(x,y),z) = h(h(x,z),y)$ which is true for $E_n(x,y) = h(x,y)$

11. Conclusion and Implementation Details

[1] showed that it is possible to compute $ABN \bmod n$ in $l(n)+1$ clock cycles. That is, a modular multiplication is performed with the same complexity (timewise and hardwarewise) as that of a standard multiplication, It is thus possible to execute all the schemes mentioned in our article in time and hardware equivalent to their execution without modular reductions at all.

Field experiments confirm this estimation since on a 68HC05 running at 3.5 MHz (internal clock), the M operation (for $l(n)$ = 512) is executed in about 135 ms (RAM usage is less than 70 bytes) and a special M^2 (squaring) version runs at 85 ms but requires a double RAM space.

12. Acknowledgment

The authors would like to thank Beni Arazi for his numerous suggestions and remarks concerning this work.

References

[1] **B. ARAZI**, *Modular multiplication is equivalent in complexity to a standard multiplication*, Fortress U&T Internal Report (1992) available from Fortress U&T Information Safeguards, P.O. Box 1350, Beer-Sheva, IL-84110, Israel.

[2] **J. BENALOH & M. de MARE**, *One-way accumulators : A decentralised alternative to digital signatures*, Advances in cryptology: Proceedings of Eurocrypt'93 , Lecture Notes in Computer Science, Springer-Verlag, to appear.

[3] **W. DIFFIE & M. HELLMAN**, *New directions in cryptography*, IEEE TIT, vol. 22, (1976), pp 644-654.

[4] **S. DUSSE & B. KALISKI**, *A cryptographic library for the Motorola DSP56000.* In Advances in Cryptology - Eurocrypt'90, pp. 230-244, Springer-Verlag, New-York, 1990.

[5] **T. EL-GAMAL**, *A public-key cryptosystem and a signature scheme based on the discrete logarithm*, IEEE TIT, vol. 31, No. 4, (1985), pp. 469-472.

[6] **S. EVEN**, *Systolic modular multiplication*, In Advances in Cryptology, Crypto'90, pages 619-624, Springer-Verlag, New-York, 1991.

[7] **A. FIAT & A. SHAMIR**, *How to prove yourself : Practical solutions of identification and signature problems*, Advances in Cryptology: Proceedings of Crypto'86, Lecture Notes In Computer Science, Springer-Verlag, Berlin, 263 (1987), pp 186-194.

[8] **P. MONTGOMERY**, *Modular multiplication without trial division*, Mathematics of Computation, vol. 44 (170), pp. 519-521 1985.

[9] **D. NACCACHE,** *Can OSS be repaired ?,* Advances in cryptology: Proceedings of Eurocrypt'93 , Lecture Notes in Computer Science, Springer-Verlag, to appear.

[10] **National Institute of Standards and Technology,** Publication XX : announcement and specifications for a digital signature standard (DSS), Federal Register, August 19, 1992.

[11] **J.J. QUISQUATER & L. GUILLOU,** *A practical zero-knowledge protocol fitted to security microprocessor minimising both transmission and memory,* Advances in cryptology: Proceedings of Eurocrypt'88 (C. Günter, ed.), Lecture Notes in Computer Science, Springer-Verlag, Berlin, 330 (1988), pp 123-128.

[12] **R. RIVEST, A. SHAMIR & L. ADLEMANN,** *A method for obtaining digital signatures and public-key cryptosystems,* CACM, vol. 21 (1978), pp. 120-126.

[13] **C. SCHNORR,** *Efficient identification and signatures for smart-cards,* Advances in cryptology: Proceedings of Eurocrypt'89 (G. Brassard ed.), Lecture Notes in computer science, Springer-Verlag, Berlin, 435 (1990), pp. 239-252.

[14] **M. SHAND & J. VUILLEMIN,** *Fast implementations of RSA cryptography,* 11th IEEE Symposium on Computer Arithmetic, 1993. To appear.

Secret Sharing Schemes with Veto Capabilities*

C. Blundo A. De Santis L. Gargano U. Vaccaro

Dipartimento di Informatica ed Applicazioni
Università di Salerno
84081 Baronissi (SA), Italy

Abstract

A secret sharing scheme permits a secret to be shared among participants in such a way that only qualified subsets of participants can recover the secret, but any non-qualified subset has absolutely no information on the secret.
In this paper we consider the problem of designing efficient secret sharing schemes having the additional feature that qualified minorities can forbid any other set of participants from reconstructing the secret key. This problem was first considered by Beutelspacher [2] who gave an algorithm, based on projective geometries, to construct threshold schemes in which qualified minorities have this "veto" capability. We show that well known tools from Error Correcting Coding Theory allow to modify the classical Shamir secret sharing algorithm [22] to handle this more general problem.

1 Introduction

A *secret sharing scheme* is a protocol to share a secret S among a set $\mathcal{P}$ of participants in such a way that participants in subsets $A \subseteq \mathcal{P}$ enabled to recover the secret, pooling together their information can reconstruct the secret; but participants in subsets $B \subseteq \mathcal{P}$, that are not enabled to recover the secret S, have no information on S.

Secret sharing schemes are useful in any important action that requires the concurrence of several designed people to be initiated, as launching a missile, opening a bank vault or even opening a safety deposit box [25]. Secret sharing schemes are also used in management of cryptographic keys and multi-party secure protocols (see [13] for example).

An *access structure* for a secret sharing scheme is the family of all subsets of participants that are enabled to reconstruct the secret. A (k,n) threshold scheme is a secret sharing scheme on a set of n participants, whose access structure consists of all subsets of participants of cardinality at least k. Threshold schemes were introduced by Blakley [4] and Shamir [22]. Other constructions for threshold schemes have been

*Partially supported by Italian Ministry of University and Research (M.U.R.S.T.) and by National Council for Research (C.N.R.) under grant 91.02326.CT12.

proposed. McEliece and Sarwate [17] showed how to construct threshold schemes using Reed-Solomon codes; Asmuth and Bloom [1] made use of the Chinese Remainder Theorem; and Stinson and Vanstone [28] realised threshold schemes based on techniques from combinatorial design theory.

Secret sharing schemes for general access structures were first defined by Ito, Saito, and Nishiseki in [16]. Given any access structure $\mathcal{A}$ they showed how to set up a secret sharing scheme for $\mathcal{A}$. Subsequently, Benaloh and Leichter [3] described a more efficient way to realise secret sharing schemes for any given access structure.

An important issue in the implementation of secret sharing schemes is the size of the shares, since the security of a system degrades as the amount of secret information increases. Thus, one of the basic problems in the field of secret sharing schemes is to derive bounds on the amount of information that must be kept secret. In all secret sharing schemes the size of the shares (i.e., of the private information given to any participant) cannot be less than the size of the secret (see [12] for a simple information-theoretic proof). Moreover, there are access structures for which any corresponding secret sharing scheme must give to some participant a share of size strictly bigger than the secret size. Lower bounds on the size of shares can be found in [12], [5], and [6]. The best lower bound on the size of shares that has been proved is $\lceil \log |S|^{2-\epsilon} \rceil$, where $|S|$ is the number of possible secrets and ϵ is any constant > 0 [5]. Upper bounds on the size of shares have been given in [10] and [24].

We also briefly mention some "extended capabilities" of secret sharing schemes that have been studied. The issue of protecting against cheating by one or more participants is addressed in [17], [29], [19], [26] and [11]. The question of how to set up a secret sharing scheme in the absence of a trusted party is solved in [15]. We refer the reader to the recent survey by Stinson [23] for a unified description of recent results in the area of secret sharing schemes. For different approaches to the study of secret sharing schemes, for schemes with "extended capabilities" as disenrollment, fault-tolerance, and pre-positioning and for a complete bibliography we recommend the survey article by Simmons [25].

In this paper we consider the problem of designing efficient secret sharing schemes having the additional feature that qualified minorities can forbid any other set of participants from reconstructing the secret key. This problem was first considered by Beutelspacher [2] who gave an algorithm, based on projective geometries, to construct threshold schemes into which qualified minorities have this "veto" capability. We show that well known tools from Error Correcting Coding Theory allow to modify the classical Shamir secret sharing algorithm [22] to handle this more general problem. Our algorithms are computationally efficient since are based on well known errors/erasures decoding algorithms of Reed–Solomon codes [21].

2 The Formal Setting

Let $\mathcal{P}$ be the set of n participants and S be the set of secret keys. Suppose that each time a subset $X \subseteq \mathcal{P}$ of participants join together, the participants in X can be partitioned into two kinds: participants of kind $\mathcal{G}$, that is participants that really want to recover the secret; and participants of kind $\mathcal{B}$, that is participants that want to forbid the reconstruction of the secret (think of $\mathcal{G}$ as the positive votes and of $\mathcal{B}$ as the negative votes).

A (t, s, n) *threshold scheme* consists of an algorithm which gives shares to participants in $\mathcal{P}$ such that the following conditions are satisfied:

- Any t (or more) participants of kind $\mathcal{G}$ together with at most $s-1$ participants of kind $\mathcal{B}$ uniquely determine the secret;
- if less than t participants of kind $\mathcal{G}$ join together, then the secret cannot be retrieved, no matter how many participants of kind $\mathcal{B}$ coalise;
- the secret cannot be reconstructed if more than $s-1$ participants of kind $\mathcal{B}$ are "active".

As in [2], we assume that the secret reconstruction phase is carried out by a trustworthy machine which keeps secret to all participants the received shares. This in not a strong assumption since it is more or less explicitly used in all usual secret sharing schemes. In fact, if the machine does not keep secret the received shares, then every one that has access to the machine could know all the shares and therefore reconstruct the secret even if he is not allowed to.

A crucial role in our algorithms is played by Reed–Solomon codes [20] [18] which can be described as follows. Let $\{\alpha_1, \ldots, \alpha_\ell\}$ be a fixed list of the non-zero elements of $GF(q^m)$ and $P_r[x]$ be the set of all $r-1$ degree polynomials with coefficients in $GF(q^m)$. Then, the Reed-Solomon $RS(\ell, r)$ code of codeword length ℓ, where $\ell = q^m - 1$, and dimension r is defined as

$$RS(\ell, r) = \{\mathbf{C} = (C_1, \ldots, C_\ell) : C_i = f(\alpha_i),\ i = 1, \ldots, \ell,\ f(x) \in P_r[x]\}. \quad (1)$$

As noticed by McEliece and Sarwate [17], Reed–Solomon codes can be used to share a secret in a way similar to that of the original Shamir scheme. Suppose that $S \in GF(q^m)$ is the secret, and let $f(x) \in P_r[x]$ be a randomly chosen polynomial such that $f(0) = S$. Then the shares given to participants are the coordinates of the codeword in $RS(\ell, r)$ associated to $f(x)$. Now, suppose that one has k values $C_{i_1} \ldots C_{i_k} \in GF(q^m)$, where $k-e$ of them are true shares and the remaining e are random elements of $GF(q^m)$. It is known that one can reconstruct $f(x)$ (and therefore the secret $S = f(0)$) if and only if the relation $k - 2e \geq r$ holds [17]. Moreover, the reconstruction algorithm is very efficient and requires $O(\ell \log^2 \ell)$ operations in $GF(q^m)$ [21]. Using above properties, we shall devise several secret sharing schemes in which minorities have veto capabilities.

3 Schemes with Veto Capabilities

In this section we show how to construct secret sharing schemes with veto capabilities. We first present a simple scheme to realise $(t, 1, n)$ threshold schemes. After, we shall describe an efficient (t, s, n) threshold schemes.

3.1 $(t, 1, n)$ Threshold Schemes

Let $\mathcal{P} = \{P_1, P_2, \ldots, P_n\}$ be the set of participants. Consider a Reed–Solomon code $RS(\ell, 2t-1)$, where $\ell = q^m - 1$, for some prime q, is an integer greater than or equal to $n[4n(n-t)+1]$. The following algorithm realises a $(t, 1, n)$ threshold scheme for the set of participants $\mathcal{P}$, $t \geq 2$

Distribution Algorithm

Input: The secret $S \in GF(q^m)$, $\mathcal{P} = \{P_1, P_2, \ldots, P_n\}$, and an integer t.

Let S_1, S_2 be randomly chosen in $GF(q^m)$ such that $S = S_1 \oplus S_2$, where $\oplus$ denote the componentwise addition modulo the prime q. Let $\{\gamma_{1,1}, \ldots, \gamma_{1,n-t+1}, \ldots, \gamma_{n,1}, \ldots, \gamma_{n,n-t+1}, \beta_1, \ldots, \beta_n\}$ be a fixed list of non-sero elements of $GF(q^m)$.
For any integer k, let $P_k[x]$ denote the set of all $k-1$ degree polynomials with coefficients in $GF(q^m)$

Choose randomly a polynomial $f(x) \in P_{2t-1}[x]$ such that $f(0) = S_1$.
For $i = 1, \ldots, n$ let

- $\mathbf{w}_i^P = (f(\gamma_{i,1})a, \cdots, f(\gamma_{i,n-t+1}), r_{i,1}, \cdots, r_{i,n-t-1})$.
- $\mathbf{w}_i^N = (r_{i,n-t}, \cdots, r_{i,3(n-t)-1})$.

where each $r_{i,j}$ is a randomly chosen element in $GF(q^m) \setminus (\{f(\gamma_{i,1}), \ldots, f(\gamma_{i,n-t+1}) \ : \ 1 \leq i \leq n\}) \cup (\{g(\beta) \ : \ \beta \in GF(q^m)\}); r_{a,b} \neq r_{c,d}$.

Choose randomly a polynomial $g(x) \in P_t[x]$ such that $g(0) = S_2$.

The two shares given to participant P_i will be the following:

- $\mathbf{s}_i^P = \mathbf{w}_i^P g(\beta_i)$ (Positive share)
- $\mathbf{s}_i^N = \mathbf{w}_i^N g(\beta_i)$ (Negative share)

Output: the shares $(\mathbf{s}_1^P, \mathbf{s}_1^N)a, \ldots, (\mathbf{s}_n^P, \mathbf{s}_n^N)$ for participants $P_1, \ldots, P_n$, respectively.

In the reconstruction phase each participant who is active gives to the reconstructing machine either his positive or his negative share (according to his intention to allow the reconstruction of the secret or his intention to assert a veto). The reconstruction machine performs a simple polynomial interpolation on the second half of the shares collected (i.e., on the pieces $g(\beta_i)$'s) in order to reconstruct $g(x)$ and then $S_2 = g(0)$, and an error and erasures decoding on the first part of the shares collected (i.e., on the pieces $\mathbf{w}_i$'s) in order to reconstruct $f(x)$ and then $f(0) = S_1$.

It is easy to see that at least t participants are needed to reconstruct $S_2 = g(0)$. Moreover, if *less* than t participants join together they do not know anything (in the information theoretic sense) about S_2 and therefore about the secret S. It is also easy to see that if the reconstructing machine receives the shares of t (or more) participants with positive votes and no share of participants with negative votes it can reconstruct the whole secret. Indeed, the machine can recover S_2 by interpolating on the $g(\beta)$'s and evaluate $g(x)$ so obtained in 0 to get S_2; moreover the machine can use the error and erasure correction Reed–Solomon decoding algorithm to recover $f(x)$. This can be done since in this case the number of $f(\gamma)$'s minus the number of r's, i.e., the *errors* in the codeword, is bigger than $2t-1$. Therefore, also S_1 can be recovered and finally $S = S_1 \oplus S_2$.

On the other hand, just a single participant can forbid the reconstruction of S. In fact, if there exists a participant asserting the veto, then, even though the remaining participants want to reconstruct the secret, the polynomial $f(x)$ cannot be reconstructed since there are more errors in the codeword associated to $f(x)$ than the Reed–Solomon code $RS(\ell, 2t-1)$ can correct. This can be seen by computing the number of "good" pieces (i.e., the $f(\gamma)$'s) minus the number of the "bad" pieces (i.e., the r's). We have that this difference is at most $2t-2 <$ degree $f(x)+1$.

Consider the following example. Let $\mathcal{P} = \{P_1, P_2, P_3, P_4, P_5\}$ be a set of 5 participants. If we want to realise a $(3,1,5)$ threshold scheme, then we have to consider a codeword C of the Reed–Solomon code $RS(63,4)$ and a polynomial $g(x) \in P_3[x]$. Let α, be the primitive root of $GF(2^6)$.[1] Suppose that the secret S is equal to $S = \alpha^{33} \in GF(2^6)$. Let $S_1 = \alpha^{25}$ and $S_2 = \alpha^3$, in this way $S = S_1 \oplus S_2$. Suppose that the polynomial $f(x)$ associated to C such that $f(0) = \alpha^{25}$ is equal to $f(x) = \alpha^{25} + \alpha^{58}x + \alpha x^2 + \alpha^{41}x^3 + \alpha^3 x^4$. Let $g(x) = \alpha^3 + \alpha^{19}x + \alpha^{12}x^2$ be a polynomial in $P_3[x]$ such that $g(0) = \alpha^3$. Suppose, for the sake of the simplicity, that $\gamma_{i,j} = \alpha^{3(i-1)+j}$ for $i = 1,2,\ldots,5$ and $j = 1,2,3$, and that $\beta_i = \alpha^{15+i}$ for $i = 1,2,\ldots,5$; in this way each $r_{i,j}$ belongs to the set $\{\alpha^0, \alpha^{11}, \alpha^{12}, \alpha^{14}, \alpha^{15}, \alpha^{24}, \alpha^{27}, \alpha^{28}, \alpha^{30}, \alpha^{34}, \alpha^{35}, \alpha^{42}, \alpha^{45}, \alpha^{50}, \alpha^{52}, \alpha^{59}, \alpha^{62}, \alpha^{63}\}$, for $i = 1,2,\ldots,5$ and $j = 1,\ldots,5$. The share distributed to participants $\{P_1, P_2, P_3, P_4, P_5\}$ are the following.

- P_1 $s_1^P = \alpha^{20}\alpha^{58}\alpha^{32}\alpha^{34}, \alpha^{55}$, $s_1^N = \alpha^{15}\alpha^{11}\alpha^{35}\alpha^{52}, \alpha^{55}$.
- P_2 $s_2^P = \alpha^{40}\alpha^{43}\alpha^{49}\alpha^{0}, \alpha^{37}$, $s_2^N = \alpha^{59}\alpha^{35}\alpha^{27}\alpha^{15}, \alpha^{37}$.
- P_3 $s_3^P = \alpha^{8}\alpha^{32}\alpha^{33}\alpha^{27}, \alpha^{23}$, $s_3^N = \alpha^{62}\alpha^{50}\alpha^{12}\alpha^{59}, \alpha^{23}$.
- P_4 $s_4^P = \alpha^{29}\alpha^{20}\alpha^{25}\alpha^{24}, \alpha^{47}$, $s_4^N = \alpha^{11}\alpha^{15}\alpha^{34}\alpha^{59}, \alpha^{47}$.
- P_5 $s_5^P = \alpha^{54}\alpha^{22}\alpha^{53}\alpha^{50}, \alpha^{51}$, $s_5^N = \alpha^{11}\alpha^{14}\alpha^{27}\alpha^{45}, \alpha^{51}$.

3.2 (t, s, n) Threshold Schemes

Now we can describe how to construct (t,s,n) threshold schemes building up on our previous construction of $(t,1,n)$ threshold schemes. The following algorithm realizes a (t,s,n) threshold scheme for a set of participants $\mathcal{P} = \{P_1, P_2, \ldots, P_n\}$.

Distribution Algorithm

Input: The secret $S \in GF(q^m)$, $\mathcal{P} = \{P_1, P_2, \ldots, P_n\}$, and integers t and s.

Let $P_k[x]$ be the set of all $k-1$ degree polynomials with coefficients in $GF(q^m)$
Choose randomly a polynomial $f(x) \in P_{n-s+1}[x]$ such that $f(0) = S$; compute the values $S_1 = f(\beta_1), \ldots, S_n = f(\beta_n)$. *[This is a classical $(n-s+1, n)$ threshold scheme, therefore S can be reconstructed if and only if at least $n-s+1$ pieces S_i are known].*

For each S_i, $1 \leq i \leq n$, perform an independent $(t,1,n)$ threshold scheme Σ_i. Let $(S_{i,1}^P, S_{i,1}^N), \ldots, (S_{i,n}^P, S_{i,n}^N)$ the shares so obtained, where $S_{i,j}^P$ and $S_{i,j}^N$ denote the positive share and the negative share, respectively.

The two shares given to participant P_i will be the following

- $\mathbf{w}_i^P = S_{1,i}^P, \ldots, S_{i,i}^P, \ldots, S_{n,i}^P$ (Positive share).
- $\mathbf{w}_i^N = S_{1,i}^P, \ldots, S_{i,i}^N, \ldots, S_{n,i}^P$ (Negative share).

Output: the shares $(\mathbf{w}_1^P, \mathbf{w}_1^N), \ldots, (\mathbf{w}_n^P, \mathbf{w}_n^N)$ for participants $P_1, \ldots, P_n$, respectively.

Suppose now that k participants are active, m of them with positive share and p of them with negative share, $k = m + p$. Let $\mathbf{w}_{i_1}^{Q_{i_1}}, \ldots, \mathbf{w}_{i_k}^{Q_{i_k}}$ be the part of shares given by them (that is, each of the above pieces is either a negative share or a positive share), with each $Q_{i_j} \in \{P, N\}$. By construction we have

$$\mathbf{w}_{i_1}^{Q_{i_1}} = S_{1,i_1}^{Q_{i_1}^1}, \ldots, S_{i_1,i_1}^{Q_{i_1}^{i_1}}, \ldots, S_{n,i_1}^{Q_{i_1}^n}$$

[1] The finite field $GF(2^6)$ is generated by the polynomial $p(x) = 1 + x + x^6$.

$$\vdots$$

$$w_{i_h}^{Q_{i_h}} = S_{1,i_h}^{Q_{i_h}^1}, \ldots, S_{i_h,i_h}^{Q_{i_h}^{i_h}}, \ldots, S_{n,i_h}^{Q_{i_h}^n}$$

where if $Q_{i_j} = P$, then $Q_{i_j}^r = P$ for all r, whereas if $Q_{i_j} = N$, then $Q_{i_j}^r = P$ for all $r \neq i_j$ and $Q_{i_j}^{i_j} = N$. Now the reconstruction machine attempts to recover S_1 by performing the reconstruction algorithm on the pieces $S_{1,i_1}^{Q_{i_1}^1}, \ldots, S_{1,i_h}^{Q_{i_h}^1}$, attempts to recover S_2 by performing the reconstruction algorithm on the pieces $S_{2,i_1}^{Q_{i_1}^2}, \ldots, S_{2,i_h}^{Q_{i_h}^2}, \ldots$, attempts to recover S_n by performing the reconstruction algorithm on the pieces $S_{n,i_1}^{Q_{i_1}^n}, \ldots, S_{n,i_h}^{Q_{i_h}^n}$. Since $S_{1,i_1}^{Q_{i_1}^1}, \ldots, S_{1,i_h}^{Q_{i_h}^1}$ were obtained from S_1 by applying a $(t, 1, n)$ threshold scheme, S_1 will be recovered if and only if at least t participants have given positive vote and no participant has given negative vote. Since any participant P_i can forbid only the reconstruction of piece S_i, (by giving $S_{i,i}^N$), and since S can be reconstructed if and only at least $n - s + 1$ pieces S_i are available, it follows that S can be reconstructed if and only if at least t participants have given positive vote and no more than $s - 1$ participants have given negative vote.

4 Conclusion

We have shown that Reed-Solomon codes can be used to set-up secret sharing schemes with veto capabilities in the sense of [2]. It should be remarked that our distribution and reconstruction algorithms are computationally efficient since they can be based on the fast decoding of Reed-Solomon codes [21].

References

[1] C. Asmuth and J. Bloom, *A Modular Approach to Key Safeguarding*, IEEE Trans. on Inform. Theory, vol. IT-29, no. 2, Mar. 1983, pp. 208-210.

[2] A. Beutelspacher, *How to Say 'No'*, in "Advances in Cryptology - EUROCRYPT 89", vol. 434 of "Lecture Notes in Computer Science", Springer-Verlag, pp. 491-496.

[3] J. C. Benaloh and J. Leichter, *Generalized Secret Sharing and Monotone Functions*, in "Advances in Cryptology - CRYPTO 88", Ed. S. Goldwasser, vol. 403 of "Lecture Notes in Computer Science", Springer-Verlag, pp. 27-35.

[4] G. R. Blakley, *Safeguarding Cryptographic Keys*, Proceedings AFIPS 1979 National Computer Conference, pp. 313-317, June 1979.

[5] C. Blundo, A. De Santis, L. Gargano, and U. Vaccaro, *On the Information Rate of Secret Sharing Schemes*, in: Advances in Cryptology – CRYPTO '92, E. Brickell (Ed.), Lectures Notes in Computer Science, vol. 740, pp. 149-169, 1993, Springer-Verlag.

[6] C. Blundo, A. De Santis, D. R. Stinson, and U. Vaccaro, *Graph Decomposition and Secret Sharing Schemes*, in "Advances in Cryptology – EUROCRYPT 92", Lecture Notes in Computer Science, Vol. 658, R. Rueppel (Ed.), Springer-Verlag, pp. 1-24, 1993. Also to appear in: Journal of Cryptology.

[7] C. Blundo, A. De Santis, A. Gaggia, and U. Vaccaro, *New Bounds on the Information Rate of Secret Sharing Schemes*, IEEE Transactions on Information Theory, to appear.

[8] C. Blundo, A. De Santis, and U. Vaccaro, *Efficient Sharing of Many Secrets*, STACS '93, 10th Annual Symposium on Theoretical Aspects of Computer Science, P. Enjalbert, A. Finkel, and K. W. Wagner (Eds.), Lecture Notes in Computer Science, vol. 665, Springer–Verlag, 1993.

[9] C. Blundo, A. Cresti, A. De Santis, and U. Vaccaro, *Fully Dynamic Secret Sharing Schemes*, in: *Advances in Cryptology - CRYPTO '93*, D. Stinson (Ed.), Lectures Notes in Computer Science, Springer-Verlag, to appear.

[10] E. F. Brickell, and D. R. Stinson, *Improved Bounds on the Information Rate of Perfect Secret Sharing Schemes*, J. Cryptology, vol. 5, No. 3, pp. 153-166, 1992.

[11] E. F. Brickell and D. R. Stinson, *The Detection of Cheaters in Threshold Schemes*, SIAM J. on Discrete Math., vol. 4, pp. 502–510, 1991.

[12] R. M. Capocelli, A. De Santis, L. Gargano, and U. Vaccaro, *On the Size of Shares for Secret Sharing Schemes*, Journal of Cryptology, vol. 6, (1993), 157–167.

[13] O. Goldreich, S. Micali, and A. Wigderson, *How to Play any Mental Game*, Proceedings of 19th ACM Symp. on Theory of Computing, pp. 218–229, 1987.

[14] E. D. Karnin, J. W. Greene, and M. E. Hellman, *On Secret Sharing Systems*, IEEE Trans. on Inform. Theory, vol. IT-29, no. 1, Jan. 1983, pp. 35-41.

[15] I. Ingemarson and G. J. Simmons, *A Protocol to Set Up Shared Secret Schemes Without the Assistance of a Mutually Trusted Party*, Lecture Notes in Computer Science, vol. 473, pp. 266–282, 1991.

[16] M. Ito, A. Saito, and T. Nishizeki, *Secret Sharing Scheme Realizing General Access Structure*, Proc. IEEE Global Telecommunications Conf., Globecom 87, Tokyo, Japan, 1987.

[17] R. J. McEliece and D. Sarwate, *On Sharing Secrets and Reed–Solomon Codes*, Communications of the ACM, vol. 24, n. 9, pp. 583–584, September 1981.

[18] F. J. MacWilliams and N.J.A. Sloane, *The Theory of Error–Correcting Codes*, North–Holland, Amsterdam, 1977.

[19] T. Rabin and M. Ben-Or, *Verifiable Secret Sharing and Multiparty Protocols with Honest Majority*, Proc. 21st ACM Symp. on Theory of Computing, pp. 73–85, 1989.

[20] I. S. Reed and G. Solomon, *Polynomial Codes over Certain Finite Fields*, SIAM J. Appl. Math., pp. 300–304, June 1960.

[21] D. Sarwate, *On the Complexity of Decoding Goppa Codes*, IEEE Trans. Inform. Theory, vol. 23, pp. 515–516, July 1977.

[22] A. Shamir, *How to Share a Secret*, Communications of the ACM, vol. 22, n. 11, pp. 612-613, Nov. 1979.

[23] D. R. Stinson, *An Explication of Secret Sharing Schemes*, Design, Codes and Cryptography, vol. 2, pp. 357–390, 1992.

[24] D. R. Stinson, *New General Lower Bounds on the Information Rate of Secret Sharing Schemes*, Proceedings of Crypto '92, Advances in Cryptology, Lecture Notes in Computer Science, E. Brickell Ed., Springer-Verlag, (to appear).

[25] G.J. Simmons, *An Introduction to Shared Secret and/or Shared Control Schemes and Their Application*, Contemporary Cryptology, IEEE Press, pp. 441–497, 1991.

[26] G.J. Simmons, *Robust Shared Secret Schemes or "How to be Sure You Have the Right Answer even though You don't Know the Question"*, Congressus Numerantium, vol. 8, pp. 215–248, 1989.

[27] G. J. Simmons, *Prepositioned Shared Secret and/or Shared Control Schemes*, Lecture Notes in Computer Science, vol. 434, pp. 436–467, 1990.

[28] D.R. Stinson and S.A. Vanstone, *A Combinatorial Approach to Threshold Schemes*, SIAM J. Disc. Math., vol. 1, No. 2, May 1988, pp. 230–236.

[29] M. Tompa and H. Woll, *How to Share a Secret with Cheaters*, J. Cryptology, vol. 1, pp. 133–138, 1988. (Also, Crypto '86, pp. 261-265.)

Group-theoretic hash functions

Jean-Pierre Tillich and Gilles Zémor

ENST

Abstract. We discuss the security of group-theoretic hash functions for cryptographic purposes. Those functions display several attractive features: they can be computed quickly, and it can be shown that local modifications of the plaintext necessarily change the hashed values. We show why the first such proposal given in [Zém91] is not secure, by giving a probabilistic algorithm for finding collisions. However, our attack is based on the special form of the matrices which were originally chosen. We propose alternative schemes which seem to be immune to such attacks.

1 Introduction

We focus on the problem of designing easily computable cryptographic hash functions. Such a function $\mathscr{H}$ should map the set of variable length texts over an alphabet $\mathscr{A}$, to a set of (short) fixed length texts.

$$\mathscr{H} : \mathscr{A}^* \longrightarrow \mathscr{A}^n$$

A hash function should have the following properties :

- It should be easily (i.e. quickly) computable.
- It should be computationally difficult to find "collisions", i.e. two texts having the same hashed value. (This is sometimes known as the strong collision criterion).

Many hashing schemes have been proposed and studied (see e.g. [Dam89]), one of those, discussed by Godlewski and Camion in [GC88], is a Knapsack-type scheme based upon error-correcting codes, with the attractive property that the modification of any set of less than d characters of text will necessarily yield a modification of the hashed value, where d is the minimum distance of an appropriately chosen code. Unfortunately, such schemes are based upon linear computations which are well-known for their cryptographic weakness. In [Zém91], a hashing scheme was proposed which retains something of the features of the coding-based scheme, while trying to enhance cryptographic strength. For that purpose, the original tool, i.e. the minimum distance of a code, was replaced by the girth of a Cayley graph.

The basic idea is as follows: choose a group G together with two generators A and B. Suppose we are working with the alphabet $\mathscr{A} = \{0, 1\}$. To any message, i.e. a string of 0's and 1's, associate a formal string of A's and B's by the substitution $0 \rightarrow A$ and $1 \rightarrow B$. The hashed value is obtained by computing in the group G the product of those A's and B's. Let us call such a function a group theoretic hash function and denote it by $\mathscr{H}_{G,A,B}$. Breaking the scheme boils down to finding short factorizations of group elements into products of A's and B's: "short" means the length of reasonably-sized texts, which should be of a different order of magnitude from the group size.

In [Zém91], the following group G and generators A,B are proposed. Take $G = SL_2(\mathbf{F}_p)$, the group of 2×2 matrices of determinant 1 over the integers modulo a prime p, and $A = A_1$, $B = B_1$ with

$$A_1 = \begin{pmatrix} 1 & 1 \\ 0 & 1 \end{pmatrix} \qquad B_1 = \begin{pmatrix} 1 & 0 \\ 1 & 1 \end{pmatrix}.$$

We will denote the hash function just defined by $\mathscr{H}_1$.

Let us restate this definition of $\mathscr{H}_{G,A,B}$ in graph-theoretic terms. Denote by $\mathscr{G}(G, \mathscr{S})$ (or simply $\mathscr{G}$ when no confusion can arise) the directed Cayley graph associated with G and $\mathscr{S} = \{A, B\}$. This means that $\mathscr{G}$ has G as its set of vertices, and there is a directed edge between vertices v and w iff $w = vs$ with s belonging to $\mathscr{S}$. In this setting, a binary text x can be considered as a directed path in the graph $\mathscr{G}$, with the identity vertex as starting point, and its endpoint is precisely the hashed value $\mathscr{H}(x)$. Now two texts yielding the same hashed value correspond to two paths with the same starting and endpoints. We would like those two paths to differ necessarily by a "minimum amount" ; this can be guaranteed, if the graph $\mathscr{G}$ is chosen without short "cycles". Let us make this slightly more formal.

Definition 1. Call the *"directed girth"* of a graph $\mathscr{G}$, the largest integer ∂ such that given any two vertices v and w, any pair of distinct directed paths joining v to w will be such that one of those paths has length (i.e. number of edges) ∂ or more.

The basic idea, initiated in [Zém91], of looking for potentially good hash functions among Cayley graphs is that their girth is a relevant parameter to hashing. Namely if the Cayley graph $\mathscr{G}$ has a large girth ∂, then the corresponding hash function will have the property that "local" modifications of a text will necessarily modify the hashed value. More precisely we have,

Proposition 2. *If a substring of k consecutive symbols of a text is replaced by a string of h consecutive symbols without modifying the hashed value, then* $\sup(k, h) \geq \partial$ *i.e. one of those strings has more than ∂ symbols.*

2 Some basic attacks

Since we compute the hash function by group multiplications, the forger may use the general algorithm for forgery of P.Camion, which may be found in [Cam87]. But using simple matrices like A_1 and B_1 enables one to choose large prime numbers p, e.g. of 150 bits, without much damage to the speed of computation. In that case the group $SL_2(\mathbf{F}_p)$ is so large (it has $O(p^3)$ elements, that is about 2^{450} elements) that the scheme is safe against this kind of attack.

The use of a group may also help the forger in the following manner : Let $\mathcal{H} = \mathcal{H}_{G,A,B}$ be a group-theoretic hash function.

- suppose that the forger finds a message M_0 which hashes to the unit element, then it is easy, for a given message M, to find another message with collides with it : the forger just needs to split M into two parts $M = P//Q$ where $//$ denotes the concatenation of two messages. Then the message $P//M_0//Q$ collides with M
- suppose also that there are many elements in the group of small order : then there will be a fast probabilistic way to obtain rather short messages whose hashed value is the unit element of the group. The forger just has to compute the order d of several outputs for the hash function $\mathcal{H}$, there is a non negligible chance to find a message M for which $\mathcal{H}(M)$ is of small order l. Thus the message M' consisting of the concatenation of l times the message M, verifies $\mathcal{H}(M') = Id$, where Id is the unit element of the group. This leads to the following attack on the hash function $\mathcal{H}$ when $G = SL_2(\mathbf{F}_p)$.

Algorithm 1 *(Attack on any $\mathcal{H}_{G,A,B}$ such that $G = SL_2(\mathbf{F}_p)$.)*

- *factorize $p-1$. This is not too long because p has less than 50 digits. This will be used to find the order of an element of $\mathbf{Z}_p$.*
- *compute the hash values of about $O(p^{0,6})$ messages.*
- *compute the order of the hash values matrices, when the corresponding matrix is diagonalizable. To perform this, find the eigenvalues of the matrix, the order of the eigenvalues will be the same as the order of the matrix.*

It can be proven that, on average, for randomly chosen p, this algorithm yields a message whose hashed value is of order less than $O(p^{0,4})$. This attack may be dangerous as long as p is not greater than 2^{60}. Actually the key of the method described above lies in the following fact : the diagonalizable matrices of the group $SL_2(\mathbf{F}_p)$ make up a non negligible part of this group. More precisely there are exactly $(p-1)(p^2-p-4)/2$ diagonalizable matrices in $SL_2(\mathbf{F}_p)$ which has $p(p-1)(p+1)$ elements. It is easy to see that the order of a diagonalizable matrix divides $p-1$. Furthermore, for any divisor d of $p-1$, there are exactly $p(p+1)\lfloor\frac{d-1}{2}\rfloor$ diagonalizable matrices ($\neq \pm Id$) of order d or a divisor of d. Therefore, using the fact that on average the greatest non trivial divisor of a number n is more or less of order $0(n^{0,63})$ it follows that our algorithm has a non negligible chance of success.

It must be added that there is an easy way to prevent this attack : instead of choosing a modulo p completely at random, it is sufficient to choose a modulo p such that the greatest non trivial divisor of $p-1$ is very large, say of the same magnitude as p. Such a p is not difficult to find.

The preceding attack makes use of the structure of the group $SL_2(\mathbf{F}_p)$, but not of the generators A and B of the group, which are used to construct the hash function. The function $\mathscr{H}_1$ has actually a drawback : A is the transpose of B. This fact may benefit the forger in the following manner : imagine that he is able to find a binary message $x_1x_2\ldots x_n$ (of length n) whose hashed value is a symmetrical matrix. Then it is straightforward to check that the message $\overline{x_n}\,\overline{x_{n-1}}\ldots\overline{x_1}$, where $\overline{x_i}$ denotes the bit $1-x_i$, has the same hashed value as the first binary message. This provides a collision when these two messages are different. The forger proceeds as follows.

Algorithm 2 *(Attack on any $\mathscr{H}_{G,A,B}$ such that $G = SL_2(\mathbf{F}_p), B = {}^tA$.)*
The forger computes the hashed value $\left(\begin{smallmatrix} a & b \\ c & d \end{smallmatrix}\right)$ of a random message. Then he tries to append this message with m "0" bits, followed by n "1" bits, such that the hashed value of this new message is a symmetrical matrix. The forger's task is to solve the following equation (in n, m) over the field of integers modulo p :

$$m = a' + \frac{b'}{c'-n} \bmod p, \quad \textit{where}\ a' = -dc^{-1},\ b' = adc^{-2} - bc^{-1} + 1,\ c' = ac^{-1}. \tag{1}$$

This attack may be performed using brute force : just try a small m and hope that the corresponding n, by (1) is small too. If the forger tries all possible values of m less than $\sqrt{p}$, he has a reasonable chance of finding a corresponding n which is of order $O(\sqrt{p})$. This gives him in $O(\sqrt{p})$ operations two messages which collide, the longest of them being about $O(\sqrt{p})$ bits long. This attack may be effective, as long as p stays rather small : $p \leq 2^{50}$. This method of finding collisions may be somewhat improved, by using tools which are described in [GTV88], but it doesn't seem to break the hashing scheme, when p becomes larger than $2^{60}, 2^{70}$.

3 An efficient attack on the hash function $\mathscr{H}_1$

To obtain collisions we look for short factorizations of the identity into a product of A_1's and B_1's in $SL_2(\mathbf{F}_p)$.

It is possible to derive an efficient attack, by using the following remark.

Proposition 3. *The matrices $A_1 = \left(\begin{smallmatrix} 1 & 1 \\ 0 & 1 \end{smallmatrix}\right)$, $B_1 = \left(\begin{smallmatrix} 1 & 0 \\ 1 & 1 \end{smallmatrix}\right)$ generate freely the monoid :*

$$\mathscr{M} = \left\{ M = \begin{pmatrix} a & b \\ c & d \end{pmatrix},\ a, b, c, d \in \mathbf{N},\ ad - bc = 1 \right\}$$

Moreover, there is an explicit algorithm which factorizes any matrix of $\mathcal{M}$ into a product of A_1's and B_1's and runs in a logarithmic number of steps (in $\max(a,b,c,d)$*).*

The latter algorithm is in fact Euclid's algorithm applied to (a,b) if $a+b > c+d$, and to (c,d) otherwise. More precisely, if we suppose for example that $a+b > c+d$, $a > b$ and n the number of steps of Euclid's algorithm applied to (a,b) is even, then

$$\begin{pmatrix} a & b \\ c & d \end{pmatrix} = A_1^{q_n} B_1^{q_{n-1}} \ldots A_1^{q_2} B_1^{q_1}$$

where $q_1, q_2, \ldots, q_n$ are the quotients that appear when Euclid's algorithm is applied to (a,b). For more details see appendix B in [Kur60].

This leads to the following strategy for a forgery.

- The forger first looks for a matrix

$$M = \begin{pmatrix} 1+k_1p & k_2p \\ k_3p & 1+k_4p \end{pmatrix}$$

  of $SL_2(\mathbf{Z})$, with non-negative coefficients, which is equal modulo p to the identity matrix.
- He then uses the above algorithm to factorize M into a product of A_1's and B_1's over $SL_2(\mathbf{Z})$. This yields a factorization of the identity in $SL_2(\mathbf{F}_p)$.

For this strategy to lead to an effective algorithm, two questions must be clarified.

1. Is it possible to find such a matrix M which is non-trivial (i.e. different from the identity) ?
2. is the factorization not too long ?

Let us first answer the first question. The forger has to solve the following: find 4 integers k_1, k_2, k_3, k_4 which are not all equal to 0, such that $(1+k_1p)(1+k_4p) - k_2pk_3p = 1$. In other words he has to solve the following diophantine equation.

$$(k_2k_3 - k_1k_4)p = k_1 + k_4. \tag{2}$$

Moreover he is looking after solutions of about the same magnitude (otherwise the factorization of the matrix M would be too long). Here is a probabilistic algorithm that gives an answer to this question.

Algorithm 3 *Choose a small integer c. Search for a solution (k_1, k_2, k_3, k_4) such that $k_1 + k_4 = cp$. To obtain this choose a random prime p' of about the same magnitude as p. Take $k_3 = p'$. Thus the forger has to solve :*

$$k_2p' = c + cpk_1 - k_1^2. \tag{3}$$

This is possible if the forger is able to solve : $k_1^2 - cpk_1 - c = 0 \bmod p'$. In other words the discriminant $c^2p^2 + 4c$ must be a quadratic residue modulo p'.

After only a small number of random tries (i.e random choices of c, p') , he shall succeed in finding such a pair (c, p') with high probability. This provides the following solution.

$$\begin{cases} k_1 = \dfrac{cp + \sqrt{c^2p^2 + 4c}}{2} \\ k_2 = \dfrac{c + cpk_1 - k_1^2}{p'} \\ k_3 = p' \\ k_4 = cp - k_1 \end{cases}$$

where $\sqrt{x}$ denotes a positive representation of a square root of x modulo p', if x is a quadratic residue modulo p'.

In this way, the forger will in general find solutions (k_1, k_2, k_3, k_4) of order $O(p)$. This gives him a matrix M with elements of order $O(p^2)$. This answers the first question. The answer to the second question will be given in the following section.

EXAMPLE.
Let us consider a small example, where we can see how the algorithm works. Let $p = 5$. We choose $c = 1$, and we want to find a small prime p', such that $c^2p^2 + 4c$ is a quadratic residue modulo p'. This is the case for $p' = 7$ for example. This gives us

$$\begin{cases} k_1 = \dfrac{5 + \sqrt{29}}{2} = 3 \text{ (the computation is done modulo 7)} \\ k_2 = \dfrac{1 + 5.3 - 3^2}{7} = 1 \text{ (here the computation is done over the integers)} \\ k_3 = p' = 7 \\ k_4 = 5 - 3 = 2. \end{cases}$$

This leads us to the matrix $M = \left(\begin{smallmatrix} 1+3.5 & 1.5 \\ 7.5 & 1+2.5 \end{smallmatrix}\right) = \left(\begin{smallmatrix} 16 & 5 \\ 35 & 11 \end{smallmatrix}\right)$.

Applying Euclid's algorithm, we obtain the factorization of M,

$$\begin{pmatrix} 16 & 5 \\ 35 & 11 \end{pmatrix} = B_1^2 A_1^5 B_1^3 \equiv \begin{pmatrix} 1 & 0 \\ 0 & 1 \end{pmatrix} \bmod 5.$$

4 Study of the average length of a factorization

In this section we assume that the output of algorithm 3 behaves essentially like a random matrix. We wish therefore to study the average length of a factorization of a random matrix into a product of $A_1 = \left(\begin{smallmatrix} 1 & 1 \\ 0 & 1 \end{smallmatrix}\right)$, $B_1 = \left(\begin{smallmatrix} 1 & 0 \\ 1 & 1 \end{smallmatrix}\right)$.

Our main result here is the following theorem.

Theorem 4. *If we choose uniformly a random matrix in the set of matrices of $SL_2(\mathbf{Z})$ that have non-negative coefficients smaller than the integer N, then, with probability tending to 1 as N tends to infinity, the length of the factorization of this matrix into a product of A_1's and B_1's is smaller than $(\ln N)^{1+\delta}$ for every constant $\delta > 0$.*

We will actually prove the equivalent theorem

Theorem 5. *If we choose uniformly a random pair of integers (a, b) smaller than N, then, with probability tending to 1 as N tends to infinity, the sum of the successive quotients q_i that appear when applying Euclid's algorithm to (a, b) is smaller than $(\ln N)^{1+\delta}$ for every constant $\delta > 0$.*

To convince oneself of the equivalence between theorems 4 and 5, call $\mathscr{M}_N$ the set of matrices of determinant 1 with nonnegative integer coefficients smaller than N that differ from the identity matrix, $\mathscr{E}_N$ the set of couples of coprime nonnegative integers smaller than N that differ from the couples $(0, 1)$ and $(1, 0)$, and notice that the mapping

$$\mathscr{M}_N \xrightarrow{\phi} \mathscr{E}_N$$

$$\begin{pmatrix} a & b \\ c & d \end{pmatrix} \mapsto \begin{cases} (a, b) & \text{if } a + b > c + d \\ (c, d) & \text{if not.} \end{cases}$$

is such that every element of $\mathscr{E}_N$ is the image of exactly two elements of $\mathscr{M}_N$. Moreover, the length of the factorization of a matrix M of $\mathscr{M}_N$ is exactly the sum $\sum_{i=1}^{n} q_i$ where $q_1, \ldots, q_n$ are the quotients of Euclid's algorithm applied to the couple $\phi(M)$.

Recall that Euclid's algorithm applied to a couple (a, b)

$$\begin{aligned} a &= bq_1 + r_1 \\ b &= r_1 q_2 + r_2 \\ r_1 &= r_2 q_3 + r_3 \\ &\vdots \\ r_{n-2} &= r_{n-1} q_n + r_n \text{with } r_n = 0 \text{ and } r_{n-1} = 1 \end{aligned}$$

is equivalent to computing the continued fraction expansion of b/a,

$$\frac{b}{a} = \cfrac{1}{q_1 + \cfrac{1}{q_2 + \cfrac{1}{\ldots + \cfrac{1}{q_n}}}} = [q_1, q_2, \ldots, q_n].$$

The formulation of the problem in terms of continued fractions is more adapted to our purposes because ergodic theory can be brought in. More precisely, following an idea due to Dixon, ([Dix70], see also [Dau93]) we shall proceed as follows.

- First obtain an upper bound for the sum $\sum_{i=1}^{n} q_i$ associated to "almost" every real number of $[0,1]$ by using the ergodic theorem.
- Deduce from it by "transfer" techniques a similar result for "almost" (in a sense to be specified later) all rational numbers of $[0,1]$.

Proof of theorem 5. We first prove:

Theorem 6. *For almost every real number $\omega \in [0,1]$ the n first denominators of its continued fraction expansion $q_1(\omega), q_2(\omega), \ldots, q_n(\omega)$ satisfy :*

$$\lim_{n\to\infty} \frac{q_1(\omega)^{1-\epsilon} + q_2(\omega)^{1-\epsilon} + \ldots + q_n(\omega)^{1-\epsilon}}{n} = \frac{1}{\ln 2}\int_0^1 \frac{q_1(x)^{1-\epsilon}}{1+x}\,dx$$

where ϵ is any real number in $]0,1[$. (for almost every real means that the set of real numbers where the property is not true has Lebesgue measure 0).

Proof. This is a straightforward application of the ergodic theorem. The interested reader may refer to [Bil65]. To use it in our case, it must be shown that

1. the transformation $\mathcal{T}$ over $[0,1[$ where

$$\begin{cases} \mathcal{T}(\omega) = \frac{1}{\omega} - \lfloor\frac{1}{\omega}\rfloor \text{if } \omega \neq 0, \\ \mathcal{T}(0) = 0 \end{cases}$$

preserves Gauss's measure $\mathcal{P}$ on the class of Borel's sets $\mathcal{F}$.

$$\mathcal{P}(A) = \frac{1}{\ln 2}\int_A \frac{dx}{1+x} \quad \text{for } A \in \mathcal{F}.$$

2. $\mathcal{T}$ has to be ergodic under $\mathcal{P}$, that is each $\mathcal{T}$-invariant Borel set A, (i.e $\mathcal{T}^{-1}A = A$), is of $\mathcal{P}$-measure 0 or 1.

A proof of these fact can be found in [Bil65] pp 43–45.

Thus we can apply the ergodic theorem which asserts that for every integrable function f on the unit interval,

$$\lim_{n\to\infty} \frac{1}{n}\sum_{k=0}^{n-1} f(\mathcal{T}^k\omega) = \frac{1}{\ln 2}\int_0^1 \frac{f(x)}{1+x}\,dx \text{ almost everywhere.}$$

The theorem follows, by taking $f(x) = q_1^{1-\epsilon}(x)$.

□

Note here that we could not choose $f(x) = q_1(x)$, which would give us direct information on the sum of denominators, because this function is not integrable. So we choose integrable functions which approach $q_1(x)$ as ϵ tends to 0.

We have the following corollary.

Corollary 1 *For every* $0 < \epsilon < 1$, *we have*

$$\lim_{n\to\infty} \frac{q_1(\omega)^{1-\epsilon} + q_2(\omega)^{1-\epsilon} + \ldots + q_n(\omega)^{1-\epsilon}}{n} \le \frac{\epsilon+1}{\epsilon \ln 2}$$

almost everywhere in $[0,1]$.

Proof. From theorem 6 we have

$$\begin{aligned}
\lim_{n\to\infty} \frac{q_1(\omega)^{1-\epsilon} + q_2(\omega)^{1-\epsilon} + \ldots + q_n(\omega)^{1-\epsilon}}{n} &= \frac{1}{\ln 2}\int_0^1 \frac{q_1(x)^{1-\epsilon}}{1+x}\,dx \\
&= \frac{1}{\ln 2}\sum_{k=1}^{\infty}\int_{1/(k+1)}^{1/k} \frac{k^{1-\epsilon}}{1+x}\,dx \\
&= \frac{1}{\ln 2}\sum_{k=1}^{\infty} k^{1-\epsilon}\ln(1+\frac{1}{k(k+2)}) \\
&\le \frac{1}{\ln 2}\sum_{k=1}^{\infty}\frac{1}{k^{\epsilon}(k+2)} \\
&\le \frac{1}{\ln 2}(1+\frac{1}{\epsilon})
\end{aligned}$$

□

>From this corollary it follows that

Corollary 2 *For almost all real numbers* $\omega \in [0,1]$, *there exists an integer* N_0 *such that for* $n \ge N_0$

$$\sum_{i=1}^{n} q_i(\omega) \le n^{1+\delta} \tag{4}$$

for every constant $\delta > 0$.

Proof. This is a consequence of the preceding corollary. It comes from the following inequality.

if $\sum_{i=1}^{n} x^{1-\epsilon} = C$, where the x_i are positive numbers, then $\sum_{i=1}^{n} x_i \le C^{1/(1-\epsilon)}$. Hence,

$$\sum_{i=1}^{n} q_i(\omega) \le \left(\frac{n}{\epsilon}\right)^{1/(1-\epsilon)} \left(\frac{\epsilon+1}{\ln 2}\right)^{1/(1-\epsilon)}$$

We upper bound the right quantity by $n^{1+\delta}$ for $\delta > 2\epsilon$, and n sufficiently large.

□

Definition 7. We will say in what follows that a non-rational real number α is *exceptional* if there is a strictly increasing sequence of positive integers (u_n) such that the sum of the u_n first denominators of the continued fraction expansion of α does not satisfy inequality (4).

We will say that a rational number is *exceptional* if the sum of the denominators of its complete continued fraction expansion does not satisfy inequality (4).

Now that we have proved that for almost every real number in $[0, 1]$ the sum of the n first denominators does not grow too fast (roughly speaking it grows at most a little bit faster than $n \ln n$), we wish to extend this result to a smaller set, namely the set E_N of rationals of [0,1] of the form a/b, with a and b coprime such that $0 \leq a \leq b \leq N$. For this purpose we will proceed as follows.

- we will suppose that a non negligible part of rationals of this form are exceptional.
- then we will show that from these exceptional rationals, we are able to construct a class of real numbers of non-zero measure, for which (4) does not apply. Hence we will arrive to a contradiction to corollary 2.

We denote

$$\alpha(N) = \frac{\#\{\text{exceptional rationals of } E_N\}}{\# E_N}$$

We suppose now that $\alpha(N)$ does not tend to zero as N tends to infinity.

This means that there is a strictly increasing sequence of integers (u_n) and a real $\epsilon > 0$ such that $\alpha(u_n) \geq \epsilon$ for every integer n. So using these exceptional rational element), we construct a "non negligible" set of exceptional real numbers. The key of our construction is the notion of fondamental interval. For a sequence of n integers $a_1, a_2, \dots, a_n$ we denote by $'_{a_1,a_2,\dots,a_n}$ the subinterval of [0,1] made up of those numbers whose continued fraction expansion starts with $a_1, a_2, \dots, a_n$. For each exceptional rational of E_N, with whole continued fraction expansion $[a_1, a_2, \dots, a_n]$, we associate the fondamental interval $'_{a_1,a_2,\dots,a_n}$ to whom it belongs. We denote by B_N the union of all intervals $'_{a_1,a_2,\dots,a_n}$, when we consider all exceptional rationals of E_N. There are two basics facts which come into play to show that this set B_N is not too small.

- the distance between two elements of E_N is at least $1/N(N-1)$.
- the length of any of these fondamental intervals (associated to an element of E_N) is at least $1/2N^2$. A proof of this last point can be found in [Khi64]

Putting these two facts together we obtain that the Lebesgue-measure of B_N is at least the number of exceptional elements of E_N multiplied by $1/2N^2$. Hence,

$$\lambda(B_N) \geq \alpha(N) \# E_N \frac{1}{2N^2}.$$

The estimation of $\#E_N$, the number of couples of coprime integers (a, b) such that $0 \leq a \leq b \leq N$ is classical. One has, see e.g. 18.5 of [HW79],

Proposition 8.

$$\#E_N = \frac{3}{\pi^2}N^2 + O(N\ln N).$$

So that, asymptotically,

$$\lambda(B_N) \geq \frac{3}{2\pi^2}\alpha(N).$$

Hence we can extract a subsequence (v_n) of (u_n) such that

$$\lambda(B_{v_n}) \geq 0.15\epsilon \text{ for every } n.$$

What have we thus obtained ? This set B_{v_n} is a subset of $[0,1]$ such that for every real number x of this set there exists a length $l(x)$ of the continued fraction expansion of x which does not obey inequality (4). This set B_{v_n} has non zero measure, but there is no contradiction yet, with corollary 2. Nevertheless using this collection of sets B_{v_n} we will now proceed to show how to construct a non-negligible set of exceptional real numbers.

We have to tackle two problems to construct such a set.

1. first we have no idea of the length l which do not obey inequality (4). This problem will be solved by restricting somehow the set B_{v_n}.
2. we have not constructed a set of reals for which an infinite number of lengths of the continued fraction process do not obey (4), whereas this is exactly what we want to do.

Problem 1 is solved is follows.

Let us recall that the length l is obtained as the total length of the continued fraction expansion of a rational whose denominator is an integer less than some v_i. Here a by now classical theorem of Dixon [Dix70] comes into play.

Theorem 9. *If we denote by $L(a/b)$ the total length of the continued fraction expansion of the rational a/b, then the following inequality holds.*

$$\#\left\{a/b \in E_N, |L(a/b) - \frac{12\ln 2}{\pi^2}\ln N| \geq (\ln N)^{1/2+\epsilon}\right\} = o(N^2) \quad \textit{for every } \epsilon > 0$$

A proof of this theorem can be found in [Dix70] or [Dau93].

The consequence of this is that we can consider that there is a non negligible part of the set of fondamental intervals which constitute B_{v_n} which are obtained by rationals whose rational fraction expansion length is in an interval of the form

$$[\frac{12\ln 2}{\pi^2}\ln v_n - \ln v_n^{3/4}, \frac{12\ln 2}{\pi^2}\ln v_n + \ln v_n^{3/4}].$$

In other words we can consider that

$$\lambda(B'_{v_n}) > 0.1\epsilon,$$

where B'_{v_n} is the union of all the fondamental intervals which contain only the exceptional rationals a/b (where a and b are coprime and $0 < a < b \leq v_i$)

for which the length of the continued fraction expansion lies in $[\frac{12\ln 2}{\pi^2}\ln v_n - \ln v_n^{3/4}, \frac{12\ln 2}{\pi^2}\ln v_n + \ln v_n^{3/4}]$.

To solve problem 2 we reason as follows. Define $F_n = \cup_{i=n}^{\infty} B'_{v_i}$. The F_n make up a decreasing sequence of measurable sets and since $\lambda(F_n) \geq \lambda(B'_{v_n}) \geq 0.1\epsilon$, we have that $\lambda(\cap_{n\geq 1} F_n) \geq 0.1\epsilon$. What does this set $\cap_{n\geq 1} F_n$ represent? In fact, it is easy to see that it is the class of reals of the unit interval [0,1], which has the following property : for each real ω of this class there exists a strictly increasing sequence (w_n) of positive integers, such that for each w_n, there exists an integer l_n in the interval $[\frac{12\ln 2}{\pi^2}\ln w_n - \ln w_n^{3/4}, \frac{12\ln 2}{\pi^2}\ln w_n + \ln w_n^{3/4}]$, for which the sum of the l_n first denominators of ω do not obey inequality (4). This clearly implies that $\cap_{n\geq 1} F_n$ is a set of reals whose measure is strictly greater than 0, and which do not satisfy (4), hence the contradiction with corollary 2. We have therefore proved

$$\lim_{N\to\infty} \alpha(N) = 0.$$

In other words, when N goes to infinity, a random pair of coprime integers a, b in $[0, N]$ satisfy with probability tending to 1,

$$\sum_{i=1}^{n} q_i \leq n^{1+\delta}$$

for every constant $\delta > 0$. where $q_1, \ldots, q_n$ is the set of quotients that appear when we apply Euclid's algorithm to (a, b). The proof of theorem 5 follows by applying once more theorem 9 to obtain that with probability tending to 1 as $N \to \infty$,

$$n \leq \ln N$$

□

To summarize the results of this section, what we have proved is that, assuming that the length of the factorization of the output M of algorithm 3 is comparable to that of a random matrix with coefficients of order $O(p^2)$, then

with probability $\to 1$ *as* $p \to \infty$ *the length of of the factorisation of* M *is smaller than* $(\ln p)^{1+\delta}$ *for every constant* $\delta > 0$.

Remark P. Liardet has pointed out to us that theorem 4 can be improved. More precisely by using the theorem given in [DV86] and Borel and Bernstein's theorem (see theorem 30 of [Khi64]) one can show that the expression $(\ln N)^{1+\delta}$ in theorem 4 may be replaced by $\ln N(\ln\ln N)^{1+\delta}$, or by any expression $\phi(\ln N)$, where $\phi(n)$ is any positive function of the positive integer n such that the series $\sum_{n=1}^{\infty} \frac{1}{\phi(n)}$ converges.

5 An example

We have tried a simplified version of the attack on the function $\mathscr{H}_1$ described in section 3 for a prime p of order 2^{100}. More precisely we have chosen the first prime exceeding 2^{100}, namely

$$p = 1267506002282294014967032056 53.$$

By modifying slightly algorithm 3 we have searched for a matrix of the form

$$\begin{pmatrix} a & b \\ b+p & c \end{pmatrix}$$

where

$$ac - b(b+p) = 1. \tag{5}$$

To find it we have chosen for c a prime of order $O(p)$ such that p^2-4 is a quadratic residue modulo c. This comes from equality (5) which implies $b^2 + bp + 1 = 0 \bmod c$, hence we choose:

$$\begin{cases} b \equiv \dfrac{-p+\sqrt{p^2-4}}{2} \bmod c, \quad 0 \le b < c-1 \\ a = \dfrac{b^2+pb+1}{c} \end{cases}$$

This has given us :

$$\begin{cases} c = 541818800100000000000000000307 \\ b = 450891106926957321832368570045 8 \\ a = 480714269218968140854238251367 7 \end{cases}$$

We have applied Euclid's algorithm simultaneously on (a, b) and $(b+p, c)$. We have obtained the following quotients.

$$15, 8, 2, 2, 2, 8, 1, 1, 6, 6, 44, 68, 2, 1, 1, 2, 1, 1, 1, 1, 1, 11, 4, 1006,$$
$$1, 2, 1, 3, 1, 1, 1, 1, 64, 55, 1, 4, 1, 1, 1, 2, 34, 1, 6, 1, 5, 3, 1, 1, 3, 1, 3, 23, 1, 4, 1.$$

Thus,

$$BA^4BA^{23}B^3AB^3ABA^3B^5AB^6AB^{34}A^2BABA^4BA^{55}B^{64}ABABA^3BA^2B$$
$$A^{1006}B^4A^{11}BABABA^2B^{68}A^{44}B^6A^6BAB^8A^2B^2A^2B^8A^{15} = \begin{pmatrix} a & b \\ b+p & c \end{pmatrix}$$

$$B^{15}A^8B^2A^2B^2A^8BAB^6A^6B44A^{68}B^2ABABAB^{11}A^4B^{1006}AB^2AB^3AB$$
$$ABA^{64}B^{55}AB^4ABAB^2A^{34}BA^6BA^5B^3ABA^3BA^3B^{23}AB^4A = \begin{pmatrix} a & b+p \\ b & c \end{pmatrix}.$$

These matrices are the same modulo p, this yields a collision for the hashing scheme. The lengths of these two colliding messages are the same and are equal to the sum of those quotients, that is 1423 bits.

6 The flaw associated to $\mathcal{H}_1$ and other choices of generators

The attack described above on the hash function $\mathcal{H}_1$ works essentially because A_1 and B_1 generate a substancial proportion of the matrices of $SL_2(\mathbf{Z})$ (proposition 3). To find collisions, it suffices therefore to obtain, by random search methods, matrices of $SL_2(\mathbf{Z})$ that reduce modulo p to a given matrix of $SL_2(\mathbf{F}_p)$. With non-negligible probability they will factorize over $SL_2(\mathbf{Z})$ into a product of A_1's and B_1's which is easily found.

If we want to prevent any attack of this sort, we should therefore choose matrices A and B that generate a very slim subset of $SL_2(\mathbf{Z})$, even though they generate modulo p the whole of $SL_2(\mathbf{F}_p)$. We should also keep A and B simple so that the speed of computation of $\mathcal{H}_{G,A,B}$ stays acceptable, and we wish to keep the local modification property, i.e. guarantee that the Cayley graph $\mathcal{G}(G,\{A,B\})$ has a large girth ∂. For reasons relevant to the next section we prefer to identify Id with $-Id$, i.e. deal with $G = PSL_2(\mathbf{F}_p)$. To guarantee a large girth, our strategy is to choose A and B that generate a free submonoid of $PSL_2(\mathbf{Z})$, so that following Margulis, [Mar82] and [Zém91],

$$\partial \geq \log_\alpha(p/2)$$

where $\alpha = \max(||A||, ||B||)$ and $||\ ||$ stands for the usual Euclidean norm. A natural choice of generators that meet those requirements are $A = A_1^i, B = B_1^i$ (with $i \geq 2$). For example,

$$A = \begin{pmatrix} 1 & 2 \\ 0 & 1 \end{pmatrix}, \quad B = \begin{pmatrix} 1 & 0 \\ 2 & 1 \end{pmatrix}$$

or

$$A = \begin{pmatrix} 1 & 4 \\ 0 & 1 \end{pmatrix}, \quad B = \begin{pmatrix} 1 & 0 \\ 4 & 1 \end{pmatrix}$$

Another choice is the following, denote by S and T the classical generators of $SL_2(\mathbf{Z})$, i.e.

$$S = \begin{pmatrix} 1 & 1 \\ 0 & 1 \end{pmatrix}, \quad T = \begin{pmatrix} 0 & -1 \\ 1 & 0 \end{pmatrix}$$

and put $A = S$, $B = S^iTS$ (with $i \geq 2$). It is not difficult to check that A and B generate a sparse free submonoid of $PSL_2(\mathbf{Z})$. This choice has the advantage over the preceding one that we can guarantee a small diameter to the graph $\mathcal{G}(G,\{A,B\})$: this will be elaborated on in the next section.

7 diameter issues

A lot of work has been devoted to the search for Cayley graphs with a small diameter, see e.g. [BKL89]. Recall that the diameter of a directed graph is the largest distance $d(v,w)$ between two vertices v and w, $d(v,w)$ being the smallest number of edges of a directed path joining v to w). This is also relevant to our hashing scheme because a relatively small diameter is necessary to ensure that every element of G is the hashed value of some reasonably-sized text (clearly a desirable feature of a hash function). Existing studies concern non-directed Cayley graphs though, (for which $\mathcal{S} = \mathcal{S}^{-1}$). This does not suit us, because we should not have both an element s and its inverse s^{-1} in $\mathcal{S}$,otherwise the factorisation $ss^{-1} = 1$ yields trivial collisions. We will therefore draw upon existing techniques for estimating the diameter of Cayley graphs, see e.g. [BKL89], [AM85], [Chu89], and adapt them to the directed case for our purposes. Our main result in this section is that the Cayley graph $\mathcal{G}_p = \mathcal{G}(G, \{A, B\})$ with $A = S$, $B = S^2TS$, has a diameter in $O(\log p)$, with an acceptable constant, so that hashed values of megabyte-texts will range over all of $G = PSL_2(\mathbf{F}_p)$. Of course the proof will be completely non constructive ; (a constructive method would be equivalent to breaking the associated hashing scheme).

7.1 Notation and plan of proof

A graph (directed or not) with vertex set V and edge set E will be denoted by (V, E). If $\mathcal{G}$ is the Cayley graph $\mathcal{G}(G, \mathcal{S})$, then $\mathcal{G}^*$ will denote the corresponding non directed graph (obtained from $\mathcal{G}$ by suppressing the orientation of the edges) i.e. the Cayley graph $\mathcal{G}(G, \mathcal{S} \cup \mathcal{S}^{-1})$. Therefore $\mathcal{G}_p^*$ will denote $\mathcal{G}(PSL_2(\mathbf{F}_p), \{A, B, A^{-1}, B^{-1}\})$.

If X is a subset of vertices of a graph, denote by $N_+(X)$ ($N_-(X)$) the set of vertices not in X, that are the endpoints of an edge with its initial point in X (that are the initial points of an edge with its endpoint in X). In the Cayley graph case, $N_+(X) = X\mathcal{S} \setminus X, N_-(X) = X\mathcal{S}^{-1} \setminus X$. Let $N(X)$ denote $N(X) = N_+(X) \cup N_-(X)$.

A method is indicated in [BKL89] to prove that $\mathcal{G}_p^*$ has a diameter in $O(\log p)$; it cannot be deduced from it directly, however, that $\mathcal{G}_p$ has a diameter in $O(\log p)$. We will need to use the "expansion" properties of those graphs. Following [AM85], call c a *magnifying coefficient* of a non directed graph with vertex set V whenever

$$\text{for all subsets } X \text{ of } V \text{ such that } |X| \leq \frac{|V|}{2}, \quad |N(X)| \geq c|X|$$

For a directed graph we will also call c a magnifying coefficient when

$$\text{for all subsets } X \text{ of } V \text{ such that } |X| \leq \frac{|V|}{2}, \begin{cases} |N_+(X)| \geq c|X| \\ |N_-(X)| \geq c|X| \end{cases}$$

It is reasonably straightforward to obtain that if the graphs $\mathcal{G}_p$ and $\mathcal{G}_p^*$ (in directed and non directed cases) have "good expansion properties" (i.e. magnifying coefficients independent of p) then they have a diameter in $O(\log p)$; it is also possible to prove that if the non oriented versions $\mathcal{G}_p^*$ of the graphs $\mathcal{G}_p$

have good expansion properties, then the oriented versions $\mathcal{G}_p$ also have good expansion properties. This is essentially the object of the next lemmas, namely to reduce the study of the diameter of $\mathcal{G}_p$ to the study of the expanding properties of a non directed graph. Actually the non directed graph we will reduce our problem to will not be $\mathcal{G}_p^*$ but rather $\mathcal{G}(PSL_2(\mathbf{F}_p), \{S, S^{-1}, T\})$ (S and T being defined as in section 6), at which point arithmetic considerations can be brought in, following [BKL89].

7.2 Reduction to the study of $\mathcal{G}(PSL_2(F_p), \{S, S^{-1}, T\})$

Lemma 10. *Let (V, E) be a directed graph, and suppose it has magnifying coefficient c. Then the diameter D verifies*

$$D \leq 2\log_{(1+c)} \frac{|V|}{2} + 1 \leq \frac{2}{c} \ln \frac{|V|}{2} + 1$$

Proof. let v and w be any two vertices of V. Denote by $N_+^{[k]}(v)$ ($N_-^{[k]}(w)$) the subset of vertices of V reachable from v by paths of length k or less (from which w can be reached by paths of length k or less). In other words define inductively

$$N_+^{[0]}(v) = v; \qquad N_+^{[k+1]}(v) = N_+^{[k]}(v) \cup N_+(N_+^{[k]}(v))$$

$$N_-^{[0]}(w) = w; \qquad N_-^{[k+1]}(w) = N_-^{[k]}(w) \cup N_-(N_-^{[k]}(w))$$

That c is a magnifying coefficient of (V, E) means that as long as

$$|N_+^{[k-1]}(v)| \leq \frac{|V|}{2} \text{ and } |N_-^{[h-1]}(v)| \leq \frac{|V|}{2},$$

we have :

$$\begin{cases} |N_+^{[k]}(v)| \geq (1+c)^k \\ |N_-^{[h]}(w)| \geq (1+c)^h \end{cases}$$

take $k > \log_{(1+c)} \frac{|V|}{2}$ and $h \geq \log_{(1+c)} \frac{|V|}{2}$, then $N_+^{[k]}(v)$ and $N_-^{[h]}(w)$ necessarily have a common vertex z, and there is therefore a path joining v to w,passing through z, with length $k+h$ or less.

□

In the non directed graph case, some other methods can be brought in to improve on the constant $\frac{2}{c}$, (see [AM85], [Chu89]) but by methods that do not seem to generalise to the directed case, (at least not for noncommutative Cayley graphs), and that do not represent a substantial improvement when c is small, both of which are the case here.

Lemma 11. *Suppose $\mathcal{G}(PSL_2(\mathbf{F}_p), \{S, S^{-1}, T\})$ has a magnifying coefficient c, then $\frac{c}{6}$ is a magnifying coefficient for $\mathcal{G}(PSL_2(\mathbf{F}_p), \{A, B\})$.*

Proof. suppose the contrary, then there is a subset $X \subset PSL_2(\mathbf{F}_p)$, with $|X| \leq |PSL_2(\mathbf{F}_p)|/2$ such that in $\mathcal{G}(PSL_2(\mathbf{F}_p), \{A, B\})$, either $N_+(X)$ or $N_-(X)$ has cardinality less than $\frac{c}{6}|X|$. Suppose, for example, that it is the case for $N_+(X)$ (the other case being analoguous), then

$$|XA \setminus X| < \frac{c}{6}|X| \text{ and } |XB \setminus X| < \frac{c}{6}|X|$$

therefore (recall that $A = S$ and $B = S^2TS$)

$$|XS \cap X| > (1 - \frac{c}{6})|X|, \text{ equivalently,} \tag{6}$$

$$|XS^2 \cap XS| > (1 - \frac{c}{6})|X|, \text{ and} \tag{7}$$

$$|XS^2TS \cap X| > (1 - \frac{c}{6})|X| \tag{8}$$

(8) is equivalent to

$$|XS^2T \cap XS^{-1}| > (1 - \frac{c}{6})|X| \tag{9}$$

but (6) means that $|XS^{-1} \cap X| > (1 - \frac{c}{6})|X|$, and applied to (9) this yields

$$|XS^2T \cap X| > (1 - \frac{2c}{6})|X|$$

or equivalently, since $T = T^{-1}$ in $PSL_2(\mathbf{F}_p)$,

$$|XS^2 \cap XT| > (1 - \frac{2c}{6})|X|$$

applying successively (7) and (6), we obtain :

$$|XS \cap XT| > (1 - \frac{3c}{6})|X|$$

$$|X \cap XT| > (1 - \frac{4c}{6})|X| \text{ i.e.}$$

$$|XT \setminus X| < \frac{4c}{6}|X| \tag{10}$$

(6) yields also

$$|XS \setminus X| < \frac{c}{6}|X| \tag{11}$$

$$|XS^{-1} \setminus X| < \frac{c}{6}|X| \tag{12}$$

and since in $\mathscr{G}(PSL_2(\mathbf{F}_p), \{S, S^{-1}, T\})$ we have

$$|N(X)| \leq |XT \setminus X| + |XS \setminus X| + |XS^{-1} \setminus X|$$

adding (10), (11), and (12) we obtain

$$|N(X)| < c|X|$$

a contradiction.

□

7.3 The expanding properties of $\mathscr{G}(PSL_2(F_p), \{S, S^{-1}, T\})$

The method we are about to describe to obtain a magnifying coefficient for the graph $\mathscr{G}(PSL_2(\mathbf{F}_p), \{S, S^{-1}, T\})$ is hinted at in [BKL89] ; we will present it here in more detail for the sake of completeness. Denote by Γ the modular group $PSL_2(\mathbf{Z})$; for more detailed information on Γ and related arithmetic, see for example [Ser73], [Kob84].

For a prime number p denote by Γ_p the congruence subgroup of Γ

$$\Gamma_p = \left\{ \begin{pmatrix} a & b \\ c & d \end{pmatrix} \in \Gamma \; ; \; \begin{pmatrix} a & b \\ c & d \end{pmatrix} \equiv \pm \begin{pmatrix} 1 & 0 \\ 0 & 1 \end{pmatrix} \bmod p \right\}$$

we have

$$\Gamma / \Gamma_p \simeq PSL_2(\mathbf{F}_p) \simeq SL_2(\mathbf{F}_p)/\{1, -1\}$$

Let us denote by $\mathscr{H}_p$ the Cayley graph $\mathscr{G}(PSL_2(\mathbf{F}_p), \{S, S^{-1}, T\})$. Let $\mathbf{H}$ denote the upper complex half-plane $\{z \mid Im z > 0\}$. Recall that Γ acts on $\mathbf{H}$ in the following way :

$$\begin{pmatrix} a & b \\ c & d \end{pmatrix} z = \frac{az + b}{cz + d}$$

A fundamental domain for this action is the region

$$\mathbf{H} = \{z = x + iy \; ; \; -\frac{1}{2} \leq x \leq \frac{1}{2}, \; x^2 + y^2 \geq 1\}$$

For any $z \in \mathbf{H}$, identify z with all Mz when $M \in \Gamma_p$. We obtain a (Riemann) surface $\Sigma_p = \mathbf{H} \setminus \Gamma_p$ on which $PSL_2(\mathbf{F}_p) = \Gamma / \Gamma_p$ acts naturally. Σ_p provides us with the following geometric representation of the Cayley graph $\mathscr{H}_p$: the vertices of $\mathscr{H}_p$ can be seen as the domains $g\mathbf{D}$ for $g \in PSL_2(\mathbf{F}_p)$, and any two domains $g_1\mathbf{D}$ and $g_2\mathbf{D}$ are adjacent in $\mathscr{H}_p$ iff they intersect in a curve in Σ_p. See figure 1 for an illustration.

The result that is brought in at this point is the following [Sel65] :

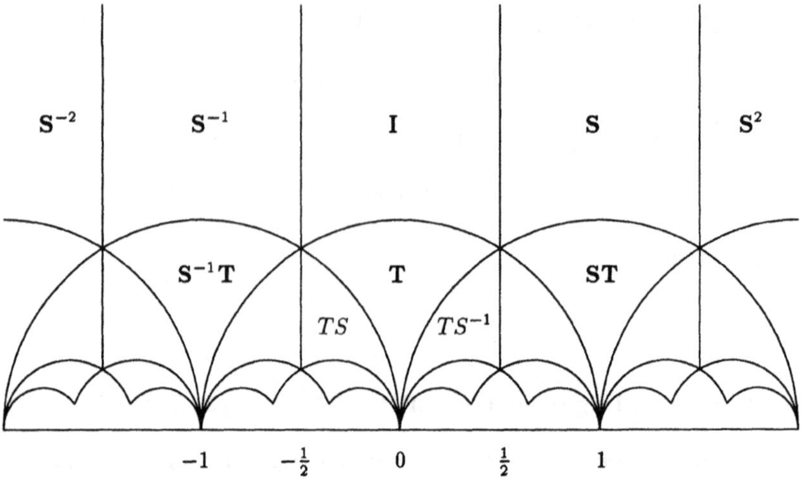

Fig. 1. A representation of the Cayley graph $\mathcal{G}(PSL_2(\mathbf{F}_p), \{S, S^{-1}, T\})$

Proposition 12. *For any real-valued function f defined on* $\mathbf{H}$ *and invariant under $_p$, and satisfying the following properties :*

$$f \text{ is continuously differentiable} \qquad (\alpha)$$

$$\iint_{D_p} f(x,y)\frac{dxdy}{y^2} = 0 \qquad (\beta)$$

$$\iint_{D_p} f^2(x,y)\frac{dxdy}{y^2} = 1 \qquad (\gamma)$$

we have

$$\frac{3}{16} \leq \iint_{D_p} \left(\left(\frac{\partial f}{\partial x}\right)^2 + \left(\frac{\partial f}{\partial y}\right)^2\right) dxdy$$

where D_p denotes a union of regions g $\mathbf{D}$ *when g describes a set of representatives of $PSL_2(\mathbf{F}_p)$ (i.e. D_p is a fundamental domain for the action of $_p$).*

Now choose any subset X of vertices of $\mathcal{H}_p$ such that $|X| \leq |PSL_2(\mathbf{F}_p)|/2$. A lower bound on $|N(X)|/|X|$ will be achieved through a lower bound on $|E(X)|/|X|$ where $E(X)$ denotes the set of edges joining vertices of X to vertices not in X. The idea is to consider the function g_X defined by $g_X(z) = 1$ if z is in the interior of $g\mathbf{D}$ with $g \in X$, and $g_X(z) = -e$ if z is in the interior of $g\mathbf{D}$

with $g \notin X$, e being an appropriatly chosen positive constant. Then the function g_X is modified so as to provide us with a function f_X satisfying conditions $\alpha), \beta), \gamma)$ above ; the point is that the function f_X will essentially be constant (with $gradf_X = 0$) except for regions corresponding to frontiers between domains $g_1\mathbf{D}$ and $g_2\mathbf{D}$ with $g_1 \in X$ and $g_2 \notin X$. Then the application of proposition 12 will give us an inequality of the form

$$\frac{3}{16} \leq \frac{1}{c_1|X|}(c_2|E(X)| + c_3|X|)$$

where c_1, c_2, c_3 are constants, and hence yield a constant lower bound on the ratio $|E(X)|/|X|$. From this, the fact that $\frac{|E(X)|}{|X|} \leq 3\frac{|N(X)|}{|X|}$, and lemmas 10 and 11 we obtain a constant κ such that

Proposition 13. *The diameter D of $\mathcal{G}_p$ satisfies $D \leq \kappa \ln p$*

With some tedious but reasonably straightforward estimations of the above constants c_1, c_2, c_3 that we wish to spare the reader, it can be grossly estimated that the constant κ is of the order 600 000 or less, which is just about satisfactory since hash functions are supposed to hash texts of several megabytes.

Acknowledgement. We wish to thank P. Liardet for careful reading of a preliminary version of this work.

References

[AM85] N. Alon and V. D. Milman. λ_1, isoperimetric inequalities for graphs, and superconcentrators. *Journal of Comb. Theory Ser. B*, 38:73–88, 1985.

[Bil65] P. Billingsley. *Ergodic theory and information.* J. Wiley and son, New York, London, Sydney, 1965.

[BKL89] L. Babai, W. M. Kantor, and A. Lubotsky. Small-diameter cayley graphs for finite simple groups. *Europ. J. of Combinatorics*, 10:507–522, 1989.

[Cam87] P. Camion. Can a fast signature scheme without secret key be secure ? In *proc. AAECC*, pages 187–196. Springer-Verlag Lec. N. Comp. Sci. 228, 1987.

[Chu89] F. R. K. Chung. Diameters and eigenvalues. *J. Am Math. Soc*, 2:187–196, 1989.

[Dam89] I. B. Damgård. Design principles for hash functions. In *Crypto*, 1989.

[Dau93] H. Daudé. *Des fractions continues, la réduction des réseaux : analyse en moyenne.* PhD thesis, Université de Caen, France, 1993.

[Dix70] J. D. Dixon. The number of steps in the euclidean algorithm. *Journal of Number Theory*, pages 414–422, 1970.

[DV86] H. Diamond and J. Vaaler. Estimates for partial sums of continued fraction partial quotients. *Pacific Journal of mathematics*, 122(1):73–82, 1986.

[GC88] P. Godlewski and P. Camion. Manipulations and errors, detection and localization. In *Advances in Cryptology, EUROCRYPT-88*, pages 96–106. LNCS 330 Springer-Verlag, 1988.

[GTV88] M. Girault, P. Toffin, and B. Vallée. How to guess l-th roots modulo n by reducing lattice points. In *First international joint conference of ISSAC-88 and AAECC-6*, july 1988.

[HW79] G. H. Hardy and E. M. Wright. *An introduction to the theory of numbers.* Oxford University Press, 1979.

[Khi64] A. Ya. Khinchin. *Continued fractions (english translation).* The University of Chicago Press, 1964.

[Kob84] N. Koblitz. *Introduction to Elliptic Curves and Modular Forms.* Springer-Verlag, 1984.

[Kur60] A. G. Kurosh. *The theory of groups.* NED, 1960.

[Mar82] G. A. Margulis. Explicit constructions of graphs without short cycles and low density codes. *COMBINATORICA*, 2(1):71–78, 1982.

[Sel65] A. Selberg. On the estimation of fourier coefficients of modular forms. *AMS Proc. Symp. Pure Math.*, 8:1–15, 1965.

[Ser73] J-P. Serre. *A course in arithmetic.* Springer-Verlag, 1973.

[Zém91] G. Zémor. Hash functions and graphs with large girths. In *EUROCRYPT 91*. LNCS 547 Springer-Verlag, 1991.

On Constructions for Optimal Optical Orthogonal Codes

Sara Bitan and Tuvi Etzion*

Computer Science Department
Technion — Israel Institute of Technology
Haifa 32000, Israel

Abstract. An optical orthogonal code, with $\lambda = 1$ is a family, of size l, of w-sets of integers modulo n in which no difference is repeated. If all the differences modulo n appear then this code coincide with the well known design called difference family and the code is called perfect. It is clear that if $l(w-1)w \leq n-1 < (l+1)(w-1)w$ then no more w-sets can be added to the code and hence the code is optimal. We give some new constructions for difference families and also constructions of optimal codes which are not difference families.

1 Introduction

An (n, w, λ) optical orthogonal code (OOC)[CSW89], C, is a family of $(0,1)$ sequences of length n and constant weight w which have cross-correlation and out of phase auto-correlation values which do not exceed λ; i.e.

$$\sum_{i=0}^{n-1} x_i x_{i+\delta} \leq \lambda$$

for $x \in C$ and $0 < \delta < n$, and

$$\sum_{i=0}^{n-1} x_i y_{i+\delta} \leq \lambda$$

* This research was supported in part by the fund of promotion of research at the Technion

for each $x \neq y \in C, 0 \leq \delta < n$, where subscripts are taken modulo n. Such codes have applications in optical code-division multiple-access communication systems [CSW89], mobile radio, frequency-hopping spread-spectrum communications radar and sonar signal design [Gol64], and constructing protocol-sequence sets for the M-active-out-of-T users collision channel without feedback [MM85].

In this paper we are concerned with the case $\lambda = 1$. In this case it is more convenient to consider a word in a $(n, w, 1)$ OOC as a w-tuple. Let C be a $(n, w, 1)$ OOC, $c \in C$ is a w-tuple $(c_1, c_2, \ldots, c_w)$, where $c_i \in Z_n$ for $1 \leq i \leq w$. $x_i = 1$ iff $c_j = i$ for some j, and all the differences of the form $c_i - c_j$ in all codewords are distinct modulo n. If each element of $Z_n - \{0\}$ appears as a difference then the code coincides with the well known combinatorial design of *difference family*[Wil72]. A *difference family* is a set of w-tuples with elements from an additive group G, such that each element of $G - \{0\}$ appears exactly once as a difference. C is an *optimal OOC*(OOOC) if there is no (n, w, λ) OOC with larger size.

Most of the known results on optical orthogonal codes are derived from *Block Designs.* [BW87] showed that an optimal (n, w, λ)-OOC, C, where w and n are relatively prime, is equivalent to a cyclic $(\lambda + 1) - (n|C|, w|C|, \lambda, 1)$ Design. In [CSW89] optimal (n, w, λ)-OOOCs for $n = \frac{q^{d+1}-1}{q-1}$, and $w = q + 1$ are presented. These codes are derived from the projective geometry $PG(q, d)$. A family of optimal orthogonal codes with $n = p^{2m-1}$, $w = p^m + 1$, where p is prime, and $\lambda = 2$ (the only known construction of optimal OOC with $\lambda = 2$ which is not derived from block design) is presented in [CK90]. Families of asymptotically optimal OOCs are presented in [AGM92] and [KMZZ93].

For $\lambda = 1$ optimal optical orthogonal codes are derived from *Difference Families*[BW87], and from *Balanced Incomplete Block Designs*[CK90]. Optimal $(n, 3, 1)$-OOCs of size $\lfloor \frac{n-1}{6} \rfloor$ are presented in [BW87]. The only unresolved cases are $n \equiv 14$ or $20 \pmod{24}$, this case is solved in [BE93] where a $(n, 3, 1)$ OOOC of size $\lfloor \frac{n-2}{6} \rfloor - 1$ is constructed. The main construction for difference families is given in [Wil72]. Other constructions are given in [Bos39] and [Mat87].

Two recursive methods for constructing a difference family are given in [CC80]. A recursive method for constructing optical orthogonal code that generalizes one of Colbourn and Colbourn's recursive construction[CC80], is given in [BW87].

In Section 2 we present new constructions for optimal optical orthogonal codes. These constructions are very similar to the constructions of Wilson [Wil72] for difference families. We modify Wilson's constructions to obtain optical orthogonal codes, and also improve his constructions to obtain new difference families.

In Section 3 we present other constructions for optimal optical orthogonal codes for which the missing differences appear in equal gaps in the range between 0 and $n-1$. These constructions as well as the constructions in Section 2 make use of the theory of numbers.

In Section 4 we present the known recursive constructions for difference families. We modify these constructions to obtain OOOCs which are not necessary difference families, and show how this construction can be applied on the known codes.

2 A Generalization of Wilson's Constructions

In this section we present four constructions for difference families, and optimal optical orthogonal codes with $\lambda = 1$. We will use the following notations, as in [Wil72]. Let q be a power of a prime and α be a primitive element in $GF(q)$; for each e, $e|q-1$ (e divides $q-1$), H^e denote the multiplicative subgroup of $GF(q)-\{0\}$ of index e. Given a subgroup H of a group G, for each $x \in G$ we define the *cosets of G modulo H* by $xH = \{xh : h \in H\}$. The cosets of H^1 modulo H^e, will be denoted by $H_0^e, H_1^e, ..., H_{e-1}^e$, where $H_m^e = \{\alpha^t : t \equiv m \pmod{e}\}$; hence, $H^e = H_0^e$.

Given a word $c = (c_1, c_2, \ldots, c_w)$ in an OOC C and $\beta \in Z_n$, let $\beta c \stackrel{\Delta}{=} (\beta c_1, \beta c_2, \ldots \beta c_w)$ where the multiplications are done modulo n. For a codeword $c \in C$ let $\Delta c \stackrel{\Delta}{=} \{c_i - c_j \ : \ 1 \leq i, j \leq n, i \neq j\}$; we define $\Delta C \stackrel{\Delta}{=} \cup_{c \in C} \Delta c$; it is easy to verify that an OOC C has $\lambda = 1$ iff $|\Delta C| = w(w-1)|C|$. For a set $\Gamma = \{\gamma_1, \gamma_2, \ldots : \gamma_r\}$ we define $\Delta \Gamma \stackrel{\Delta}{=} \{\gamma_i - \gamma_j \ : \ 1 \leq i, j \leq r, i \neq j\}$.

We now describe two constructions (A and B) which generate difference families $(q, w, 1)$, where q is a power of a prime, and $(p, w, 1)$ OOOC, where p is prime.

Construction A: Let $w = 2t$, α be a primitive element in $GF(q)$, $q = 2(w-1)m + 1$, and ξ be a primitive $(w-1)^{th}$ root of unity in $GF(q)$. Let e be the largest integer $e|q-1$ such that $\{1, \xi - 1, \xi^2 - 1, ..., \xi^{t-1} - 1\} \subseteq H^e$. Let $g = g.c.d.(\frac{q-1}{e}, 2(w-1))$, $d = \frac{q-1}{ge}$, and $s = \frac{\lfloor \frac{m}{t} \rfloor}{\frac{m}{d}}$. If s is an integer such that $s|d$ and the elements of $\{1, \xi - 1, \xi^2 - 1, ..., \xi^{t-1} - 1\}$ belong to distinct cosets $\{H_{k_l}^{e\frac{d}{s}} : 1 \leq l \leq t\}$ then let

$$C = \left\{ \alpha^{i\frac{m}{s}+j}(0, 1, \xi, \xi^2, ..., \xi^{w-2}) : \begin{matrix} 0 \leq i < s \\ 0 \leq j < \frac{m}{d} \end{matrix} \right\}$$

Construction B: Let $w = 2t + 1$, α be a primitive element in $GF(q)$, $q = 2wm + 1$, and ξ be a primitive wth root of unity in $GF(q)$. Let e be the largest integer $e|q-1$ such that $\{\xi - 1, \xi^2 - 1, ..., \xi^t - 1\} \subseteq H_z^e$, for some z, $0 \leq z < e$. Let $g = g.c.d.(\frac{q-1}{e}, 2w)$, $d = \frac{q-1}{ge}$, and $s = \frac{\lfloor \frac{m}{t} \rfloor}{\frac{m}{d}}$. If s is an integer such that $s|d$ and the elements of $\{\xi - 1, \xi^2 - 1, ..., \xi^t - 1\}$ belong to distinct cosets $\{H_{k_l}^{e\frac{d}{s}} : 1 \leq l \leq t\}$ then let

$$C = \left\{ \alpha^{i\frac{m}{s}+j}(1, \xi, \xi^2, ..., \xi^{w-1}) : \begin{matrix} 0 \leq i < s \\ 0 \leq j < \frac{m}{d} \end{matrix} \right\}$$

Note, that in both constructions, $\frac{m}{d}$ is an integer by the definitions of q, g and d.

Constructions A and B produce $(q, w, 1)$ OOOCs and the proof is based on the following lemmas and theorems, whose proof can be found in [BE93]. The lemmas analyze the structure of a graph $G_w(q) = (V_w(q), E_w(q))$, defined below, whose vertices are words of weight w.

Definition 1. Let w, q, t, α and ξ be as defined in constructions A and B. We define the graph $G_w(q) \stackrel{\Delta}{=} (V_w(q), E_w(q))$ whose vertices are w-sets from $GF(q)$, $V_w(q) \stackrel{\Delta}{=} \{v_i : 0 \leq i < m\}$, where

$$v_0 = \begin{cases} (0, 1, \xi, ..., \xi^{w-2}) & \text{if } w = 2t \\ (1, \xi, ..., \xi^{w-1}) & \text{if } w = 2t + 1 \end{cases}$$

and $v_i = \alpha^i v_0$.

$$E_w(q) \stackrel{\Delta}{=} \{[v_i, v_j] : \Delta v_i \cap \Delta v_j \neq \phi, 0 \leq i < j < m\}$$

Note that $e\frac{d}{s}$ divides m; in Lemmas 2 through 4 which follow the computation of the k_l's of constructions A and B is performed modulo m, and not modulo $e\frac{d}{s}$.

Lemma 2. *For $w = 2t$, if $\forall l$, $1 \leq l < t$, $\xi^l - 1 \in H^m_{k_l}$, then $\forall i$, $0 \leq i < m$, $\Delta v_i = H^m_i \cup (\bigcup_{l=1}^{t-1} H^m_{k_l+i})$.*

Lemma 3. *For $w = 2t + 1$, if $\forall l$, $1 \leq l \leq t$, $\xi^l - 1 \in H^m_{k_l}$, then $\forall i$, $0 \leq i < m$, $\Delta v_i = \bigcup_{l=1}^{t} H^m_{k_l+i}$.*

Lemma 4. *$[v_i, v_{i+\delta}] \in E_w(q)$ iff $\delta \in \Delta\{k_l : 1 \leq l \leq t\}$, where the differences and subscripts are taken modulo m.*

Corollary 5.

1. *For $0 \leq i < m$, the degree of $v_i \in V_w(q)$ is at most $t(t-1)$.*
2. *For $0 \leq i < m$, $f_i : v \rightarrow \alpha^i v$, is an automorphism of $G_w(q)$.*

Lemma 6. *If $[v_i, v_j] \in E_w(q)$ then $i \equiv j \pmod{\frac{m}{d}}$.*

Following Lemma 6, we define subgraphs of $V_w(q)$. $G^i_w(q)$, $0 \leq i < \frac{m}{d}$, is the subgraph induced by $V^i_w(q) = \{v_r : v_r \in V_w(q) \text{ and } r \equiv i \pmod{\frac{m}{d}}\}$. Note that by Lemma 6, there is no edge which connects $G^i_w(q)$ with $G^j_w(q)$ for $i \neq j$.

Lemma 7. *$G_w^i(q)$, $0 \leq i < \frac{m}{d}$ are isomorphic connected components of $G_w(q)$.*

Corollary 8. *If $G_w^0(p)$, p prime, contains an independent set of size s, then the words corresponding to the vertices of the independent set of size $s\frac{m}{d} = \lfloor \frac{m}{t} \rfloor$ in $G_w(p)$ form a $(p, w, 1)$ OOOC; if $\frac{m}{t}$ is an integer and $G_w^0(q)$ contains an independent set of size s, then we form a $(q, w, 1)$ difference family.*

Lemma 9. *Let q be a power of a prime which satisfies the conditions in Construction A for even w, or the conditions of construction B for odd w, then the set $S_i = \{v_i, v_{\frac{m}{s}+i}, v_{2\frac{m}{s}+i}, \ldots, v_{(s-1)\frac{m}{s}+i}\}$ is an independent set of size s in $G_w^i(q)$.*

Lemma 10. *If q is a power of a prime satisfying the conditions of construction A for even w, or the conditions of construction B for odd w, then the elements of ΔC are all distinct.*

Lemma 11. *s and d of constructions A and B are independent of the choice of ξ.*

Theorem 12. *If a power of a prime $q \equiv 1 \pmod{w(w-1)}$ satisfies the conditions of construction A for even w, or the conditions of construction B for odd w, then C is a $(q, w, 1)$ difference family.*

Theorem 13. *If a prime $p \equiv 1 \pmod{2(w-1)}$, w even, satisfies the conditions of construction A, then C is a $(p, w, 1)$ OOOC.*

Theorem 14. *If a prime $p \equiv 1 \pmod{2w}$, w odd, satisfies the conditions of construction B, then C is a $(p, w, 1)$ OOOC.*

As an example, by using constructions A and B we found some OOOC, e.g. (113,7,1), (211,8,1), (181,9,1), (163,10,1), (433,10,1) OOOCs.

Definition 15. Let $q = w(w-1)m + 1$ be a power of a prime. We will say that q satisfies condition R_w[Wil72] iff

- for $w = 2t$, $\{1, \xi - 1, ..., \xi^{t-1} - 1\}$ belong to distinct cosets modulo H^t, where ξ is a primitive $(w-1)^{th}$ root of unity in $GF(q)$.
- for $w = 2t + 1$, $\{\xi - 1, ..., \xi^t - 1\}$ belong to different cosets modulo H^t, where ξ is a primitive wth root of unity in $GF(q)$.

If a power of prime q satisfies condition R_w then there exists a $(q, w, 1)$ DF. Conditions R_4 and R_5 defined by Wilson[Wil72] are equivalent to the conditions in Bose's constructions for $(q, 4, 1)$ and $(q, 5, 1)$ difference families[Bos39].

Theorem 16. *If a power of a prime $q \equiv 1 \pmod{w(w-1)}$ satisfies the condition R_w, presented by Wilson ([Wil72]) for even w, then it satisfies the conditions of construction A. If q satisfies condition R_w for odd w, then it satisfies the conditions of construction B.*

Proof. Since q satisfies R_w it follows that for $w = 2t$, $\{1, \xi - 1, \xi^2 - 1, ..., \xi^{t-1} - 1\}$ belong to distinct cosets modulo H^t; for $w = 2t + 1$, $\{\xi - 1, \xi^2 - 1, ..., \xi^t - 1\}$ belong to distinct cosets modulo H^t. Thus, $e = 1$, $d = m$, in both constructions A and B, and $s = \frac{m}{t}$ is an integer which divides d. □

Thus, difference families for all the powers of primes which satisfy condition R_w are generated by constructions A and B. Note that the converse theorem is false, and indeed we found some new difference families for prime values that don't satisfy R_w, e.g., (4231,6,1), (8821,6,1), (13681,6,1), (5419,7,1), (35533,7,1), (36919,7,1), (35393,8,1), (23761,9,1), (45361,9,1), (54001,9,1) difference families.

We will now present two constructions C and D, which are in some sense specific cases of constructions A and B, for $w = 4$ and $w = 5$. We given these constructions for two reasons:

1. These constructions produce the same difference families as constructions A and B, but they produce some OOOCs which are not produced by constructions A and B.
2. The conditions which have to be checked in these constructions are easier to compute then the conditions in constructions A and B.

Construction C: Let α be a primitive element in $GF(q)$, where $q = 6m + 1$ is a prime power. Let ξ be a primitive third root of unity in $GF(q)$, $\xi - 1 \in H_l^m$, and $k = g.c.d.(l, m)$. If $k = 1$ or $2k|m$ then let

$$C = \left\{ \alpha^{2il+j}(0, 1, \xi, \xi^2) : \begin{matrix} 0 \leq i < \lfloor \frac{m}{2k} \rfloor \\ 0 \leq j < k \end{matrix} \right\}$$

Lemma 17. *$q = 6m + 1$, m even, satisfies the conditions of construction C, iff it satisfies the conditions of construction A.*

Theorem 18. *If $2k|m$ in construction C, then C is a $(q, 4, 1)$ difference family.*

Theorem 19. *If $p = 6m + 1$ is a prime, m odd, $k = 1$, then C is a $(p, 4, 1)$ OOOC.*

Proof. The graph $G_4(p)$ in this case, is a cycle of length m, where v_i is connected to v_{i+l} and v_{i-l}. Clearly the words of C form an independent set in $G_4(p)$ in this case, C is optimal, since $|C| = \lfloor \frac{m}{2} \rfloor$. □

As in the general case, we found using construction C, difference families with $w = 4$ for primes that do not satisfy R_4; e.g., $(193, 4, 1)$, $(313, 4, 1)$ difference family.

Theorem 20. *If $p = 6m + 1$ is a prime, where $m > 1$ is prime, then $k = 1$ in construction C.*

Theorem 20 implies that the code generated by construction C, for $p = 6m + 1$, p and $m > 1$ are primes, is an $(p, 4, 1)$ OOOC.

Construction D: Let α be a primitive element in $GF(q)$, where $q = 10m + 1$ is a prime power. Let ξ be a primitive fifth root of unity in $GF(q)$, l is the least integer such that $\alpha^l = \xi + 1$, and $d = \frac{o(\xi+1)}{g.c.d.(o(\xi+1),10)}$. If d is even or $d = m$ then let

$$C = \left\{ \alpha^{2il+j}(1, \xi, \xi^2, \xi^3, \xi^4) : \begin{matrix} 0 \leq i < \lfloor \frac{d}{2} \rfloor \\ 0 \leq j < \frac{m}{d} \end{matrix} \right\}$$

Lemma 21. *For $q = 10m+1$, m even, d in construction D is even, iff it satisfies the conditions of construction B.*

Theorem 22. *If d from construction D is even, then C is a $(q, 5, 1)$ difference family.*

Using construction D, we found some difference families with $w = 5$, for primes that do no satisfy R_5; e.g, $(401, 5, 1)$, $(761, 5, 1)$ difference family.

Theorem 23. *Let $p = 10m + 1$, m odd, be a prime; if d from construction C is equal to m, then C is a $(p, 5, 1)$ OOOC.*

Theorem 24. *For a prime $p = 10m + 1$, where $m > 1$ is prime, the condition $d = m$ in construction D is satisfied.*

Theorem 24 implies that the code generated by construction D, for $p = 10m + 1$, p and $m > 1$ are primes, is an $(p, 5, 1)$ OOOC.

3 Other Constructions

Definition 25. An $(kn, w, 1)$ optical orthogonal code C is *uniform* $\Delta C = Z_{kn} - \{i \cdot n \ : \ 0 \leq i < k\}$.

Note that an $(n, w, 1)$ difference family is uniform and $\Delta C = Z_n - \{0\}$. For Construction E which follows let $p = 4q + 1 \equiv 5 \pmod 8$ be a prime and α a residue modulo $3p$ such that $\alpha \equiv 2 \pmod 3$, and $\alpha' \equiv \alpha \pmod p$ is a primitive root modulo p. Also, let $H = \{\alpha^i \ : \ 0 \leq i < p-1\}$ where the computation is performed modulo $3p$.

Lemma 26.

(1) $\alpha^q - 1 \equiv 1 \pmod 3$.

(2) $\alpha^{2q} - 1 \equiv 0 \pmod 3$.

Proof. Since $p \equiv 5 \pmod 8$, q is odd. If $\alpha \equiv 2 \pmod 3$ then $\alpha^i \equiv 2 \pmod 3$ iff i is odd, and $\alpha^i \equiv 1 \pmod 3$ iff i is even. □

Lemma 27. $Z_{3p} - \{0, p, 2p\} = H \cup (-1)H \cup 3H$.

Proof. Since the order of α modulo p is $p-1$ it follows that the order of α modulo $3p$ is at least $p-1$. By lemma 26 $\alpha^{p-1} = \alpha^{4q} \equiv 1 \pmod 3$ and since $\alpha^{p-1} \equiv 1 \pmod p$ it follows that $\alpha^{p-1} \equiv 1 \pmod{3p}$ and $|H| = p - 1$. Therefore, H is a subgroup of the multiplicative group modulo $3p$, M_{3p}. By lemma 26 $\alpha^{2q} \equiv 1 \pmod 3$ and hence $-1 \notin H$ and therefore $M_{3p} = H \cup (-1)H$.

Clearly, $3H \cap M_{3p} = \phi$ and since α is primitive it follows that $3H$ contains $p-1$ distinct elements. To complete the proof we only have to note that $\{p, 2p\} \cap (H \cup (-1)H \cup 3H) = \phi$. □

Lemma 28. *If $p \equiv 5 \pmod 8$ is a prime then there exists α, $0 < \alpha < 3p$ such that $\alpha \equiv 2 \pmod 3$, and $\alpha' \equiv \alpha \pmod p$ is a primitive root modulo p.*

Proof. Let $p \equiv j \pmod 3$; if $\alpha' \equiv j - 1 \pmod 3$ then $\alpha' + 2p \equiv 2 \pmod 3$. If $\alpha' \equiv 2 - j \pmod 3$ then $\alpha' + p \equiv 2 \pmod 3$. □

Construction E: Let $p = 4q + 1 \equiv 5 \pmod 8$ be a prime and α a residue modulo $3p$ such that $\alpha \equiv 2 \pmod 3$, and $\alpha' \equiv \alpha \pmod p$ is a primitive root modulo p, let

$$C = \left\{(\alpha^i, \alpha^{q+i}, \alpha^{2q+i}, \alpha^{3q+i}) : 0 \leq i < q\right\}$$

where the computation is performed modulo $3p$.

Theorem 29. *C is an uniform $(3p, 4, 1)$ OOOC.*

Proof. $\Delta C = \Delta\left\{(\alpha^i, \alpha^{q+i}, \alpha^{2q+i}, \alpha^{3q+i}) : 0 \leq i < q\right\} = \left\{\alpha^i : 0 \leq i < q\right\} \cdot \left\{\alpha^q - 1, \alpha^{2q} - 1, 1 - \alpha^q\right\} \cdot \left\{\alpha^{jq} : 0 \leq j < 4\right\} = \left\{\alpha^q - 1, \alpha^{2q} - 1, 1 - \alpha^q\right\} \cdot H$. By Lemma 26 and by the proof of Lemma 27, $\alpha^{2q} - 1 \in 3H$, and either $\alpha^q - 1 \in H$ and $1 - \alpha^q \in (-1)H$ or $\alpha^q - 1 \in (-1)H$ and $1 - \alpha^q \in H$. Since H and $(-1)H$ are cosets of M_{3p}, $hH = H$ for $h \in H$, and $hH = -H$ for $h \in -H$. It is also easy to verify that $hH = 3H$ for $h \in 3H$. Thus, by lemma 27, $\Delta C = H \cup (-1)H \cup 3H = Z_{3p} - \{0, p, 2p\}$ and C is an uniform $(3p, 4, 1)$ OOOC. □

Construction F: For a prime $p = 6q + 1$, α an odd residue modulo $2p$ such that $\alpha' \equiv \alpha$ (mod p) is a primitive root modulo p, let

$$C = \left\{(0, \alpha^i, \alpha^{2q+i}, \alpha^{4q+i}) : 0 \leq i < q\right\}$$

where the computation are performed modulo $2p$.

Theorem 30. *C is an uniform $(2p, 4, 1)$ OOOC.*

Proof. Let M_{2p} denote the multiplicative group of Z_{2p}. It is easy to verify that $\{\alpha^i : 0 \leq i < 6q\}$, $|M_{2p}| = p-1$, and for even $\beta \in Z_{2p}$ $|\beta M_{2p}| = p-1$

$\Delta(0,1,\alpha^{2q},\alpha^{4q}) = \{1,\alpha^{2q},\alpha^{4q}\} \cdot \{1,-1,\alpha^{2q}-1,\alpha^{4q}-1\}$. Since $\alpha^{3q} \equiv -1$ (mod $2p$) and $\alpha^q(\alpha^{2q}-1) = \alpha^{4q}-1$, it follows that $\Delta(0,1,\alpha^{2q},\alpha^{4q}) = \{\alpha^{jq} : 0 \leq j < 6\} \cdot \{1, \alpha^{2q}-1\}$, and $\Delta C = M_{2p} \cdot \{1,\alpha^{2q}-1\}$. Since $\alpha^{2q}-1$ is even, it follows that all the elements in ΔC are distinct. Finally note that $p \notin \Delta C$. □

4 Recursive Constructions

Colbourn and Colbourn[CC80] gave two recursive constructions for a difference family from two other difference families. Brickell and Wei[BW87] gave a generalization of these constructions which produces also OOOCs. Their construction uses a combinatorial structure called difference array. An $n \times k$ *difference array* A is an array of integers from 0 to $n-1$ such that for any two columns i, j and any integer l, there exists a unique row s, such that $A(s,j) - A(s,i) \equiv l \pmod{n}$, where $B(i,j)$ denote the entry in row i and column j of the array B.

We now give a new construction which generalizes the constructions of Colbourn and Colbourn [CC80] and Brickell and Wei[BW87]. Our construction will use one uniform OOC and another OOOC to construct a new OOOC.

Let C_1 be a uniform $(rn, w, 1)$ OOC with r missing differences, C_2 a $(rm, w, 1)$ OOOC, and A a $m \times w$ difference array. For any given word $c = (c_1, c_2, ..., c_w) \in C_1$ we form the words $(nA(i,1) + c_1, nA(i,2) + c_2, ..., nA(i,w) + c_w)$ for each i, $1 \leq i \leq m$. For any given word $c = (c_1, c_2, ..., c_w) \in C_2$ we form the word $(nc_1, nc_2, ..., nc_w)$. It is easy to verify that the constructed words form an $(rnm, w, 1)$ OOOC. If C_1 and C_2 are difference families then also C is a difference family, if C_2 is uniform then also C is uniform.

Unfortunately, not many methods to construct difference arrays are known. The classic methods was mentioned in[BW87] and it obtains $n \times r$ difference array, where r is the smallest divisor of n. Another construction of [BW87] uses $(n, k, 1)$ difference family with l codewords, where $n = 1 + lk(k-1)$ and k is a power of a prime to obtain an $n \times k$ difference array. For small values of n we can find difference arrays by computer search. An 15×5 difference array is given here

0	0	0	0	0
0	1	2	5	12
0	2	4	12	3
0	3	6	10	8
0	4	9	1	13
0	5	11	8	9
0	6	13	11	5
0	7	1	3	2
0	8	3	9	14
0	9	8	7	11
0	10	14	4	7
0	11	7	2	4
0	12	5	14	10
0	13	10	6	1
0	14	12	13	6

Another recursive construction is given below.

Theorem 31. *If there exist $n_1 \times k$ and $n_2 \times k$ difference arrays then there exist an $(n_1 n_2) \times k$ difference array.*

Proof. Let A and B be the $n_1 \times k$ and $n_2 \times k$ difference arrays, respectively. We form an $(n_1 n_2) \times k$ array C whose entries are defined by $C[(r-1)n_1 + i, j] = B[r, j]n_1 + A[i, j]$, $1 \leq r \leq n_2$, $1 \leq i \leq n_1$, $1 \leq j \leq k$.

One can easily verify that C is an $(n_1 n_2) \times k$ difference array. □

The problem of constructing difference arrays is an interesting combinatorial problem. It is equivalent to some other combinatorial problems and it has some more applications in coding theory[BE93].

5 Summary

Table A contains a list of all the values of between 13 and 99, for which an $(n, 4, 1)$ OOOC was found. Note that the entries marked by h, are either value of n where a uniform OOOC cannot exist (i.e. n modulo $w(w-1)$ does not divide n), or when this condition is satisfied, but such a code does not exist (e.g. $n = 16, 18, 24, 27, 28, 30, 32$ and 45).

13^a	-	37^s	49^s	61^s	73^a		97^a
14^f	26^f	38^f	50^s	62^f	74^f	86^f	98^g
15^e	27^h	39^e	51^s	63^s	75^g	87^e	
16^h	28^h	40^s	52^d	64^s			
17^h	29^h	41^h	53^h	65^s			
18^h	30^h	42^s	54^s	66^s	78		
19^c	31^c	43^c	55^h	67^c	79^c	91^g	
20^h	32^h	44^h	56^s	68^h	80^s		
21^h	33^h	45^h	57^h	69^h	81^s		
22^h	34^h	46^h	58^h	70^g			
23^h	35^h	47^h	59^h	71^h			
24^h	36^h	48^h	60^h	72^s			

a - Wilson
b - construction B - DF
c - Construction B - OOC
d - Colbourn & Mathon
e - construction E
f - construction F
g - recursive construction
s - computer search - uniform
h - computer search

References

[AGM92] N. Q. A, L. Gyorfi, and J. L. Massey. Constructions of binary constant-weight cyclic codes and cyclically permutable codes. *IEEE Trans. on Inform. Theory*, IT-38:940–949, May 1992.

[BE93] S. Bitan and T. Etzion. Constructions for optimal constant weight cyclically permutable codes and difference families. preprint, 1993.

[Bos39] R.C. Bose. On the construction of balanced incomplete block designs. *Ann. Eugenics*, 9:353–399, 1939.

[BW87] E. F. Brickell and V.K. Wei. Optical orthogonal codes and difference families. In *Proc. of the Southeastern Conference on Combinatorics Graph Theory and Algorithms.*, 1987.

[CC80] M.J. Colbourn and C.J. Colbourn. On cyclic block designs. *Math. Report of Canadian Academy of Science*, 2:21–26, 1980.

[CK90] H. Chung and P.V. Kumar. Optical orthogonal codes - new bounds and an optimal construction. *IEEE Transactions on Information Theory*, 36(4):866–873, July 1990.

[CSW89] F. R.K. Chung, J.A. Salehi, and V.K. Wei. Optical orthogonal codes:design, analysis, and applications. *IEEE Transactions on Information Theory*, 35(3):595–604, May 1989.

[Gol64] S.W. Golomb. *Digital communication with space application.* Prentice Hall, Englewood Cliff, N.J., Penisula Publishing, Los Altos, CA. 1982, second edition, 1964.

[Mat87] R. Mathon. Constructions for cyclic steiner 2-designs. *Annals of Discrete Mathematics*, 34:353–362, 1987.

[MM85] J. L. Massey and P. Mathys. The collision channel without feedback. *IEEE Trans. on Inform. Theory*, IT-31:192–204, March 1985.

[KMZZ93] P.V. Kumar, O. Moreno, Z. Zhang and V.A. Zionviev. New constructions of optimal cylically permutable constant weight codes. to appear in IEEE Trans. on Information Theory, 1993.

[Wil72] R.M. Wilson. Cyclotomy and difference families in elementary abelian groups. *J. Number Theory*, 4:17–47, 1972.

On Complementary Sequences

Amnon Gavish [*] and Abraham Lempel [**]

Technion - Israel Institute of Technology
Haifa 32000, Israel

Abstract. A set of complex-valued sequences $\{S_j\}_{j=1}^{k}$, where $S_j = (s_{j1}, s_{j2}, \ldots, s_{jN})$ is called complementary if the sum $R(\cdot)$ of their autocorrelation functions $\{R_{S_j}(\cdot)\}_{j=1}^{k}$ satisfies

$$R(\tau) = \sum_{j=1}^{k} \sum_{i=1}^{N-\tau} s_{ji} s_{ji+\tau}^{*} = 0 \quad , \quad \forall \tau \neq 0.$$

In this paper, we introduce a new family of complementary pairs of sequences over the alphabet $\alpha_3 = \{+1, -1, 0\}$. The inclusion of zero in the alphabet, which may correspond to a pause in transmission, leads both to a better understanding of the conventional binary case, where the alphabet is $\alpha_2 = \{+1, -1\}$, and to new nontrivial constructions over the ternary alphabet α_3. For every length N, we derive restrictions on the location of the zero elements and on the form of the member sequences of the pair. We also derive a bound on the minimum number of zeros, necessary for the existence of a complementary pair of length N over α_3. The bound is tight, as it is met by some of the proposed constructions, for infinitely many lengths.

Keywords: Complementary sequences, autocorrelation, Golay pairs.

1 Introduction

Let $S = (s_1, s_2, \ldots, s_N)$ be a sequence of length N over the complex numbers. The aperiodic autocorrelation function, $R_S(\cdot)$, of the sequence S is defined as

$$R_S(\tau) = \sum_{i=1}^{N-\tau} s_i s_{i+\tau}^{*} \quad , \quad \forall \tau,$$

where s^* denotes the complex conjugate of s.

A set of k complex sequences $\{S_i\}_{i=1}^{k}$ is called a *complementary* set of sequences, if the sum of the aperiodic autocorrelation functions of the k sequences

---

[*] Department of Electrical Engineering, Technion – Israel Institute of Technology.
[**] Department of Computer Science, Technion – Israel Institute of Technology.

vanishes for every $\tau \neq 0$, i.e.,

$$R(\tau) = \sum_{j=1}^{k} R_{S_j}(\tau) = \sum_{j=1}^{k} \sum_{i=1}^{N-\tau} s_{j\,i} s^*_{j\,i+\tau} = 0 \quad , \quad \forall \tau \neq 0.$$

Complementary sets of sequences were first introduced by Golay in [1],[2], where he also presented methods for the construction of complementary pairs of sequences over the binary alphabet $\alpha_2 = \{+1, -1\}$. Given binary complementary pairs (in short, BCPs) of lengths M and N, known as seeds, one can use these construction methods to construct BCPs of lengths $2M$, $2N$, and MN. In a computer search, done in [2] and [10], seeds of lengths 2, 10, and 26 were found. Golay [2], also, showed that the complementary property of a BCP is invariant under negation and/or reversal of pair members. The same holds under negation of every other element in each member of a pair, e.g., under $s_i \rightarrow (-1)^i s_i$. Complementary pairs (CPs), that can be transformed into one another, using the above transformations, are said to be *equivalent*.

Golay has shown [2], that the length of a BCP must be of the form $N = 2(q^2 + p^2)$, where p and q are integers (except for the trivial case of $N = 1$). Additional constraints on the possible lengths of BCPs were introduced in [3], [4], and [5].

There are two obvious generalizations of the BCP: increasing the cardinality of the set and increasing the size of the alphabet. Tseng and Liu [6] presented binary complementary sets of cardinality greater than 2. In this case, there are no constraints on the possible lengths, and the members of a set do not have to be all of the same length. Tseng and Liu also generalized the construction method of BCPs, to obtain longer binary complementary sets from known seeds.

Several authors, [7]–[11], investigated *polyphase* complementary sets (in short, PCSs) over the complex roots of unity. This generalization of the binary alphabet leads to new construction methods of complementary sets of various lengths and cardinalities. In the next section we present some properties of the PCSs.

Throughout the rest of this paper, we consider another generalization of the alphabet, and introduce ternary CPs (in short, TCPs) over the alphabet $\alpha_3 = \{+1, -1, 0\}$. The inclusion of zero in the alphabet corresponds to a pause in transmission. Throughout this paper, we use the term *ternary*, for the alphabet α_3, and the term *binary* for α_2.

The *deficiency* $\delta(S)$ of a ternary sequence S is defined as the number of zero elements in S. The deficiency $\delta(A, B)$ of a TCP A, B is defined as the sum $\delta(A, B) = \delta(A) + \delta(B)$. Clearly, TCPs exist for every length. Given a length N, let $\delta[N]$ denote the minimum feasible deficiency of a TCP of length N. A TCP A, B of length N with $\delta(A, B) = \delta[N]$ will be referred to as an *optimal* ternary pair. In the sequel we show that, for $N > 3$, there are no pairs with $\delta = 1$, and we derive constraints on the possible locations of the zero elements in the members of a pair.

Next, we derive a necessary and sufficient condition for the insertion of two zeros into a given BCP, so that the resulting TCP is optimal with $\delta = 2$. The *penultimate pair* of a BCP, as defined in [12], is actually a TCP. We use the penultimate pair of a given BCP to construct new TCPs, some of which turn out to be optimal.

It also turns out that the methods to obtain longer BCPs from known seeds can be generalized to the ternary case at hand. Finally, we present a table, containing optimal TCPs up to length 14.

2 Polyphase Complementary Sequences

Let $q = p^m$, where p is prime. Let $\omega_q = e^{j2\pi/q}$ denote the $q-th$ complex root of unity. We define a set B of k sequences $\{B_j\}_{j=1}^k$, where $B_j = \{\omega_q^{a_{ji}}\}_{i=1}^N$, $a_{ji} \in \{0, 1, \ldots, q-1\}$. The set B is complementary if

$$R(\tau) = \sum_{j=1}^{k} \sum_{i=1}^{N-\tau} \omega^{a_{ji}-a_{j\ i+\tau}} = 0 \quad , \quad \forall \tau \neq 0. \tag{1}$$

A set B satisfying (1) is called a *q-ary* polyphase complementary set (in short, QPCS). The matrix $A = [a_{ji}]$ is called the *exponent* matrix of B.

Theorem 1. *For every prime p and $q = p^m$, the cardinality of a QPCS B is divisible by p.*

Proof:

$$x^q - 1 = x^{p^m} - 1 = (x^{p^{m-1}} - 1)m_q(x)$$

where

$$m_q(x) = (x^{(p-1)p^{m-1}} + x^{(p-2)p^{m-1}} + \ldots + x^{p^{m-1}} + 1).$$

$m_q(x)$ is a cyclotomic polynomial of order q over the rationals, so it is irreducible and thus it is the minimal polynomial of ω_q over the rationals. Let $f(x)$ be a polynomial with integer coefficients, such that $f(\omega_q) = 0$. Then $f(x) = k(x)m_q(x)$, where the coefficients of $k(x) = \sum_{i=0}^{d} k_i x^i$ are also integers. So, the sum of the coefficients of $f(x)$ is $p\sum_{i=1}^{d} k_i$. Now, the complementary property of B implies

$$R(\tau = N-1) = \sum_{j=1}^{k} \omega_q^{a_{j1}-a_{jN}} = 0,$$

which, in turn, implies that k is divisible by p. □

For every $1 \leq \tau \leq N-1$, we define the componentwise products matrix $E_\tau = [e_{ji}]$, where $e_{ji} \equiv a_{ji} - a_{j\ i+\tau}\ mod\ q$.

Lemma 2. *A q-ary set B is complementary if and only if for every τ, any two exponents, that are congruent modulu p^{m-1}, appear the same number of times in E_τ.*

Proof: Let $f(x)$ be a polynomial of degree $< q$ with integer coefficients, such that $f(\omega_q) = 0$. Then $f(x) = k(x)m_q(x)$ for some $k(x) = \sum k_i x^i$. Hence, each monomial of $f(x)$, whose exponent is congruent to i modulu p^{m-1}, appears exactly k_i times in $f(x)$. □

Corollary 1 *bf : When $q = p$, E_τ contains each of the exponents $0, 1, \ldots, p-1$ exactly $\frac{(N-\tau)k}{p}$ times.*

Corollary 2 *bf : Let A denote the exponent matrix of a QPCS. Then for every $1 \le i \le N$, the difference between the sum of elements in culomn A_i and the sum of elements in column A_{N+1-i} is divisible by p.*

Let M be a $k \times N$ matrix over the powers of ω_q. M is called *orthogonal* if

$$M^* \cdot M = NI_N,$$

where M^* is the conjugate transpose of M, and I_N is the $N \times N$ unit matrix.

Lemma 3. *Every $k \times N$ orthogonal matrix $M = [m_{ji}]$ over the powers of ω_q forms a QPCS.*

Proof: For every $\tau \neq 0$

$$R(\tau) = \sum_{j=1}^{k} \sum_{i=1}^{N-\tau} m_{ji} m^*_{j\ i+\tau} = \sum_{i=1}^{N-\tau} \sum_{j=1}^{k} m_{ji} m^*_{j\ i+\tau} = \begin{cases} NK & for\ \tau = 0 \\ 0 & othewise \end{cases}.$$

□

As orthogonality is a much stronger property than the complementarity, the complementary sets obtained from orthogonal matrices are not very good because their length cannot exceed their cardinality. But orthogonal matrices can be used as tools to construct longer QPCSs, as will be shown later.

Let $A \odot B$ denote the *Kronecker* product of matrices $A = [a_{ji}]$, $1 \le j \le l$ $1 \le j \le m$ and $B = [b_{ji}]$, $1 \le j \le k$ $1 \le i \le n$. That is

$$A \odot B = \begin{bmatrix} a_{11}B & a_{12}B & \ldots & a_{1m}B \\ a_{21}B & a_{22}B & \ldots & \cdot \\ \cdot & \cdot & \cdot & \cdot \\ a_{l1}B & a_{l2}B & \ldots & a_{lm}B \end{bmatrix}.$$

Construction 1 [8]: Let $B_{l \times m}$ and $D_{k \times n}$ denote two matrices, each of whose rows form a QPCS. Then, the $k \cdot l$ rows of the matrix $C = B \odot D$ form a QPCS of length $m \cdot n$.

Construction 2 [8]: Let $(B_1, B_2, \ldots, B_k)$ denote a QPCS of length N. Let M be a $m \times k$ q-ary orthogonal matrix, with the columns $M_1, M_2, \ldots, M_k$. Then, the rows of the $m \times kN$ matrix

$$C = [\ B_1 \odot M_1\ B_2 \odot M_2\ \ldots\ B_k \odot M_k]$$

form a QPCS of length kN and cardinlity m.

There exist transformations which, when applied to the exponent matrix A of a QPCS, preserve its dimensions and the complementary property:
(T1) Addition of a constanct to all elements of A [8]
(T2) Addition of $r \cdot i$ to the column i for all $1 \leq i \leq N$ [8]
(T3) If $q = p$: Multiplication of all elements in A by an integer z, where $g.c.d(z,p) = 1$
Proof of T3: Multiplication of all elements of A by z causes the multiplication of all elements of E_τ by z. As E_τ contains each residue modulo p an eqaul number times, multiplication by z will preserve the complementary property. □

3 Some Properties of Ternary Complementary Pairs

Lemma 4. *Let A and B be a TCP. Then, for every $\tau \neq 0$, the total number of zero componentwise products is even.*

Proof: The complementary property of A and B implies

$$\sum_{i=1}^{N-\tau} a_i a_{i+\tau} + \sum_{j=1}^{N-\tau} b_j b_{j+\tau} = 0 \quad , \quad \tau \neq 0 \tag{2}$$

The total number of products in (2) is $2(N-\tau)$.

Let $I = \{\, i \,|\, a_i \neq 0 \,\}$ and $J = \{\, j \,|\, b_j \neq 0 \,\}$. Then (2) can be rewritten as

$$\sum_{i;\, i, i+\tau \in I} a_i a_{i+\tau} + \sum_{j;\, j, j+\tau \in J} b_j b_{j+\tau} = 0 \quad , \quad \tau \neq 0 \tag{3}$$

Since each product in (3) equals either $+1$ or -1, their sum can vanish only if the total number of products in (3) is also even. Therefore, the vanishing number of products in (2) is even, as claimed. □

It is clear that $\delta[N] \geq 0$, with equality holding only when there exists a complementary pair over α_2. Lemmas 4, 5, and 6 below, present necessary conditions on the form of TCPs with deficiency of 3 or less.

Lemma 5. *Let A,B be a TCP of length N. Then*
(a) If $\delta(A,B) = 2$ and $a_i = 0$, $1 \leq i \leq N$, then either $b_i = 0$ or $b_{N+1-i} = 0$.
(b) If N is odd, i.e. $N = 2L+1$, and $\delta(A,B) = 1$, then either a_{L+1} or $b_{L+1} = 0$.
(c) If N is even, then $\delta(A,B) \neq 1$.

Proof: Let A,B be a TCP of length N and deficiency 2, and let z be the largest integer such that for all $1 \leq i < z$ $a_i \neq 0$, $a_{N+1-i} \neq 0$, $b_i \neq 0$, and $b_{N+1-i} \neq 0$. We can assume, without loss of generality, that $a_z = 0$. If $z = \frac{N+1}{2}$, then we must also have $b_z = 0$. If $z < \frac{N+1}{2}$, then for $\tau = N - z$, there are four potential vanishing products: $a_1 a_{N+1-z}$, $a_z a_N$, $b_1 b_{N+1-z}$, and $b_z b_N$. By assumption, $a_z a_N = 0$. Since $\delta(A,B) = 2$, at most one of a_{N+1-z}, b_z, b_{N+1-z}

can be zero. Hence, the total number of zero componentswise products in $R(\tau)$ is either 1, if $a_{N+1-z}b_zb_{N+1-z} \neq 0$, or 2, if $a_{N+1-z}b_zb_{N+1-z} = 0$. By Lemma 3, the total number of zero products in $R(\tau)$ must be even, so exactly one of a_{N+1-z}, b_z, and b_{N+1-z} must equal zero.

Suppose, contrary to statement (a) of the lemma, that $a_{N+1-z} = 0$. Then for $\tau = N+1-2z$, the two zero elements a_z and a_{N+1-z} form the single zero product $a_z a_{N-z+1}$, which contradicts Lemma 3. This concludes the proof of (a).

Parts (b) and (c) of the Lemma also follow from Lemma 3 : A TCP of odd length N and deficiency 1 must have its only zero element located at the exact middle ($i = L+1$) of one of its member sequences. And, if N is even, we must have $\delta(A,B) \neq 1$, as none of the members of the complementary pair A, B have an exact middle. □

Note: The fact that for complementary pairs with deficiency 2, each member of the pair must contain one zero and that the zeros can be aligned at the same coordinate, makes it rather convenient from a practical point of view, as a zero element corresponds to a pause in transmission.

The restrictions on the zero locations, when $\delta = 3$ are stated, without proof, in the following lemma.

Lemma 6. *Let A,B be a TCP of length N and deficiency 3. Then N must be of the form* $N = 4m+2$, *for some integer m, and either* $a_{2m+1} = b_{m+1} = b_{3m+2} = 0$ *or* $b_{2m+1} = a_{m+1} = a_{3m+2} = 0$.

Lemma 7. *Every TCP A,B of length N and deficiency 1 must satisfy*

$$A = C \; y \; \bar{C}'$$
$$B = D \; 0 \; \bar{D}$$

where C and D are binary sequences of length L over α_2, $\bar{X}$ *denotes the reversed image of* X, *and* X' *denotes the negation of* X; y *can be either +1 or -1.*

Proof: By Lemma 4, there exists no TCP of even length and deficiency 1. So, it remains to consider only pairs of odd length. Consider a TCP A, B of length $N = 2L+1$ with $\delta(A) = 0$ and $\delta(B) = 1$. By Lemma 4, we must have $b_{L+1} = 0$. For $\tau = L+1$, the complementary property of A and B implies

$$\prod_{i=1}^{L} a_i a_{L+1+i} b_i b_{L+1+i} = \prod_{i=1}^{L} a_i \prod_{j=L+2}^{2L+1} a_j \prod_{k=1}^{L} b_k \prod_{l=L+1}^{2L+1} b_l = (-1)^L. \tag{4}$$

Similarly for $\tau = L-u$, $0 \leq u \leq L-1$ we must have

$$\prod_{i=1}^{L+u+1} a_i a_{i+L-u} \prod_{\substack{i=1 \\ i\neq L+1 \\ i\neq u+1}}^{L+u+1} b_i b_{i+L-u} = \prod_{i=1}^{L+u+1} a_i \prod_{j=L-u+1}^{2L+1} a_j \prod_{\substack{k=1 \\ k\neq L+1 \\ k\neq u+1}}^{L+u+1} b_k \prod_{\substack{l=L-u+1 \\ l\neq L+1 \\ l\neq 2L-u+1}}^{2L+1} b_l$$
$$= (-1)^{L-u}. \tag{5}$$

Notice that the products $b_{u+1}b_{L+1}$ and $b_{L+1}b_{2L-u+1}$ vanish.
Dividing (4) by (5), and letting $u = 0$, yields

$$\frac{b_1 b_{2L+1}}{a_{L+1}^2} = b_1 b_{2L+1} = 1. \tag{6}$$

Dividing (5) with $u = x + 1$ by (5) with $u = x$, $0 \leq x \leq L - 2$, yields

$$\frac{a_{L+x+2}a_{L-x}b_{L+x+2}b_{L-x}b_{x+1}b_{2L-x+1}}{b_{x+2}b_{2L-x}} = -1. \tag{7}$$

As is well known [2], a BCP A, B over α_2 satisfies

$$a_r a_{N-r+1} b_r b_{N-r+1} = -1 \quad , \quad 0 < r < \frac{N}{2}. \tag{8}$$

It is easy to verify that (8) also holds for the case considered here. Dividing (7) by (8), with $r = L - x$ and $N = 2L + 1$, yields

$$b_{x+1}b_{2L-x+1}b_{x+2}b_{2L-x} = 1 \quad , \quad 0 \leq x < L - 1. \tag{9}$$

Applying (6) to (9), we obtain

$$b_{x+2}b_{2L-x} = 1 \quad , \quad 0 \leq x < L - 1. \tag{10}$$

Dividing (8) by (10), with $r = x + 2$, yields

$$a_{x+2}a_{2L-x} = -1 \quad , \quad 0 \leq x < L - 1. \tag{11}$$

Finally, dividing (8), with $r = 1$, by (6), yields

$$a_1 a_{2L+1} = -1. \tag{12}$$

By (6), (10), (11), and (12), every TCP A, B of length N and deficiency 1 must satisfy

$$\begin{aligned} A &= C\ y\ \bar{C}' \\ B &= D\ 0\ \bar{D}. \end{aligned}$$
□

Lemma 6 restricts the form of potential candidates for TCPs of length N and deficiency 1. It turns out, as shown in Theorem 2 below, that the restriction of Lemma 6 can be satisfied only by a TCP of length 3 and deficiency 1, e.g. , the pair

$$\begin{bmatrix} + & + & - \\ + & 0 & + \end{bmatrix}.$$

Theorem 8. *For $N > 3$, if $\delta[N] > 0$, then $\delta[N] \geq 2$.*

Proof: Suppose there is a TCP A, B of length $N > 3$ with $\delta(A) = 0$ and $\delta(B) = 1$. By Lemma 6, N must be odd, i.e. $N = 2L + 1$, and we can assume without loss of generality, that A and B are of the form

$$A = C\ 1\ \bar{C}'$$
$$B = D\ 0\ \bar{D}$$

where $C = (c_1, c_2, c_3, \ldots, c_L)$ and $D = (d_1, d_2, d_3, \ldots, d_L)$ are binary sequences of length L over α_2, with $d_1 = c_1$. First, we show, by induction, that C and D must be identical. The complementary property of A and B for $\tau = N - 2$ implies

$$R(N-2) = 2c_1d_2 + (-2c_1c_2) = 0$$

which, in turn, implies $d_2 = c_2$.

Suppose $d_i = c_i$ for $1 \leq i \leq m < L$. Then, by the complementary property,

$$0 = R(N-m-1) = \sum_{i=1}^{m+1} a_i a_{i+N-m-1} + \sum_{i=1}^{m+1} b_i b_{i+N-m-1}$$
$$= \sum_{i=1}^{m+1} c_i(-c_{m+2-i}) + \sum_{i=1}^{m+1} d_i d_{m+2-i} = 2c_1(d_{m+1} - c_{m+1}).$$

It follows that $d_{m+1} = c_{m+1}$, and by induction, $D = C$.

Now, it is easy to verify that, with $D = C$ and $L > 1$,

$$R(L-1) = 4c_1c_L.$$

Thus, with $L > 1$, or $N > 3$, setting $R(L-1) = 0$ contradicts the fact that C has no zero components. For the unique case of $N = 3$, $R(L-1) = R(0) = 2N - 1 = 5$, and $R(1) = R(2) = 0$. □

Now, with the case of $\delta[N] = 1$ out of the way, we shall focus our attention on optimal TCPs of length N and deficiency 2. The class of TCPs with $\delta[N] = 2$ can be partitioned into subclasses: the subclass of pairs over α_3 derived from BCPs and a subclass of new seed pairs with no binary parents. The first subclass is obtained by inserting a zero at the same location in each member of a BCP of length N, to obtain a TCP of length $N + 1$, in accordance with Lemma 4(a). Theorem 3 below, specifies the necessary and sufficient conditions on the binary complementary pair, for the derived pair to be complementary.

For $n \geq 0$, we define $A_n = (C\,0^n\,D)$ and $B_n = (E\,0^n\,F)$, where $C = (c_1, c_2, \ldots, c_{l_1})$, $D = (d_1, d_2, \ldots, d_{l_2})$, $E = (e_1, e_2, \ldots, e_{l_1})$, and $F = (f_1, f_2, \ldots, f_{l_2})$ are binary sequences over α_2, and $l_1 \geq l_2$.

The *aperiodic crosscorrelation* of C and D, $R_{CD}(\tau)$, is defined as

$$R_{CD}(\tau) = \sum_{i=1}^{l_1} c_i d_{i+\tau} \quad , \quad \forall \tau,$$

where $d_i \equiv 0$ for all $i \leq 0$ or $i > l_2$.

Theorem 9. *Let A_n and B_n be a pair of sequences, as defined above. Let $R_C(\tau)$, $R_D(\tau)$, $R_E(\tau)$, and $R_F(\tau)$ denote the aperiodic autocorrelation functions of C, D, E, and F, respectively. Let $R_{CD}(\tau)$ and $R_{EF}(\tau)$ denote the aperiodic crosscorrelations of C, D and E, F, respectively. Then, any two of the following four properties imply the other two.*
(P1) A_0, B_0 form a BCP.
(P2) A_1, B_1 form a TCP.
(P3) C, D, E, and F form a complementary set.
(P4) $R_{CD}(\tau) + R_{EF}(\tau) = 0$, for every τ.

Proof: Let $R_0(\tau) = R_{A_0}(\tau) + R_{B_0}(\tau)$. Then, for $\tau \neq 0$,

$$R_0(\tau) = \sum_{i=1}^{l_1-\tau} c_i c_{i+\tau} + \sum_{i=l_1-\tau+1}^{l_1} c_i d_{i-l_1+\tau} + \sum_{i=1}^{l_2-\tau} d_i d_{i+\tau} + \sum_{i=1}^{l_1-\tau} e_i e_{i+\tau}$$
$$+ \sum_{i=l_1-\tau+1}^{l_1} e_i f_{i-l_1+\tau} + \sum_{i=1}^{l_2-\tau} f_i f_{i+\tau}.$$

Similarly, for $R_1(\tau) = R_{A_1}(\tau) + R_{B_1}(\tau)$ and $\tau \neq 0$,we obtain

$$R_1(\tau) = \sum_{i=1}^{l_1-\tau} c_i c_{i+\tau} + \sum_{i=l_1-\tau+1}^{l_1-1} c_{i+1} d_{i-l_1+\tau} + \sum_{i=1}^{l_2-\tau} d_i d_{i+\tau} + \sum_{i=1}^{l_1-\tau} e_i e_{i+\tau}$$
$$+ \sum_{i=l_1-\tau+1}^{l_1-1} e_{i+1} f_{i-l_1+\tau} + \sum_{i=1}^{l_2-\tau} f_i f_{i+\tau}.$$

It follows that, for $\tau \neq 0$,

$$R_0(\tau) = R_C(\tau) + R_{CD}(\tilde{\tau}) + R_D(\tau) + R_E(\tau) + R_{EF}(\tilde{\tau}) + R_F(\tau)$$

and

$$R_1(\tau) = R_C(\tau) + R_{CD}(\tilde{\tau}+1) + R_D(\tau) + R_E(\tau) + R_{EF}(\tilde{\tau}+1) + R_F(\tau)$$

where $\tilde{\tau} = \tau - l_1$.

One can readily verify now that any pair of properties that includes either (P3) or (P4) implies the other two. It remains to be shown that (P1) and (P2) imply (P3) and (P4). Since $R_0(0) = R_1(0) = 2(l_1 + l_2)$, the validity of (P1) and (P2) implies

$$R_{EF}(\tilde{\tau}) + R_{CD}(\tilde{\tau}) = R_{EF}(\tilde{\tau}+1) + R_{CD}(\tilde{\tau}+1) \quad , \quad \forall \tilde{\tau}$$

which, in turn, implies

$$R_{EF}(\tau) + R_{CD}(\tau) = 0 \quad , \quad \forall \tau.$$

Hence, (P4) and, therefore, (P3). □

The following corollaries are immediate deductions from Theorem 3.

Corollary 3: *Any two of (P1) ... (P4) imply the complementary property of the pair A_n, B_n for all $n \geq 0$.*

Corollary 4: *If the pair A_n, B_n is complementary for some $n \geq l_1 - 1$, then it is such for all $n \geq 0$.*

Corollary 2 follows from the observation that, when $n \geq l_1 - 1$, the complementary property of A_n and B_n implies (P3) (for $1 \leq \tau \leq l_1 - 1$) and (P4) (for $n + 1 \leq \tau \leq l_1 + l_2 + n - 1$).

4 Construction of Ternary Complementary Pairs

Theorem 10. *Let C and D be a BCP of length L. Then, for every positive integer n, the ternary sequences $A_n = (C\,0^n\,D)$ and $B_n = (C\,0^n\,D')$ of length $2L + n$, form a complementary pair.*

Proof: Let $R(\tau) = R_A(\tau) + R_B(\tau)$. For $0 < \tau \leq L$, we have

$$R(\tau) = \sum_{i=1}^{2L+n-\tau} a_i a_{i+\tau} + \sum_{i=1}^{2L+n-\tau} b_i b_{i+\tau} = \sum_{i=1}^{L-\tau} c_i c_{i+\tau} + \sum_{i=1}^{L-\tau} d_i d_{i+\tau}$$
$$+ \sum_{i=1}^{\tau-n} c_{L-\tau+n+i} d_i + \sum_{i=1}^{L-\tau} c_i c_{i+\tau} + \sum_{i=1}^{L-\tau} (-d_i)(-d_{i+\tau}) + \sum_{i=1}^{\tau-n} c_{L-\tau+n+i}(-d_i)$$
$$= 2\sum_{i=1}^{L-\tau} c_i c_{i+\tau} + 2\sum_{i=1}^{L-\tau} d_i d_{i+\tau}.$$

By the complementary property of C and D

$$R(\tau) = 0 \quad , \quad 0 < \tau \leq L.$$

For $L + 1 \leq \tau \leq L + n$, we have

$$R(\tau) = \sum_{i=L+n-\tau+1}^{L} c_i d_{i-L-n+\tau} + \sum_{i=L+n-\tau+1}^{L} c_i(-d_{i-L-n+\tau}) = 0$$

And, finally, for $\tau > L + n$ we obtain

$$R(\tau) = \sum_{i=1}^{2L+n-\tau} c_i d_{i+\tau-L-n} + \sum_{i=1}^{2L+n-\tau} c_i(-d_{i+\tau-L-n}) = 0.$$

□

Note that, by Theorem 2, the construction of Theorem 4 is optimal, when $n = 1$ and $L > 1$.

In Theorem 5 below, we apply the well known Golay constructions [2] to obtain new TCPs from known ones. These constructions were originally presented by Golay for BCPs.

Theorem 11. *Let A and B be a TCP of length N. Let C and D be a BCP of length M. Then, the pairs*
(1) AB *and* AB'
(2) $(a_1b_1a_2b_2\ldots a_Nb_N)$ *and* $(a_1b'_1a_2b'_2\ldots a_Nb'_N)$
(3) $(A^{c_1}A^{c_2}\ldots A^{c_M}B^{d_1}B^{d_2}\ldots B^{d_M})$ *and*
$(A^{d_M}A^{d_{M-1}}\ldots A^{d_1}B^{-c_M}B^{-c_{M-1}}\ldots B^{-c_1})$
are CPs. Here $A^1 = A$ *and* $A^{-1} = A'$.

The proof is identical to Golay's proof for the binary case.

For every complementary pair A, B, we define the *penultimate pair* as:

$$P = \frac{1}{2}(A + B) \quad , \quad Q = \frac{1}{2}(\bar{A} - \bar{B}),$$

where $P = (p_1p_2\ldots p_N)$ with $p_i = \frac{1}{2}(a_i + b_i)$, and $Q = (q_1q_2\ldots q_N)$, with $q_i = \frac{1}{2}(a_{N-i+1} - b_{N-i+1})$. The *reduced* penultimate pair P_r, Q_r is obtained, by removing the run of zeros, if any, at the beginning and at the end of P and Q. It can be easily shown, that P, Q, as well as P_r, Q_r form a CP. Note, that if A, B are binary, or if they are ternary with zeros occupying the same positions in both, then P, Q are ternary sequences over α_3.

For every CP A, B of length N, we define the *generating sequence* ϵ as the sequence $\epsilon = (\epsilon_1, \epsilon_2, \ldots, \epsilon_N)$ where $\epsilon_i = a_ib_i$. If $\epsilon_1 = 1$, then A, B is called a *normalized pair*. Every complementary pair can be easily normalized, by negating one of the sequences.

Theorem 12. *Let* A, B *be a normalized BCP of length* N*, with generating sequence* ϵ*. Let* P_r, Q_r *be the reduced penultimate pair of* A, B*. Let* m *be the largest positive integer, such that,* $\epsilon_i = 1$ *for all* $1 \le i \le m$*. Then*
(1) P_r, Q_r *form a TCP of length* $l = N - m$*.*
(2) $p_i = 0$ *if and only if* $q_i = 0$*.*
(3) $\delta(P_r) = \delta(Q_r) = g$*, where* $g = \frac{N-2m}{2} = \frac{l-m}{2}$*.*
(4) Each of the subsequences $P_r(m+1, l) = (p_{m+1}, \ldots, p_l)$*, and* $Q_r(m+1, l) = (q_{m+1}, \ldots, q_l)$ *contains* $g = \frac{l-m}{2}$ *zeros and* g *nonzeros.*

Proof: We have $p_i = \frac{1}{2}a_i(1 + \epsilon_i)$ and $q_i = \frac{1}{2}a_{N+1-i}(1 - \epsilon_{N+1-i})$ for $1 \le i \le N$. It is well known that the BCP A, B satisfies the symmetry property

$$a_ia_{N-i+1}b_ib_{N-i+1} = \epsilon_i\epsilon_{N-i+1} = -1 \quad , \quad 1 \le i \le N. \tag{13}$$

Thus, by the definition of m, $\epsilon_{N-i+1} = -1$ for $1 \le i \le m$, and $q_i = \frac{1}{2}a_{N+1-i}(1 + \epsilon_i)$ for $1 \le i \le N$. Hence, $l = N - m$ and $p_i = 0$ if and only if $q_i = 0$. By (13), the number of negative ϵ_i is exactly $\frac{N}{2}$, which is also the number of zeros in each of P and Q. Hence, the number of zeros in each of P_r and Q_r equals $\frac{N}{2} - m = \frac{1}{2}(l - m)$. □

The case of $g = 1$, has been thoroughly discussed in [12]. In the present context, the main result of [12] can be reformulated as follows.

Theorem 13. [12] *Let A, B be a normalized BCP, and let P, Q be the corresponding penultimate pair. If $g = 1$, then, either*
$A = (+ - - - + - + - - +)$ *and* $B = (+ - - - - - - + + -)$, *or*
$A = (+ + + + - + + - - + - + - + - - + - + + + - - + + +)$ *and*
$B = (+ + + + - + + - - + - + + + + + - + - - - + + - - -)$.

Thus, up to normalization, the only reduced penultimate CPs of deficiency $\delta = 2$ over α_3, are the pairs $P_1 = (+ - - - 0 -)$, $Q_1 = (+ - - + 0 +)$ of length 6 and
$P_2 = (+ + + + - + + - - + - + 0 +)$, $Q_2 = (+ + + - - + + + - + - - 0 -)$ of length 14.

It can be readily verified that, every normalized TCP U, V of length l and deficiency $2g$, which satisfies

(1) The zero elements occupy the same locations in U and in V,

(2) $\delta(U) = \delta(V) = g$,

(3) U (and, therefore, V) can be padded with, say m, zeros at the end, so that the resulting sequence $\tilde{U}$ of length $l + m$, satisfies $\tilde{u}_i = 0$ if and only if $\tilde{u}_{l+m+1-i} \neq 0$,

must be the reduced penultimate pair of some BCP. In particular, every TCP of deficiency $\delta = 2$, with zeros located at the second (or next to the last) coordinate of each of its members, is a reduced penultimate pair.

The case of $g = 2$, produces TCPs of deficiency 4. All possibilities for this case are derived in [12]. The only new optimal penultimate pair with $g = 2$ obtained via Theorem 6 is the one of length 7 and deficiency 4, resulting from a BCP of length 10 (See Table 1).

The next theorem introduces an alternate construction, using the penultimate pair.

Theorem 14. *Let A, B denote a BCP of length L, and let P, Q be the corresponding penultimate pair. Then the sequences $H = (\, A\, P\, B' \,)$ and $I = (\, A\, \bar{Q}'\, B \,)$ form a TCP of length $3L$.*

Proof: Let $R(\tau) = R_H(\tau) + R_I(\tau)$. Then, for $1 \leq \tau \leq L$, the complementary property of A, B yields

$$R(\tau) = \sum_{i=1}^{L-\tau} a_i a_{i+\tau} + \frac{1}{2} \sum_{i=L-\tau+1}^{L} a_i (a_{i-L+\tau} + b_{i-L+\tau})$$

$$+\frac{1}{4}\sum_{i=\tau+1}^{L}(a_i+b_i)(a_{i-\tau}+b_{i-\tau})-\frac{1}{2}\sum_{i=1}^{\tau}b_i(a_{L-\tau+i}+b_{L-\tau+i})+\sum_{i=1}^{L-\tau}b_ib_{i+\tau}$$
$$+\sum_{i=1}^{L-\tau}a_ia_{i+\tau}+\frac{1}{2}\sum_{i=L-\tau+1}^{L}a_i(b_{i-L+\tau}-a_{i-L+\tau})+\frac{1}{4}\sum_{i=\tau+1}^{L}(b_i-a_i)(b_{i-\tau}-a_{i-\tau})$$
$$+\frac{1}{2}\sum_{i=1}^{\tau}b_i(b_{L-\tau+i}-a_{L-\tau+i})+\sum_{i=1}^{L-\tau}b_ib_{i+\tau}=\frac{5}{2}(\sum_{i=1}^{L-\tau}a_ia_{i+\tau}+\sum_{i=1}^{L-\tau}b_ib_{i+\tau})=0.$$

For $L+1\leq\tau\leq 2L$, we have

$$R(\tau)=\frac{1}{2}\sum_{i=1}^{2L-\tau}a_i(a_{i+\tau-L}+b_{i+\tau-L})-\sum_{i=2L-\tau+1}^{L}a_ib_{i-2L+\tau}$$
$$-\frac{1}{2}\sum_{i=\tau-L+1}^{L}b_i(a_{i+L-\tau}+b_{i+L-\tau})+\frac{1}{2}\sum_{i=1}^{2L-\tau}a_i(b_{i+\tau-L}-a_{i+\tau-L})$$
$$+\sum_{i=2L-\tau+1}^{L}a_ib_{i-2L+\tau}+\frac{1}{2}\sum_{i=\tau-L+1}^{L}b_i(b_{i+L-\tau}-a_{i+L-\tau})=0.$$

Finally, for $2L+1\leq\tau\leq 3L-1$,

$$R(\tau)=\sum_{i=1}^{3L-\tau}a_i(-b_{i-2L+\tau})+\sum_{i=1}^{3L-\tau}a_ib_{i-2L+\tau}=0.$$

□

The construction of Theorem 8 produces TCPs of length $N=3L$, from BCPs of length L. The resulting pairs have deficiency L. As the deficiency grows linearly with the length, this construction seems to produce non-optimal pairs for large lengths. However, it produces some short optimal pairs, such as the optimal pair of length 6 $\begin{pmatrix}+-+0--\\+-0+++\end{pmatrix}$, obtained from the BCP $\begin{pmatrix}+-\\++\end{pmatrix}$, and the optimal pair of length 12 $\begin{pmatrix}+++-++00--+-\\+++-00-+++-+\end{pmatrix}$, obtained from the BCP $\begin{pmatrix}+++-\\++-+\end{pmatrix}$.

The known optimal complementary pairs of length $N\leq 14$, are listed in Table 1. The optimality of the given pairs is guaranteed either by Theorem 2 (lenghts 5,6,9,14) or by an exhaustive computer search (lengths 7,11,12).

N	$\delta[N]$	Pair	Remarks
2	0	++ +−	Binary
3	1	++− +0+	The only example of $\delta = 1$
4	0	+++− ++−+	Binary
5	2	++0+− ++0−+	Standard construction with zero insertion
6	2	++−+0+ ++−−0−	Standard construction and penultimate pair
		−−−0+− +−+0−−	Theorem 8
7	4	−−+0−0− −−+0+0+	Penultimate pair
		−−−++−+ −000−0−	Seed
8	0	−−−−−++− −−++−+−+	Binary
9	2	+++−0++−+ +++−0−−+−	Standard construction with zero insertion
10	0	+−−+−+−−−+ +−−−−−−++−	Binary
11	6	+++−000++−+ +++−000−−+−	As in N=9 with insertion of 3 zeros
12	4	+++−++00−−+− +++−00−+++−+	Theorem 8
13		$\delta[N] \geq 4$	
14	2	++++−++−−+−+0+ +++−−+++−+−−0−	Penultimate pair

Table 1. Optimal TCPs up to length 14.

References

[1] M. J. E. Golay, "Static multislit spectrometry and its application to the panoramic display of infrared spectra", J. Opt. Soc. Amer. 41 (1951), pp. 468-472.

[2] M. J. E. Golay, "Complementary Series", IRE Trans. of Information Theory IT-7, pp. 82-87, April 1961.

[3] M. Griffin, "There are no Golay Complementary Sequences of Length $2 \cdot 9^t$" Aequationes Math. 15 (1977), pp. 73-77.

[4] T. H. Andres and R. G. Stanton, "Golay Sequences", Combinatorial Mathematics, Vol. 5, pp. 44-54, New York: Spring-verlag, 1977.

[5] Shalom Eliahou, Michel Kervaire, and Bahman Safari, "A New Restriction on the length of Golay Complementary Sequences",Journal of Combinatorial Theory, Series A 55, (1990), pp. 49-59.

[6] C. -C. Tseng and C. L. Liu, "Complementary Sets of Sequences", IEEE Trans. on Information Theory, Vol. IT-18, No. 5, pp. 644-651, Sept. 1972.

[7] R. Sivaswamy, "Multiphase Complementary Codes", IEEE Trans. on Information Theory, Vol. IT-24, No. 5, pp. 546-552, Sept. 1978.

[8] Robert L. Frank, "Polyphase Complementary Codes", IEEE Trans. on Information Theory, Vol. IT-26, No. 6, pp. 641-647, November 1980.

[9] Bi Guangguo, "Methods of Constructing Orthogonal Complementary Pairs and Sets of Sequences", IEEE International Conference on Communications 1985, pp. 839-843, June 1985.

[10] Stephen Jauregui Jr., "Complementary Sequences of Length 26", IRE Trans. on Information Theory, Vol. IT-7, p. 323, July 1962.

[11] D. V. Sarwate, "Sets of Complementary Sequences", Electronics Letters, Vol. 19, No. 18, pp. 711-712, Sept. 1983.

[12] Shalom Eliahou, Michel Kervaire, and Bahman Safari, "On Golay Polynomial Pairs", Advances In Applied Mathematics 12 (1991), pp. 235-292.

[13] Amnon Gavish, "On Complementary Sequences", Msc. research thesis, Technoin -IIT, January 1993.

Spectral-Null Codes and Null Spaces of Hadamard Submatrices

Ron M. Roth

Computer Science Department
Technion — Israel Institute of Technology
Haifa 32000, Israel.
e-mail: ronny@cs.technion.ac.il

Abstract. Codes $\mathcal{C}(m,r)$ of length 2^m over $\{1,-1\}$ are defined as null spaces of certain submatrices of Hadamard matrices. It is shown that the codewords of $\mathcal{C}(m,r)$ all have an rth order spectral null at zero frequency. Establishing the connection between $\mathcal{C}(m,r)$ and the parity-check matrix of Reed-Muller codes, the minimum distance of $\mathcal{C}(m,r)$ is obtained along with upper bounds on the redundancy of $\mathcal{C}(m,r)$.

1 Introduction

Let Φ denote the alphabet $\{1,-1\}$. A word $x = [x_0\, x_1 \ldots x_{n-1}]$ over Φ is said to have an rth order spectral null at zero frequency if $\sum_{j=0}^{n-1} j^i\, x_j = 0$ for $i = 0, 1, \ldots, r-1$, where operations are taken over the integers. Codes consisting of words with prescribed spectral-null properties have appeared in the literature in several applications, e.g., in reducing the notch width of the spectrum of DC-free words at zero frequency [12],[13] or in enhancing the error-correction capability of codes used in partial-response channels [6],[14].

Let $\mathcal{C}$ be a nonempty subset of Φ^n. We refer to n as the length of $\mathcal{C}$ and to $n - \log_2 |\mathcal{C}|$ as the redundancy of $\mathcal{C}$, denoted $red(\mathcal{C})$. The minimum distance $dist(\mathcal{C})$ of $\mathcal{C}$ is the minimum Hamming distance between any two distinct words in $\mathcal{C}$.

Let $H(n,r)$ denote the integer matrix

$$H(n,r) = \begin{bmatrix} 1 & 1 & 1 & \ldots & 1 \\ 0 & 1 & 2 & \ldots & n-1 \\ 0^2 & 1^2 & 2^2 & \ldots & (n-1)^2 \\ \vdots & \vdots & \vdots & \ldots & \vdots \\ 0^{r-1} & 1^{r-1} & 2^{r-1} & \ldots & (n-1)^{r-1} \end{bmatrix}$$

and let $\mathcal{S}_{IR}(n,r)$ denote the $(n-r)$-dimensional space over the real field IR consisting of all words $x \in IR^n$ satisfying $H(n,r)\,x' = 0$. The set $\mathcal{S}(n,r) = \mathcal{S}_{\Phi}(n,r)$ is defined by $\mathcal{S}_{IR}(n,r) \cap \Phi^n$. That is, $\mathcal{S}(n,r)$ is the set of all words in Φ^n with an rth order spectral null at zero frequency.

The case $r = 0$ corresponds to unconstrained words and, therefore, $\mathcal{S}(n,0) = \Phi^n$ with redundancy zero and minimum distance 1. The set $\mathcal{S}(n,1)$ consists of all balanced (or DC-free) words, with redundancy $\frac{1}{2}\log_2 n + O(1)$ and minimum distance 2 [12],[15]. See also [1],[2],[10]. Subsets of $\mathcal{S}(n,1)$ with certain error-correcting capability have been described in [3],[4],[5],[7],[8],[20], among others.

The problem of analyzing and synthesizing codes $\mathcal{C} \subseteq \mathcal{S}(n,r)$ for values of r greater than 1 has also been dealt with in a number of papers recently, e.g., [13],[14],[17],[19]. Some of the constructions, as in [14] and [17], are based on approaching the set $\mathcal{S}(n,r)$, when n goes to infinity, by a sequence of sets of words that are generated by finite labeled directed graphs. It also follows from [13] and [14] that the minimum distance of $\mathcal{S}(n,r)$ is at least $2r$. The case of combining such constructions with prescribed error-correcting capability, and the design of efficient encoders and decoders for such codes, seem to be problems that have yet to be explored.

In this work we aim at finding relatively large subsets of $\mathcal{S}(n,r)$ that have minimum distance which is much larger than $2r$. To this end, we define null spaces $\mathcal{C}(m,r)$ of certain submatrices of Hadamard matrices (Section 2). In Section 3 we show that $\mathcal{C}(m,r)$ is a subset of $\mathcal{S}(2^m,r)$. Using the connection between $\mathcal{C}(m,r)$ and the parity-check matrix of Reed-Muller codes, we then show that the minimum distance of $\mathcal{C}(m,r)$ is 2^r (Section 4). Finally, in Section 5, we obtain upper bounds on the redundancy of $\mathcal{C}(m,r)$. We point out that there exists an efficient encoding algorithm of unconstrained binary sequences into $\mathcal{C}(m,2)$ with redundancy $O(m^2)$. Such an algorithm will be described in a full version of this paper.

2 Null spaces of Hadamard submatrices

Let F be the alphabet $\{0,1\}$, regarded as a subset of the integers. Denote by $\mathcal{H} = \mathcal{H}(m)$ the $2^m \times 2^m$ Sylvester-type Hadamard matrix whose rows and columns are indexed by elements of F^m and whose entries are given by

$$(\mathcal{H})_{u,v} = (-1)^{u \cdot v} , \quad u,v \in F^m .$$

For example, letting "+" stand for 1 and "−" for −1, we have, $\mathcal{H}(0) = [+]$,

$$\mathcal{H}(1) = \begin{bmatrix} + & + \\ + & - \end{bmatrix} , \quad \text{and} \quad \mathcal{H}(2) = \begin{bmatrix} + & + & + & + \\ + & - & + & - \\ + & + & - & - \\ + & - & - & + \end{bmatrix} .$$

For a vector $u \in F^m$, we denote by $wt(u)$ the Hamming weight of u and by $\mathcal{B}(m,r)$ the set of all vectors $u \in F^m$ with $wt(u) < r$. Clearly, $|\mathcal{B}(m,r)| = V(m,r) = \sum_{i=0}^{r-1} \binom{m}{i}$.

Let $\mathcal{H}(m,r)$ be the $V(m,r) \times 2^m$ submatrix of $\mathcal{H}(m)$ consisting of all rows of $\mathcal{H}(m)$ indexed by $u \in \mathcal{B}(m,r)$. For example,

$$\mathcal{H}(3,3) = \begin{bmatrix} + & + & + & + & + & + & + & + \\ + & - & + & - & + & - & + & - \\ + & + & - & - & + & + & - & - \\ + & + & + & + & - & - & - & - \\ + & - & - & + & + & - & - & + \\ + & - & + & - & - & + & - & + \\ + & + & - & - & - & - & + & + \end{bmatrix},$$

where the rows here are indexed in $\mathcal{H}(m)$ by $u = 000, 001, 010, 100, 011, 101, 110$. The matrix $\mathcal{H}(3,2)$ is obtained by taking the first four rows in $\mathcal{H}(3,3)$.

We now define the set $\mathcal{C}_{IR}(m,r)$ as the right null space of $\mathcal{H}(m,r)$ in IR^{2^m} i.e.,

$$\mathcal{C}_{IR}(m,r) = \left\{ x \in IR^{2^m} \mid \mathcal{H}(m,r)\, x' = 0 \right\}, \quad 0 \le r \le m,$$

where hereafter we interchange integer indices of entries of vectors (such as x) with their binary representations.

The code $\mathcal{C}(m,r) = \mathcal{C}_{\Phi}(m,r)$ is defined as the restriction of $\mathcal{C}_{IR}(m,r)$ to Φ^{2^m}, namely,

$$\mathcal{C}(m,r) = \mathcal{C}_{IR}(m,r) \cap \Phi^{2^m}.$$

Example 1. The code $\mathcal{C}(m,0)$ corresponds to an 'empty' matrix $\mathcal{H}(m,0)$ and, hence, $\mathcal{C}(m,0) = \Phi^{2^m} = \mathcal{S}(2^m,0)$. Since $\mathcal{H}(m,1)$ is the all-one vector $\mathbf{1}$, we have $\mathcal{C}(m,1) = \mathcal{S}(2^m,1)$, namely, $\mathcal{C}(m,1)$ is the set of all balanced words in Φ^{2^m}. •

Let $\mathcal{G}(m,r)$ denote the $V(m,m-r+1) \times 2^m$ submatrix of $\mathcal{H}(m)$ consisting of all rows of $\mathcal{H}(m)$ indexed by $u \in F^m - \mathcal{B}(m,r)$. Since the rows of $\mathcal{H}(m)$ are orthogonal, the rows of $\mathcal{G}(m,r)$ form a basis of $\mathcal{C}_{IR}(m,r)$. Therefore,

$$\mathcal{C}(m,r) = \left\{ x \in \Phi^{2^m} \mid x = y\,\mathcal{G}(m,r) \text{ for some } y \in IR^{V(m,m-r+1)} \right\}.$$

Note that, in particular, the rows of $\mathcal{G}(m,r)$ are words of $\mathcal{C}(m,r)$.

Example 2. For $r = m$, the matrix $\mathcal{G}(m,r)$ contains one row, namely, the last row of $\mathcal{H}(m)$. This row vector is also known as a *Morse sequence* [9] and will be denoted hereafter by $\mu(m)$. The vth entry in $\mu(m)$ is given by $= (-1)^{wt(v)}$ for every $v \in F^m$. Therefore, the code $\mathcal{C}(m,m)$ contains two words, $\mu(m)$ and $-\mu(m)$, and, as such, it has redundancy $2^m - 1$. The minimum distance of $\mathcal{C}(m,m)$ is 2^m. •

An equivalent definition for $\mathcal{C}_{IR}(m,r)$ and $\mathcal{C}(m,r)$ can be obtained by using the parity-check matrix of Reed-Muller (RM) codes, as we now show. This alternative definition will turn out to be useful while analyzing the spectral properties and the minimum distance of $\mathcal{C}(m,r)$.

Let u and v be two integer vectors in F^m. We write $u \le v$ if the inequality holds component-by-component. Similarly, we write $u < v$ if $u \le v$ and $u \ne v$. Define the $2^m \times 2^m$ matrix $H_{RM} = H_{RM}(m)$ by

$$(H_{RM})_{u,v} = \begin{cases} 1 & \text{if } u \le v \\ 0 & \text{otherwise} \end{cases}, \quad u, v \in F^m .$$

For example, $H_{RM}(0) = [\,1\,]$,

$$H_{RM}(1) = \begin{bmatrix} 1\,1 \\ 0\,1 \end{bmatrix}, \quad \text{and} \quad H_{RM}(2) = \begin{bmatrix} 1\,1\,1\,1 \\ 0\,1\,0\,1 \\ 0\,0\,1\,1 \\ 0\,0\,0\,1 \end{bmatrix} .$$

The parity-check matrix of the binary $(m-r)$th order Reed-Muller (RM) code of length 2^m is defined as the $V(m,r) \times 2^m$ submatrix of $H_{RM}(m)$ consisting of all rows of $H_{RM}(m)$ indexed by $u \in \mathcal{B}(m,r)$. For example,

$$H_{RM}(3,3) = \begin{bmatrix} 1\,1\,1\,1\,1\,1\,1\,1 \\ 0\,1\,0\,1\,0\,1\,0\,1 \\ 0\,0\,1\,1\,0\,0\,1\,1 \\ 0\,0\,0\,0\,1\,1\,1\,1 \\ 0\,0\,0\,1\,0\,0\,0\,1 \\ 0\,0\,0\,0\,0\,1\,0\,1 \\ 0\,0\,0\,0\,0\,0\,1\,1 \end{bmatrix} ,$$

where the rows here are indexed in $H_{RM}(m)$ by $u = 000, 001, 010, 100, 011, 101, 110$. The matrix $H_{RM}(3,2)$ is obtained by taking the first four rows of $H_{RM}(3,3)$. Note that we regard the matrices $H_{RM}(m,r)$ as *integer* matrices with entries in F, rather than their common definition as parity-check matrices over $GF(2)$ [16, Ch. 13].

For integer vectors $x = [x_0\, x_1\, \ldots\, x_{\ell-1}]$ and $y = [y_0\, y_1\, \ldots\, y_{\ell-1}]$, we denote by $x \oplus y$ the (unique) vector in F^ℓ whose sth entry is congruent to $x_s + y_s$ modulo 2 for every $s = 0, 1, \ldots, \ell-1$. Also, we denote by $x * y$ the integer vector whose sth entry is given by $x_s y_s$.

It is easy to verify that the rows of $\mathcal{H}$ that are indexed by u, v, and $u \oplus v$ are related by

$$(\mathcal{H})_{u \oplus v} = (\mathcal{H})_u * (\mathcal{H})_v .$$

If, in addition, we have $u * v = 0$, then a similar relation holds also for the rows of H_{RM}:

$$(H_{RM})_{u \oplus v} = (H_{RM})_u * (H_{RM})_v .$$

The following lemma establishes a connection between $\mathcal{H}$ and H_{RM}.

Lemma 1. *$\mathcal{H} = H_{RM}{}' D H_{RM}$, where D is the $2^m \times 2^m$ diagonal matrix $(D)_{u,u} = (-2)^{wt(u)}$, $u \in F^m$.*

Proof. We show by induction on $wt(u)$ that

$$(\mathcal{H})_u = \sum_{v \le u} (-2)^{wt(v)} (H_{RM})_v .$$

For $u = 0$ we have $(\mathcal{H})_0 = (H_{RM})_0$ and for every unit vector $e \in F^m$ we have $(\mathcal{H})_e = (H_{RM})_0 - 2\,(H_{RM})_e$.

Assume now that $wt(u) > 1$ and write $u = e \oplus z$ for some unit vector $e \le u$. Since $wt(z) = wt(u) - 1$, we can apply the induction hypothesis to $(\mathcal{H})_e$ and $(\mathcal{H})_z$ to obtain

$$\begin{aligned}
(\mathcal{H})_u &= (\mathcal{H})_{e\oplus z} = (\mathcal{H})_e * (\mathcal{H})_z \\
&= \Big((H_{RM})_0 - 2\,(H_{RM})_e\Big) * \Big(\sum_{v \le z} (-2)^{wt(v)} (H_{RM})_v\Big) \\
&= \sum_{v \le z} (-2)^{wt(v)} (H_{RM})_v \;-\; 2 \sum_{v \le z} (-2)^{wt(v)} \Big((H_{RM})_e * (H_{RM})_v\Big) \\
&= \sum_{v \le z} (-2)^{wt(v)} (H_{RM})_v \;+\; \sum_{v \le z} (-2)^{wt(e\oplus v)} (H_{RM})_{e\oplus v} \\
&= \sum_{v \le u} (-2)^{wt(v)} (H_{RM})_v ,
\end{aligned}$$

as claimed. □

Proposition 2. *$C_{IR}(m,r) = \Big\{ x \in IR^{2^m} \mid H_{RM}(m,r)\, x' = 0 \Big\}$.*

Proof. We show that the rows of $\mathcal{H}(m,r)$ and $H_{RM}(m,r)$ span the same real vector space. Indeed, since H_{RM} is upper-triangular, then, by Lemma 1, each row $(\mathcal{H})_u$ is a linear combination of rows $(H_{RM})_v$ for $v \le u$. Conversely, the inverse of H_{RM} is also upper-triangular (in fact, $(H_{RM}{}^{-1})_{u,v} = (-1)^{wt(u)+wt(v)} (H_{RM})_{u,v}$). Writing $H_{RM} = D^{-1}(H_{RM}{}')^{-1}\mathcal{H}$, each row $(H_{RM})_u$ is a linear combination of rows $(\mathcal{H})_v$ for $v \le u$. □

Let $G_{RM}(m,r)$ be the $V(m, m{-}r{+}1) \times 2^m$ matrix whose rows are given by

$$(G_{RM}(m,r))_u = (H_{RM}(m, m{-}r{+}1))_u * \mu(m) , \quad u \in \mathcal{B}(m, m{-}r{+}1) ,$$

where $\mu(m)$ is the Morse sequence of length 2^m (see Example 2). It is easy to verify that $H_{RM}(m,r)\, G_{RM}(m,r)' = 0$ and, therefore, the rows of $G_{RM}(m,r)$ form a basis of $C_{IR}(m,r)$. In other words, the codes $\mathcal{C}(m,r)$ and $\mathcal{C}(m, m{-}r{+}1)$ are dual codes, up to a fixed sign-flipping of components of the words in one of the codes according to a Morse sequence.

3 Spectral properties of $\mathcal{C}(m, r)$

The next theorem, following the lemma, exhibits the spectral properties of $\mathcal{C}_{IR}(m, r)$ (and of $\mathcal{C}(m, r)$).

Lemma 3. *Let $\ell \leq m$ be a positive integer. Then, there exist integers α_u, $u \in \mathcal{B}(m, \ell)$, such that for every $v = [v_0\, v_1\, \dots\, v_{\ell-1}] \in F^m$,*

$$\left(\sum_{s=0}^{m-1} v_s\, 2^s\right)^{\ell-1} = \sum_{u \in \mathcal{B}(m,\ell)} \alpha_u \, (H_{RM})_{u,v} \,. \tag{1}$$

Proof. Expanding the left-hand side of (1) we have, for every $v \in F^m$,

$$\left(\sum_{s=0}^{m-1} v_s\, 2^s\right)^{\ell-1} = \sum_U \alpha_U \prod_{s \in U} v_s \,,$$

where U ranges over all subsets of $\{0, 1, \dots, m-1\}$ of size smaller than ℓ and the α_U's are integers that do not depend on v. Note that since v is restricted to F^m, the degree of each v_s in every term of the expansion can be assumed to be 1 at most. Let $u = [u_0\, u_1\, \dots\, u_{m-1}]$ be the characteristic vector of U, that is, $u_s = 1$ if $s \in U$ and $u_s = 0$ otherwise. The lemma now follows by the easily-verified equality $\prod_{s \in U} v_s = (H_{RM})_{u,v}$. □

Theorem 4. *$\mathcal{C}_{IR}(m, r) \subseteq \mathcal{S}_{IR}(2^m, r)$.*

Proof. Let $x = [x_0\, x_1\, \dots\, x_{2^m-1}] = [\, x_v\,]_{v \in F^m}$ be a word in $\mathcal{C}_{IR}(m, r)$. We now compute the first r entries of $H(2^m, r)\, x'$. By Lemma 3 we have, for every $1 \leq \ell \leq r$,

$$\begin{aligned}
(H(2^m, r)\, x')_{\ell-1} &= \sum_{j=0}^{2^m-1} j^{\ell-1} x_j = \sum_{v=[v_0 v_1 \dots v_{m-1}] \in F^m} \left(\sum_{s=0}^{m-1} v_s\, 2^s\right)^{\ell-1} x_v \\
&= \sum_{v \in F^m} \sum_{u \in \mathcal{B}(m,\ell)} \alpha_u \, (H_{RM})_{u,v}\, x_v \\
&= \sum_{u \in \mathcal{B}(m,\ell)} \alpha_u \, (H_{RM}\, x')_u \,.
\end{aligned}$$

However, $(H_{RM}\, x')_u = 0$ for every $u \in \mathcal{B}(m, r)$. Hence, $H(2^m, r)\, x' = 0$. □

Remark. By Theorem 4 we have $\mathcal{C}(m,r) \subseteq \mathcal{S}(2^m, r)$, with equality for $r = 0, 1$. It is also known that $|\mathcal{S}(2^m, m)| = 2$ for $m \leq 5$ [12, p. 243]. Hence, $\mathcal{C}(m,m) = \mathcal{S}(2^m, m) = \{\mu(m), -\mu(m)\}$ for $m \leq 5$. However, for $m = 6$, the set $\mathcal{S}(64, 6)$ contains words other than a Morse sequence, e.g., the word $++----++-+--+-+++++--+--+--+--++$, concatenated with its reflection. •

4 Distance properties of $\mathcal{C}(m,r)$

Next we obtain the minimum distance of $\mathcal{C}(m,r)$.

Lemma 5. *[16, p. 374].*

$$H_{RM}(m,r) = \begin{bmatrix} H_{RM}(m-1,r) & H_{RM}(m-1,r) \\ 0 & H_{RM}(m-1,r-1) \end{bmatrix} .$$

Theorem 6. *The minimum distance of $\mathcal{C}(m,r)$ is 2^r.*

Proof. The proof is very similar to that used for RM codes. We show by induction on m that the nonzero words of $\mathcal{C}_{IR}(m,r)$ all have Hamming weight at least 2^r. This will imply a lower bound of 2^r on $dist(\mathcal{C}(m,r))$.

The claim is trivial for $m = 0$. As for $m > 0$, let x be a nonzero word in $\mathcal{C}_{IR}(m,r)$, and write $x = [y \;\, z]$, where y and z are words of length 2^{m-1}. By Lemma 5 we have $H_{RM}(m-1,r-1)\,z' = 0$ and $H_{RM}(m-1,r)\,(y+z)' = 0$. Therefore, $z \in \mathcal{C}_{IR}(m-1,r-1)$ and $y + z \in \mathcal{C}_{IR}(m-1,r)$. Now, if $y \neq -z$, then, by the induction hypothesis, the Hamming weight of $y+z$, and therefore that of x, must be at least 2^r. On the other hand, if $y = -z$, then $z \neq 0$ and, so, the Hamming weight of z is at least 2^{r-1}. Hence, the Hamming weight of x, being in this case twice that of z, must be at least 2^r.

To show that the 2^r lower bound is tight, note that if y and z are two words in $\mathcal{C}(m-1,r-1)$ at distance 2^{r-1}, then $[-y \;\; y]$ and $[-z \;\; z]$ are two words in $\mathcal{C}(m,r)$ at distance 2^r. The tightness now follows by induction on m, starting with $\mathcal{C}(0,0) = \Phi$. □

Example 3. The matrix $\mathcal{G}(m,m-1)$ has $m+1$ rows and, hence, $|\mathcal{C}(m,m-1)| \geq 2m+2$. We now show by induction on m that this bound is tight. This is obviously true for $m = 1$. Assume now that $m > 1$ and let $[y \;\, z]$ be a word in $\mathcal{C}(m,m-1)$, where y and z are words of length 2^{m-1}. By Lemma 5 we have $z \in \mathcal{C}(m-1,m-2)$ and $y+z \in \mathcal{C}_{IR}(m-1,m-1)$. However, since $\mathcal{C}_{IR}(m-1,m-1)$ consists of scalar multiplies of $\mu(m-1)$, all the nonzero words in $\mathcal{C}_{IR}(m-1,m-1)$

do not contain any zero components. Therefore, we must have either $y = -z$ or $y = z$. Moreover, when $y = z$ we have $z = \pm\mu(m-1)$. Hence,

$$|\mathcal{C}(m, m-1)| \le |\mathcal{C}(m-1, m-2)| + 2 \le 2m + 2 .$$

We thus conclude that $red(\mathcal{C}(m, m-1)) = 2^m - 1 - \log_2 m$. The minimum distance of $\mathcal{C}(m, m-1)$ is 2^{m-1}. •

The syndrome-based decoding algorithm for RM codes described in [18] can be adapted easily to correct up to $2^{r-1} - 1$ errors (and detect 2^{r-1} errors) in any word of $\mathcal{C}_{IR}(m, r)$ and, hence, in any word of $\mathcal{C}(m, r)$. The outline of the algorithm is as follows. Let $x = [y\ z] \in \mathcal{C}_{IR}(m, r)$ be the 'transmitted' word, where, by Lemma 5, $z \in \mathcal{C}_{IR}(m-1, r-1)$ and $y + z \in \mathcal{C}_{IR}(m-1, r)$. In fact, since $H_{RM}(m-1, r-1)$ is a submatrix of $H_{RM}(m-1, r)$, we also have $y \in \mathcal{C}_{IR}(m-1, r-1)$. Assume that at most 2^{r-1} errors have occurred, resulting in a 'received' word $\tilde{x} = [\tilde{y}\ \tilde{z}]$. We first correct the errors in $\tilde{y} + \tilde{z}$ to obtain $y + z$ by applying recursively the decoding algorithm for $\mathcal{C}_{IR}(m-1, r)$. Next, we attemp to correct each of the words $\tilde{y}$ and $\tilde{z}$ by applying the decoding algorithm for $\mathcal{C}_{IR}(m-1, r-1)$. Since one of these words contains no more than 2^{r-1} errors, at least one of the decoding attempts will end up with the right detection indication or (if the number of errors in $\tilde{x}$ is less than 2^{r-1}) with the correct error locations. Knowing $y + z$, we can then reconstruct the other half of x. The complexity analysis in [18] shows that the running time of such a decoding algorithm is $O(2^r\, V(m, r))$ i.e., it is polynomial in the minimum distance and the number of rows of $H_{RM}(m, r)$.

5 Bounds on the redundancy of $\mathcal{C}(m, r)$

We now turn to bounding the redundancy of $\mathcal{C}(m, r)$ from above, starting with the following theorem.

Theorem 7. $red(\mathcal{C}(m, r)) \le \log_2(2^{m-r} + 1) \cdot V(m, r)$.

Proof. The proof is carried out by induction on r. The claim obviously holds for $r = 0$, so we assume from now on that $m \ge r > 0$.

For an integer vector $s \in \mathcal{Z}^{V(m-1,r)}$, denote by $\mathcal{C}(m-1, r; s)$ the set of all words $x \in \mathcal{C}(m-1, r-1)$ such that $H_{RM}(m-1, r)\, x' = s$. Also, let $A(m-1, r)$ denote the set of all vectors s for which $\mathcal{C}(m-1, r; s)$ is nonempty.

Consider the sets

$$\Delta(m, r; s) = \left\{ [x\ y] \in \Phi^{2^m} \mid x \in \mathcal{C}(m-1, r; s) \text{ and } y \in \mathcal{C}(m-1, r; -s) \right\}$$

which are defined for every $s \in A(m-1,r)$. By Lemma 5 we have

$$\mathcal{C}(m,r) = \bigcup_{s \in A(m-1,r)} \Delta(m,r;s) .$$

Furthermore, by definition, the sets $\Delta(m,r;s)$ are disjoint for distinct vectors s and $|\Delta(m,r;s)| = |\mathcal{C}(m-1,r;s)|^2$. Since $\sum_{s \in A(m-1,r)} |\mathcal{C}(m-1,r;s)| = |\mathcal{C}(m-1,r-1)|$, we thus have

$$|\mathcal{C}(m,r)| \geq \frac{|\mathcal{C}(m-1,r-1)|^2}{|A(m-1,r)|} , \tag{2}$$

by which we obtain,

$$red(\mathcal{C}(m,r)) \leq 2\, red(\mathcal{C}(m-1,r-1)) + \log_2 |A(m-1,r)| . \tag{3}$$

Now, for every $s \in A(m-1,r)$, the entries that correspond to the rows of $H_{RM}(m-1,r)$ indexed by $v \in \mathcal{B}(m-1,r-1)$ are all zero. The remaining $\binom{m-1}{r-1}$ entries of s correspond to rows in $H_{RM}(m-1,r)$ whose Hamming weight is 2^{m-r} and, as such, each of these entries in s is an even integer that may take at most $2^{m-r}+1$ values. Hence, we can bound $|A(m-1,r)|$ from above by $(2^{m-r}+1)^{\binom{m-1}{r-1}}$. Applying the induction hypothesis on (3) we thus obtain,

$$\begin{aligned} red(\mathcal{C}(m,r)) &\leq \log_2(2^{m-r}+1) \cdot \left(2\, V(m-1,r-1) + \tbinom{m-1}{r-1}\right) \\ &= \log_2(2^{m-r}+1) \cdot V(m,r) , \end{aligned}$$

as desired. □

The bound of Theorem 7 is tight for the extreme cases $r = 0$ and $r = m$, but not for values of r in between. We can improve the bound by taking into account the fact that the sizes of the sets $\mathcal{C}(m-1,r;s)$ vary with $s \in A(m-1,r)$. We now elaborate on this.

Let the real function $\eta : [0,1] \to [0,1]$ be defined by $\eta(t) = ((1-t)\log_2(1-t) + (1+t)\log_2(1+t))/2$. The following Lemma is the well-known Chernoff bound (see, for instance, [16, p. 310]).

Lemma 8. *Let δ be a real number in the interval $[0,1]$. The number of words $x = [x_0\, x_1\, \ldots\, x_{n-1}]$ over Φ with $\sum_{j=0}^{n-1} x_j \geq \delta n$ is at most $2^{n(1-\eta(\delta))}$.*

Lemma 9. *For $m \geq r > 0$,*

$$red(\mathcal{C}(m,r)) \leq 2\, red(\mathcal{C}(m-1,r-1)) + \tfrac{1}{2}\tbinom{m-1}{r-1}(m-2r+c) + T(m,r) ,$$

where $c = 2 - \log_2 \log_2 e$ and

$$\begin{aligned} T(m,r) = \max_{\alpha \in (0,1)} \Big\{ & -2\log_2(1-\alpha) \\ & + \tfrac{1}{2}\tbinom{m-1}{r-1} \log_2\left(red(\mathcal{C}(m-1,r-1)) + \log_2 \tbinom{m-1}{r-1} + 1 - \log_2 \alpha\right)\Big\} . \end{aligned}$$

Proof. Fix U to be a subset $\{i_1, i_2, \ldots, i_{r-1}\}$ of $\{0, 1, \ldots, m-2\}$ where $i_1 < i_2 < \cdots < i_{r-1}$. For $x \in \Phi^{2^{m-1}}$ and $v \in F^{r-1}$, we denote by $x^{(v)}$ the subvector of x consisting of all 2^{m-r} entries of x that are indexed by vectors $[z_0\, z_1\, \ldots\, z_{m-2}] \in F^{m-1}$ such that $[z_{i_1}\, z_{i_2}\, \ldots\, z_{i_{r-1}}] = v$. Also, denote by $\sigma(x)$ the vector in $\mathcal{Z}^{2^{r-1}}$ whose vth component, $(\sigma(x))_v$, equals the sum of elements in $x^{(v)}$ for every $v \in F^{r-1}$.

Now, assume that x is a word in $\mathcal{C}(m-1, r-1)$ and let $u \in F^{m-1}$ be the characteristic vector of U. It is easy to verify that $(H_{RM}(m-1, r)\, x')_u = (\sigma(x))_1$. On the other hand, we have $\mathcal{H}(m-1, r-1)\, x' = 0$, from which it is fairly easy to deduce the equality $\mathcal{H}(r-1, r-1)\, \sigma(x)' = 0$. Hence, $\sigma(x)$ is a multiple of $\mu(r-1)$ and, as such,

$$|(\sigma(x))_v| \;=\; |(\sigma(x))_1| \;=\; |(H_{RM}(m-1, r)\, x')_u| \quad \text{for every } v \in F^{r-1} .$$

For the fixed set U and for $\delta \in [0, 1]$, let $Q(u, \delta)$ be the set defined by

$$Q(u, \delta) = \left\{ \pm x \in \Phi^{2^{m-1}} \mid \sigma(x)_v\, (-1)^{wt(v)} \geq \delta\, 2^{m-r} \quad \text{for every } v \in F^{r-1} \right\} .$$

In particular, if $x \in \mathcal{C}(m-1, r-1) \cap Q(u, \delta)$, then

$$|(\sigma(x))_v| \;=\; |(H_{RM}(m-1, r)\, x')_u| \geq \delta\, 2^{m-r}$$

for every $v \in F^{r-1}$.

Let α be such that $|Q(u, \delta)| \leq \alpha\, |\mathcal{C}(m-1, r-1)| / \binom{m-1}{r-1}$ and suppose that $\alpha < 1$. We now re-iterate the proof of Theorem 7, except that we ignore sets $\mathcal{C}(m-1, r; s)$ which correspond to vectors s containing at least one entry with absolute value $\geq \delta\, 2^{m-r}$. This allows us effectively to reduce the set $A(m-1, r)$ into a set $\tilde{A}(m-1, r)$ which contains only vectors s whose entries all have absolute values less than $\delta\, 2^{m-r}$. In such a case we have $|\tilde{A}(m-1, r)| \leq (\delta\, 2^{m-r})^{\binom{m-1}{r-1}}$ and the inequality (2) becomes

$$|\mathcal{C}(m, r)| \;\geq\; \frac{|\mathcal{C}(m-1, r-1) - \cup_u Q(u, \delta)|^2}{|\tilde{A}(m-1, r)|} \;\geq\; \frac{(1-\alpha)^2\, |\mathcal{C}(m-1, r-1)|^2}{(\delta\, 2^{m-r})^{\binom{m-1}{r-1}}} ,$$

which translates into

$$red(\mathcal{C}(m, r)) \leq 2\, red(\mathcal{C}(m-1, r-1)) + \tbinom{m-1}{r-1}\, (m-r+\log_2 \delta) - 2\, \log_2(1-\alpha) . \quad (4)$$

We now obtain a bound that relates α and δ in terms of m and r. By Lemma 8,

$$|Q(u, \delta)| \;\leq\; 2 \cdot \left(2^{2^{m-r}(1-\eta(\delta))}\right)^{2^{r-1}} \;=\; 2 \cdot 2^{2^{m-1}(1-\eta(\delta))} .$$

Hence, α and δ can be chosen to be any real numbers in $(0, 1)$ satisfying the inequality

$$\tbinom{m-1}{r-1} \cdot 2 \cdot 2^{2^{m-1}(1-\eta(\delta))} \;\leq\; \alpha\, |\mathcal{C}(m-1, r-1)|$$

i.e.,

$$2^{m-1}\eta(\delta) \geq red(\mathcal{C}(m-1,r-1)) + \log_2\binom{m-1}{r-1} + 1 - \log_2\alpha .$$

Noting that $\eta(\delta) \geq \delta^2(\log_2 e)/2$, we can take α and δ to be such that

$$2\log_2\delta = \log_2\Big(red(\mathcal{C}(m-1,r-1))+1+\log_2\binom{m-1}{r-1} - \log_2\alpha\Big) - m+2-\log_2\log_2 e .$$

The lemma now follows by plugging this value of $\log_2\delta$ into (4) and minimizing over α. □

Corollary 10. *For $m \geq r > 0$,*

$$red(\mathcal{C}(m,r)) \leq 2\,red(\mathcal{C}(m-1,r-1)) + \tfrac{1}{2}\binom{m-1}{r-1}(m + c_0\, r\log_2(m+1))$$

for some absolute constant c_0.

Proof. Fix α to some constant (say, $\alpha = \frac{1}{2}$) and let c_0 be a constant such that $T(m,r) \leq (c_0/2)\binom{m-1}{r-1} r \log_2(m+1)$ for every $m \geq r > 0$; by Theorem 7, such a constant indeed exists. The corollary now follows from Lemma 9. □

The following theorem is an improvement on Theorem 7 for small values of r, namely, when $r = o(m/\log m)$.

Theorem 11. $red(\mathcal{C}(m,r)) \leq (m/2)\,V(m,r)\Big(1 + O((r/m)\log(m+1))\Big)$.

Proof. Let c_0 be as in Corollary 10 and set $\epsilon(m,r)$ to be the expression $(c_0\, r/m)\log_2(m+1)$. We prove the inequality

$$red(\mathcal{C}(m,r)) \leq (m/2)\,V(m,r)\,(1+\epsilon(m,r))$$

by induction on r. The inequality obviously holds when $r = 0$, so we assume from now on that $m \geq r > 0$.

By Corollary 10 and the induction hypothesis we have

$$\begin{aligned} red(\mathcal{C}(m,r)) &\leq 2\,(m/2)\,V(m-1,r-1)\,(1+\epsilon(m-1,r-1)) \\ &\quad + \tfrac{1}{2}\binom{m-1}{r-1}(m + c_0\, r\log_2(m+1)) . \end{aligned}$$

Now, $\epsilon(m,r) \geq \epsilon(m-1,r-1)$ and, therefore,

$$\begin{aligned} red(\mathcal{C}(m,r)) &\leq (m/2)\cdot 2\,V(m-1,r-1)\,(1+\epsilon(m,r)) + (m/2)\binom{m-1}{r-1}(1+\epsilon(m,r)) \\ &= (m/2)\,V(m,r)\,(1+\epsilon(m,r)) , \end{aligned}$$

as claimed. □

We therefore conclude that when m is large and r is fairly small, the redundancy of $\mathcal{C}(m,r)$ is not greater than approximately $m/2$ times the redundancy of the respective $(m-r)$th order RM code of length 2^m. In particular, for $r=1$ we obtain the well-known fact that the redundancy of $\mathcal{S}(2^m,1) = \mathcal{C}(m,1)$ is approximately $m/2$.

It is worthwhile comparing the bound of Theorem 11 to the known bounds on $red(\mathcal{S}(n,r))$. For $n=2^m$ we have [19]

$$red(\mathcal{S}(2^m,r)) = O(2^r m) , \tag{5}$$

compared to the bound

$$red(\mathcal{C}(m,r)) = O(m^r)$$

implied by Theorem 11. Recall, however, that the minimum distance of $\mathcal{C}(m,r)$ is *exponential* in r, whereas the minimum distance of $\mathcal{S}(2^m,r)$ is guaranteed to be only *linear* in r [13],[14] and is bounded from above by $r(r-1)+2$ [11, p. 506],[19]. We point out, however, that it is not yet known as to what extent the bound of (5) is tight: the lower bound on $red(\mathcal{S}(2^m,r))$ shown in [19] is $(r-1)(m-r+1)$.

The next table presents the values of $red(\mathcal{C}(m,r))$ for $m \le 5$.

$r \backslash m$	1	2	3	4	5
1	1	1.42	1.87	2.35	2.84
2		3	5	8.21	12.38
3			7	12.68	22.89
4				15	28.42
5					31

Values of $red(\mathcal{C}(m,r))$ for $m \le 5$.

Acknowledgment

The author thanks Tuvi Etzion, Paul Siegel, and Alexander Vardy for helpful discussions.

References

1. S. Al-Bassam, B. Bose, *On balanced codes, IEEE Trans. Inform. Theory,* IT-36 (1990), 406–408.
2. N. Alon, E.E. Bergmann, D. Coppersmith, A.M. Odlyzko, *Balancing sets of vectors, IEEE Trans. Inform. Theory,* IT-34 (1988), 128–130.

3. A.M. Barg, *Incomplete sums, DC-constrained codes, and codes that maintain synchronization, Designs, Codes, and Cryptography,* 3 (1993), 105–116.
4. A.M. Barg, S.N. Lytsin, *DC-constrained codes from Hadamard matrices, IEEE Trans. Inform. Theory,* IT-37 (1991), 801–807.
5. M. Blaum, *A (16,9,6,5,4) error-correcting DC-free block code, IEEE Trans. Inform. Theory,* IT-34 (1988), 138–141.
6. E. Eleftheriou, R. Cideciyan, *On codes satisfying Mth order running digital sum constraints, IEEE Trans. Inform. Theory,* IT-37 (1991), 1294–1313.
7. T. Etzion, *Constructions of error-correcting DC-free block codes, IEEE Trans. Inform. Theory,* IT-36 (1990), 899–905.
8. H.C. Ferreira, *Lower bounds on the minimum Hamming distance achievable with runlength constrained or DC-free block codes and the synthesis of a (16,8), $D_{\min} = 4$, DC-free block code, IEEE Trans. Magn.,* MAG-20 (1984), 881–883.
9. W.H. Gottschalk, G.A. Hedlung, *Topological Dynamics, Colloquium Publications of the AMS,* 36, American Math. Society, Providence, Rhode Island, 1955.
10. H.D.L. Hollman, K.A.S. Immink, *Performance of efficient balanced codes, IEEE Trans. Inform. Theory,* IT-37 (1991), 913–918.
11. L.K. Hua, *Introduction to Number Theory,* Springer, Berlin, 1982.
12. K.A.S. Immink, *Coding Techniques for Digital Recorders,* Prentice-Hall, London, 1991.
13. K.A.S. Immink, G. Beenker, *Binary transmission codes with higher order spectral zeros at zero frequency, IEEE Trans. Inform. Theory,* IT-33 (1987), 452–454.
14. R. Karabed, P.H. Siegel, *Matched spectral-null codes for partial-response channels, IEEE Trans. Inform. Theory,* IT-37 (1991), 818–855.
15. D.E. Knuth, *Efficient balanced codes, IEEE Trans. Inform. Theory,* IT-32 (1986), 51–53.
16. F.J. MacWilliams, N.J.A. Sloane, *The Theory of Error-Correcting Codes,* North-Holland, Amsterdam, 1977.
17. C.M. Monti, G.L. Pierobon, *Codes with a multiple spectral null at zero frequency, IEEE Trans. Inform. Theory,* IT-35 (1989), 463–472.
18. R.M. Roth, G.M. Benedek, *Interpolation and approximation of sparse multivariate polynomials over $GF(2)$, SIAM J. Comput.,* 20 (1991), 291–314.
19. R.M. Roth, P.H. Siegel, A. Vardy, *High-order spectral-null codes: Constructions and bounds,* submitted to *IEEE Trans. Inform. Theory.*
20. H. van Tilborg, M. Blaum, *On error-correcting balanced codes, IEEE Trans. Inform. Theory,* IT-35 (1989), 1091–1095.

On small families of sequences with low periodic correlation

Sascha Barg

IPPI, Moscow
E-mail: aBarg@ippi.msk.su

Abstract. We survey families of binary sequences with good correlation properties of period n and size of order n and n^2.

1. Introduction.

Good families of low-correlated sequences are mostly constructed from the so-called trace codes. In other words, they are formed by cyclically distinct vectors of some well-studied cyclic codes of BCH type. Recent progress in the theory of cyclic codes and maximum period sequences over finite rings [1]-[4] allows to construct some very good sequence families.

2. Bounds.

In this correspondence we restrict ourselves to families of binary n-periodic sequences of size M of order from n to n^2 (the summary of families of lesser size is given in [5]). For any two sequences $a, b \in \{0,1\}^*$, let

$$\theta_{a,b}(\tau) = \sum_{i=0}^{n-1} (-1)^{a(i)+b(i+\tau)}$$

denote their periodic correlation. By $\theta(S)$ let us denote the maximum modulo of the correlation of a family S. Denote by $L(n,\theta)$ the maximum possible cardinality of a family S with period n and $\theta(S) = \theta$ and put $\theta(n,M) = \min\{\theta(S) \,|\, S$ has period n and size $M\}$. These two quantities were studied in quite a few papers, [6]-[8] among them. In particular, it is known that for $\theta^2 \leq 3n-8$,

$$L(n,\theta) \leq \frac{n^2-\theta^2}{3n-2-\theta^2} \qquad \text{(the Sidelnikov bound [6, p.21]).} \qquad (1)$$

From this it follows that for any positive constants a, b with $(\sqrt{n+a}+b)^2 \leq 3n-8$

$$L(n, \sqrt{n+a}+b) \leq n/2 + \text{const}\,\sqrt{n}. \qquad (2)$$

Further, from [7, p.87] it follows that

$$L(n, 2\sqrt{n+a}+b) \leq 3n^2/10 + \text{const}\, n, \tag{3}$$

where $a, b > 0$ and $(2\sqrt{n+a}+b)^2 \leq 3n - 10 + \sqrt{6n^2 - 42n + 76}$.

3. Families.

3.1. $M = \Theta(n)$. Consider a code of length $n = 2^{2l+1} - 1$ with the generator idempotent $\theta_1^* + \theta_{2^t+1}^*$, $gcd(t, 2l+1) = 1$. Its cyclically distinct vectors form the set of Gold sequences (in particular, for $l = 1$, the "dual 2-error-correcting BCH family" [6]) with $M = n + 2$ and $\theta = 1 + \sqrt{2n}$ which is optimal by (1).

While these results are rather old, it is only recently that it became possible to construct good families for $n = 2^{2l}$ (even exponents). It was discovered by a number of authors [1], [4] that the binary Kerdock codes are just binary images of quaternary maximum period sequences. In particular, Nečaev [1] has proved the following fact.

For any $u \in Z_4$, denote the coordinates of its 2-adic expansion by u_0, u_1: $u = u_0 + 2u_1$. Let $F(x) \in Z_4[x]$ be a primitive basic irreducible polynomial of degree $2l - 1$ and define $F_0(x), F_1(x)$ by $F(x) = F_0(x^2) + xF_1(x^2)$. Put

$$\begin{aligned} G(x) &= -(F_0(x)^2 - xF_1(x)^2), \\ H(x) &= G(3x). \end{aligned}$$

Consider the set $L(H)$ of quaternary linear recurrent sequences $u(i)$ with characteristic polynomial $H(x)$. Form the set of binary vectors

$$\mathcal{K}(2l) = \{(a_0, a_1, u_1(0) + a_1, u_1(1) + a_0, u_1(2) + a_0, \ldots, u_1(2^{2l} - 3) + a_1)\},$$

$$a_0, a_1 \in \{0, 1\}.$$

Theorem 1 *(Nečaev). The set $\mathcal{K}(2l), l \geq 2$, forms a $(2^{2l}, 2^{4l}, 2^{2l-1} - 2^{l-1})$-code equivalent to the Kerdock code. The twice shortened code $\mathcal{K}^*(2l)$ is a nonlinear cyclic $(2^{2l} - 2, 2^{4l}, 2^{2l-1} - 2^{l-1} - 2)$-code.*

The weight spectrum of $\mathcal{K}(2l)$ is known from [9]. Take $2^{2l}(2^{2l-1} - 1)$ words in the code $\mathcal{K}^*(2l)$ that correspond to the words of minimum weight in $\mathcal{K}(2l)$. A straightforward argument shows that they are generated by 2^{2l-1} different cyclic representatives, each of period $2^{2l} - 2$.

Proposition 1 *The family $S(\mathcal{K}^*)$ of sequences has the parameters $n = 2^{2l} - 2, M = 2^{2l-1}, \theta(S) = 2 + \sqrt{n+2}$ and is asymptotically optimal according to (1).*

3.2 $M = \Theta(n^2)$. Based on the Nečaev's results, it is natural to try to construct larger families. Namely, consider the binary Delsarte–Goethals codes $\mathcal{DG}(m, (m-2)/2), m = 2l + 2, l \geq 1$ [10, Eq.15.37]. Again starting from a theorem in [1] about the trace description of coordinates of the 2-adic representation of elements of quaternary linear recurrent sequences, it is possible to prove

Corollary 1 *There exists a nonlinear cyclic code with the parameters* $(2^m - 2, 2^{3m-1}, 2^{m-1} - 2^{m/2} - 2)$ *which is equivalent to the twice shortened code* $\mathcal{DG}(m, (m-2)/2)$.

The proof follows from the fact that the code $\mathcal{DG}(m, (m-2)/2)$ consists of the vectors of two types: (1) vectors of the Kerdock code $K(m)$ and (2) vectors of the type $|c|c| + a$, where $a \in K(m)$ and $c \in C$, an extended cyclic code generated by the idempotent $\theta_0 + \theta_1^* + \theta_3^*$ (see [10, Ch.15]).

From this we get the following

Proposition 2 *There exists a family S of sequences of period* $n = 2^{2l+2} - 2$, *size of order* $n^2/4$ *and* $\theta(S) = 2^{l+2} + 2 \sim 2\sqrt{n}$.

Note that this is only slightly worse than the upper bound (3).

A family of somewhat greater size $M \sim n^2, n = 2^{2l+2} - 1, \theta(S) \leq 4\sqrt{n}$, is formed by cyclically distinct vectors of the dual triple-error-correcting BCH codes. As noted in [11], [12], the "very large Kasami set" has the same order of θ and the size asymptotically greater by $\sqrt{n}$ (see the table below). The best known bound [7] asserts that $\theta(n, n^2) \gtrsim \sqrt{5n}$.

Finally, for $n = 2^{2l+1} - 1$ a family of size $M \sim n^2$ can be formed by cyclically distinct vectors of the cyclic code generated by $\theta_1^* + \theta_{2^{j-1}+1}^* + \theta_{2^j+1}^*$, where $j = 2$ (dual 3-error-correcting BCH codes) or $j = l$ ([10, p. 439]) or by a proper enlargement of the set of Gold sequences [12].

In the table below we write that a family is asymptotically optimal if it asymptotically meets Sidelnikov bound (1),(2) and put a question mark if it does not meet the bounds quoted above and no better judgement can be made.

References

[1] A.A. Nečaev, "The cyclic form of the Kerdock code," *Discr. Math. Appl.*, **1** 4 (1989), 123–139, in Russian.

[2] A.S. Kuz'min and A.A. Nečaev, "Construction of error-correcting codes using linear recurrences over Galois rings," *Uspekhi Mat. Nauk*, **47** 5 (1992), 183–184, in Russian.

[3] A.S. Kuz'min and A.A. Nečaev, "Linear recurrent sequences over Galois rings," Uspekhi Mat. Nauk, **48** 1(1993), 176–168, in Russian.

[4] A.R. Hammons, P.V. Kumar, A.R. Calderbank, N.J.A. Sloane, and P. Solé, "The Z_4-linearity of Kerdock, Preparata, Goethals, and related codes," manuscript (1993).

[5] P.V. Kumar and O. Moreno, "Prime-phase sequences with periodic correlation properties better than binary sequences," *IEEE Trans. Inf. Theory*, **IT-37** (May 1991), 603–616.

[6] V.M. Sidelnikov, "On mutual correlation of sequences", *Problemy Kibernetiki*, **24** (1971), 15–42, in Russian.

[7] V.I. Levenshtein, "Bounds for packings of metric spaces and certain applications," *Problemy Kibernetiki*, **40** (1983), 47–110, in Russian.

[8] A. Tietäväinen, "On the cardinality of sets of sequences with given maximum correlation," *Discrete Math.*, **106/107** (1992), 471–477.

[9] E.R. Berlekamp, "The weight enumerators for certain subcodes of the second order binary Reed–Muller codes," *Information and Control*, **17** (1970), 485–500.

[10] F.J. MacWilliams and N.J.A. Sloane, *The Theory of Error-Correcting Codes* [Russian Translation], Moscow, Svyaz, 1979.

[11] H. Tarnanen and A. Tietäväinen, "A simple method to estimate the maximum nontrivial correlation of some sets of sequences," manuscript.

[12] O.S. Rothaus, "Modified Gold codes," *IEEE Trans. Inf. Theory*, **IT-39** (1993), 654–656.

[13] A. Barg, "A large family of sequences with low periodic correlation," submitted for publication.

Small binary families

Period	Size	$\theta_{\max}$	Asymptotic optimality	Notes
$n = 2^{2l+1} - 1$	$n + 2$	$1 + \sqrt{2n+2}$	optimal	Gold families dual 2-error-corecting BCH codes [6]
$n = 2^{2l} - 2$ $l \geq 2$	$(n/2) + 1$	$2 + \sqrt{n+2}$	optimal	twice shortened cyclic Kerdock codes [1]
$n = 2^{2l} - 2$ $l \geq 2$	$\sim n^2/4$	$\sim 2\sqrt{n}$	?	twice shortened Delsarte-Goethals codes $\mathcal{DG}(2l, l-1)$
$n = 2^{2l} - 1$	$(n+1)(n+2)+1$	$4\sqrt{n}$	?	dual 3-error-correcting BCH codes
$n = 2^{2l} - 1$	$(n+1)^{5/2}$	$4\sqrt{n}$ (upper bound)	?	"very large Kasami set" [11]
$n = 2^{2l+1} - 1$	$(n+1)(n+2)+1$	$2\sqrt{2n+2}$	?	dual 3-error correcting BCH codes, $\theta_1^* + \theta_{2^{l-1}+1}^* + \theta_{2^l+1}^*$-codes.
$n = p^l$	$n^2 - n$	$4 + 3\sqrt{n}$ (upper bound)	?	sequences from multiplicative characters [13]

Disjoint Systems (Extended Abstract)

Noga Alon [1] **Benny Sudakov**
Department of Mathematics
Raymond and Beverly Sackler Faculty of Exact Sciences
Tel Aviv University, Tel Aviv, Israel

Abstract. A *disjoint system of type* $(\forall, \exists, k, n)$ is a collection $\mathcal{C} = \{\mathcal{A}_1, \ldots, \mathcal{A}_m\}$ of pairwise disjoint families of k-subsets of an n-element set satisfying the following condition. For every ordered pair $\mathcal{A}_i$ and $\mathcal{A}_j$ of distinct members of $\mathcal{C}$ and for every $A \in \mathcal{A}_i$ there exists a $B \in \mathcal{A}_j$ that does not intersect A. Let $D_n(\forall, \exists, k)$ denote the maximum possible cardinality of a disjoint system of type $(\forall, \exists, k, n)$. It is shown that for every fixed $k \geq 2$,

$$lim_{n \to \infty} D_n(\forall, \exists, k) \binom{n}{k}^{-1} = \frac{1}{2}.$$

This settles a problem of Ahlswede, Cai and Zhang. Several related problems are considered as well.

1 Introduction

In Extremal Finite Set Theory one is usually interested in determining or estimating the maximum or minimum possible cardinality of a family of subsets of an n element set that satisfies certain properties. See [5], [7] and [9] for a comprehensive study of problems of this type. In several recent papers (see [3], [1],[2]), Ahlswede, Cai and Zhang considered various extremal problems that study the maximum or minimum possible cardinality of a collection of families of subsets of an n-set, that satisfies certain properties. They observed that many of the classical extremal problems dealing with families of sets suggest numerous intriguing questions when one replaces the notion of a family of sets by the more complicated one of a collection of families of sets.

In the present note we consider several problems of this type that deal with disjoint systems. Let $N = \{1, 2, \ldots, n\}$ be an n element set, and let

[1] Research supported in part by a United States - Israel BSF Grant

$\mathcal{C} = \{\mathcal{A}_1, \ldots, \mathcal{A}_m\}$ be a collection of pairwise disjoint families of k-subsets of N. $\mathcal{C}$ is a ***disjoint system of type*** $(\exists, \forall, k, n)$ if for every ordered pair $\mathcal{A}_i$ and $\mathcal{A}_j$ of distinct members of $\mathcal{C}$ there exists an $A \in \mathcal{A}_i$ which does not intersect any member of $\mathcal{A}_j$. Similarly, $\mathcal{C}$ is a ***disjoint system of type*** $(\forall, \exists, k, n)$ if for every ordered pair $\mathcal{A}_i$ and $\mathcal{A}_j$ of distinct members of $\mathcal{C}$ and for every $A \in \mathcal{A}_i$ there exists a $B \in \mathcal{A}_j$ that does not intersect A. Finally, $\mathcal{C}$ is a ***disjoint system of type*** $(\exists, \exists, k, n)$ if for every ordered pair $\mathcal{A}_i$ and $\mathcal{A}_j$ of distinct members of $\mathcal{C}$ there exists an $A \in \mathcal{A}_i$ and a $B \in \mathcal{A}_j$ that does not intersect A.

Let $D_n(\exists, \forall, k)$ denote the maximum possible cardinality of a disjoint system of type $(\exists, \forall, k, n)$. Let $D_n(\forall, \exists, k)$ denote the maximum possible cardinality of a disjoint system of type $(\forall, \exists, k, n)$ and let $D_n(\exists, \exists, k)$ denote the maximum possible cardinality of a disjoint system of type $(\exists, \exists, k, n)$. Trivially, for every n,

$$D_n(\exists, \forall, 1) = D_n(\forall, \exists, 1) = D_n(\exists, \exists, 1) = n.$$

It is easy to see that every disjoint system of type $(\exists, \forall, k, n)$ is also a system of type $(\forall, \exists, k, n)$, and every system of type $(\forall, \exists, k, n)$ is also of type $(\exists, \exists, k, n)$. Therefore, for every $n \geq k$

$$D_n(\exists, \forall, k) \leq D_n(\forall, \exists, k) \leq D_n(\exists, \exists, k).$$

In this note we determine the asymptotic behaviour of these three functions for every fixed k, as n tends to infinity.

Theorem 1.1 *For every $k \geq 2$*

$$lim_{n \to \infty} D_n(\exists, \forall, k) \binom{n}{k}^{-1} = \frac{1}{k+1}.$$

Theorem 1.2 *For every $k \geq 2$*

$$lim_{n \to \infty} D_n(\forall, \exists, k) \binom{n}{k}^{-1} = \frac{1}{2}.$$

Corollary 1.3 *For every $k \geq 2$*

$$lim_{n \to \infty} D_n(\exists, \exists, k) \binom{n}{k}^{-1} = \frac{1}{2}.$$

Theorem 1.1 settles a conjecture of Ahlswede, Cai and Zhang [2], who proved it for $k = 2$ [1]. The main tool in its proof is a result of Frankl and Füredi [8]. The proof of Theorem 1.2, which settles another question raised in [2] and proved for

$k = 2$ in [1], is more complicated and combines combinatorial and probabilistic arguments. A sketch of this proof and the simple derivation of Corollary 1.3 from its assertion are presented in Section 2. The proof of Theorem 1.1 and the full proof of Theorem 1.2 will appear in the full version of this paper.

2 Random graphs and disjoint systems

In this section we give a sketch of the proof of Theorem 1.2. We need the following two probabilistic lemmas.

Lemma 1 (Chernoff, see e.g. [4], Appendix A). *Let X be a random variable with the binomial distribution $B(n,p)$. Then for every $a > 0$ we have*

$$Pr(|x - np| > a) < 2e^{-2a^2/n}.$$

Let L be a graph-theoretic function. L satisfies ***the Lipschitz condition*** if for any two graphs H, H' on the same set of vertices that differ only in one edge we have $|L(H) - L(H')| \leq 1$. Let $G(n,p)$ denote, as usual, the random graph on n labeled vertices in which every pair, randomly and independently, is chosen to be an edge with probability p. (See, e.g., [6].)

Lemma 2 ([4], Chapter 7). *Let L be a graph-theoretic function satisfying the Lipschitz condition and let $\mu = E[L(G)]$ be the expectation of $L(G)$, where $G = G(n,p)$. Then for any $\lambda > 0$*

$$Pr(|L(G) - \mu| > \lambda\sqrt{m}] < 2e^{-\lambda^2/2}$$

where $m = \binom{n}{2}$.

Sketch of the proof of Theorem 1.2 Let n_1 be the number of families containing only one element in a disjoint system of type $(\forall, \exists, k, n)$. Since sets in any two one-element families are disjoint we have $n_1 \leq n/k$. This settles the required upper bound for $D_n(\forall, \exists, k)$, since all other families contain at least 2 sets.

We prove the lower bound using probabilistic arguments. We show that for any $\varepsilon > 0$ there are at least $\frac{1}{2}(1-\varepsilon)\binom{n}{k}$ families which form a disjoint system of type $(\forall, \exists, k, n)$, provided n is sufficiently large (as a function of ε and k). Let $G = G(n,p)$ be a random graph, where p is a constant, to be specified later, which is very close to 1. We use this graph to build another random graph G_1, whose vertices are all k-cliques in G. Two vertices of G_1 are adjacent if and only if the induced subgraph on the corresponding k-cliques in G is the union of two vertex disjoint k-cliques with no edges between them. We prove that almost surely (i.e.,with probability that tends to 1 as n tends to infinity) the following two events happen. First, the number of vertices in G_1 is greater than $(1-\varepsilon/2)\binom{n}{k}$. Second, G_1 is almost regular, i.e., for every (small) $\delta > 0$ there exists a (large) number d such that the degree $d(x)$ of any vertex x of G_1 satisfies $(1-\delta)d < d(x) < (1+\delta)d$, provided n is sufficiently large.

Suppose $G_1 = (V, E)$ satisfies these properties. By Vizing's Theorem [10], the chromatic index $\chi'(G_1)$ of G_1 satisfies $\chi'(G_1) \leq (1+\delta)d+1$. Since for any $x \in G_1$ we have $d(x) \geq (1-\delta)d$, the number of edges $|E|$ of G_1 is at least $\frac{(1-\delta)d|V|}{2}$. Hence there exists a matching in G_1 which contains at least $\frac{(1-\delta)d|V|}{2}/\chi'(G_1) \sim \frac{(1-\delta)|V|}{2(1+\delta)}$ edges. This matching covers almost all vertices of G_1, as δ is small, providing a system of pairs of k-sets covering almost all the $\binom{n}{k}$ k-sets. Taking each pair as a family we have a disjoint system of size at least $\frac{1}{2}(1-\varepsilon)\binom{n}{k}$ and ε can be made arbitrarily small for all n sufficiently large.

We next show that the resulting system is a disjoint system of type $(\forall, \exists, k, n)$. Assume this is false and let $\mathcal{A} = \{A_1, A_2\}$ and $\mathcal{B} = \{B_1, B_2\}$ be two pairs where $A_1 \cap B_i \neq \emptyset$ for $i = 1, 2$. Choose $x_1 \in A_1 \cap B_1$ and $x_2 \in A_1 \cap B_2$. Since x_1 and x_2 belong to A_1 they are adjacent in $G = G(n,p)$. However, $x_1 \in B_1$, $x_2 \in B_2$ and this contradicts the fact that the subgraph of G induced on $B_1 \cup B_2$ has no edges between B_1 and B_2. Thus the system is indeed of type $(\forall, \exists, k, n)$ and

$$D_n(\forall, \exists, k) > \frac{1}{2}(1-\varepsilon)\binom{n}{k}$$

for every $\varepsilon > 0$, provided $n > n_0(k, \varepsilon_1)$, as needed.

The proof that indeed G_1 has the required properties stated above almost surely can be established using Lemmas 1 and 2. We omit the details.

Proof of Corollary 1.3 Let n_1 be the number of one element families in a disjoint system of type $(\exists, \exists, k, n)$. The trivial argument used in the proof of Theorem 1.2 shows that $n_1 \leq n/k$. Since each other family contains at least two elements

$$D_n(\exists, \exists, k) \leq \frac{n}{k} + \frac{1}{2}\binom{n}{k}.$$

As observed in Section 1, $D_n(\forall, \exists, k) \leq D_n(\exists, \exists, k)$ and hence, by Theorem 1.2, the desired result follows.

References

1. R. Ahlswede, N. Cai and Z. Zhang, *A new direction in extremal theory for graphs*, to appear in JCISS.
2. R. Ahlswede, N. Cai and Z. Zhang, *Higher level extremal problems*, to appear.
3. R. Ahlswede and Z. Zhang, *On cloud-antichains and related configurations*, Discrete Math. 85 (1990), 225-245.
4. N. Alon and J. H. Spencer, **The Probabilistic Method**, Wiley, 1991.
5. I. Anderson, **Combinatorics of Finite Sets**, Clarendon Press, 1987.
6. B. Bollobás, **Random Graphs**, Academic Press, 1985.
7. B. Bollobás, **Combinatorics**, Cambridge University Press, 1986.
8. P. Frankl and Z. Füredi, *Colored packing of sets in combinatorial desgin theory*, Annals of Discrete Math. 34 (1987), 165-178.
9. Z. Füredi, *Matchings and covers in hypergraphs*, Graphs and Combinatorics 4 (1988), 115-206.
10. V. G. Vizing, *Coloring the vertices of a graph in prescribed colors* (in Russian), Diskret. Analiz. No. 29, Metody Diskret. Anal. v. Teorii Kodov i Shem 101 (1976), 3-10.

Some sufficient conditions for 4-regular graphs to have 3-regular subgraphs *

Oscar Moreno[1] and Victor A. Zinoviev[1,2]

[1] Department of Mathematics, University of Puerto Rico, Rio Piedras, P.R. 00931 USA
Email: o_moreno@uprenet.bitnet,
moreno@sun386-gauss.uprr.pr
[2] Institute for Problems of Information Transmission Russian Academy of Sciences, Ermolova str.19, GSP-4, Moscow, 101447, Russia.
Email: zinov@sun386-gauss.uprr.pr
zinov@cmnet.uprr.pr

Abstract. We are interested in the following problem: when would a 4-regular graph (with multiple edges) have a 3-regular subgraph. We give several sufficient conditions for 4-regular graph to have a 3-regular subgraph.

1. Introduction.

All graphs considered are finite, undirected and without loops. Note that we allow multiple edges. A simple graph is a graph without multiple edges.

A graph G has a set of vertices V and a set of edges E. A graph G is 4-regular if the degree of every vertex of G is equal to four.

A well-known conjecture of Berge and Sauer (see [4] and [5]), asserts that every 4-regular simple graph has a 3-regular subgraph. This conjecture was proved by Taŝkinov [9].

In [2] (see also [3] and [8]) the Chevalley-Warning theorem was used to extend this result to graphs with multiples edges, that are 4-regular plus an edge.

Theorem A [2]. Every 4-regular graph plus one edge contains a 3-regular subgraph.

In a previous paper [7] the authors using similar techniques as in [2] presented a sufficient condition for 4-regular graphs (possibly with multiple edges) to have a 3-regular subgraph.

This sufficient condition looks as follows: let G be a 4-regular graph with n vertices: $v_1, v_2, \cdots, v_n$. A 2-decomposition will be a disjoint partition

* Work partially supported by NSF Grants RII-9014056, the Component IV of the EPSCoR of Puerto Rico Grant, U.S. Army Center of Excellence for Symbolic Methods in Algorithmic Mathematics (ACSyAM), of Cornell MSI. Contract DAAL03-91-C-0027, and the Office of Naval Research under grant number N00014-90-F-1301.

$A_{v_1}, A_{v_2}, \cdots, A_{v_n}$ of the set E of edges into sets with two edges, both incident with one corresponding vertex. In other words,

$$E = \bigcup_{i=1}^{n} A_{v_i},$$

where

$$A_{v_i} \bigcap A_{v_j} = \emptyset \text{ for any } i \neq j,$$

and for any i

$$A_{v_i} = \{e_1^{(i)}, e_2^{(i)}\},$$

where both edges $e_1^{(i)}$ and $e_2^{(i)}$ are incident to vertex v_i. Denote N_G the total number of distinct 2-decompositions of G. It has been pointed out by N. Alon [1] that the above notion of a 2-decomposition coincides with that of an Euler orientation (see[6]). This makes easy to see that there are such decompositions, since it is well known that a connected graph, where every vertex has even degree, must have an Euler orientation (see [6]).

Our main result in [7] is the following theorem:

Theorem B [7]. A 4-regular graph G (possibly with multiple edges) has a 3-regular subgraph, if the number N_G of distinct 2-decompositions (Euler orientations) of G is such that $N_G \not\equiv 1$ (modulo 3).

Here we give several new sufficient conditions for a 4-regular graph to have a 3-regular subgraph. Two of them are natural generalizations of Theorem A, and the third one gives a sufficient condition that works even in cases when Theorem B does not work. We illustrate these results with examples.

2. Sufficient Conditions

We need new concepts. Different from the case of an (usual) edge $e = e(x, y)$, when the edge e is incident with two vertices x and y, we shall use the concept of a half-edge $e = e(x)$, when the edge e is incident with only one vertex x. We will say that we cut the edge $e(x, y)$ on two half-edges $e(x)$ and $e(y)$, if we change the edge $e(x, y)$ by two half-edges $e(x)$ and $e(y)$. See Fig.1 for illustration of these concepts.

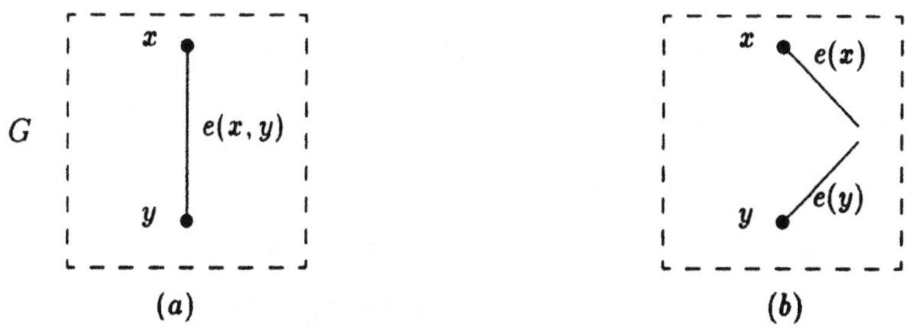

Fig.1. Cutting of edge $e(x, y)$ in graph G

The following theorem is a modification of Theorem A of Alon, Friedland and Kalai [2,3].

Theorem 2.1. A 4-regular graph G has a 3-regular subgraph if we:
(i) add one half-edge to any vertices of G;
(ii) cut any one edge of G on two half-edges.

Proof. The proof is similar to the proof of Theorem A in [2].

Definition 2.2. Let G be a 4-regular graph minus one edge. Any partition

$$A(1,1) = (A_{v_1}, \cdots, A_{v_n})$$

of the set $E(G)$, where A_{v_i} has size two for all i, except exactly in one value, when it has size one, and A_{v_i}, $i = 1, \cdots, n$, consists of edges which are incident with vertex v_i, will be called a partition of type (1,1) (one subset has one edge). Denote by $N(1,1)$ the number of distinct partitions of G of type $(1,1)$.

Definition 2.3. Let G be a 4-regular graph minus one edge. Any partition

$$A(2,1) = (A_{v_1}, \cdots, A_{v_n})$$

of the set $E(G)$, where A_{v_i} has size two for all i, $i = 1, \cdots, n$, and A_{v_i} consists of two different edges, which are incident with vertex v_i, but there is exactly one edge which belongs to two different subsets A_{v_i} and A_{v_j} will be called a partition of type (2,1) (two subsets have one edge in common). Denote by $N(2,1)$ the number of distinct partitions of G of type $(2,1)$.

Example 2.4. The definitions of the numbers $N(1,1)$ and $N(2,1)$ are illustrated by graph the G, which is 4-regular without one edge (see Fig.2).

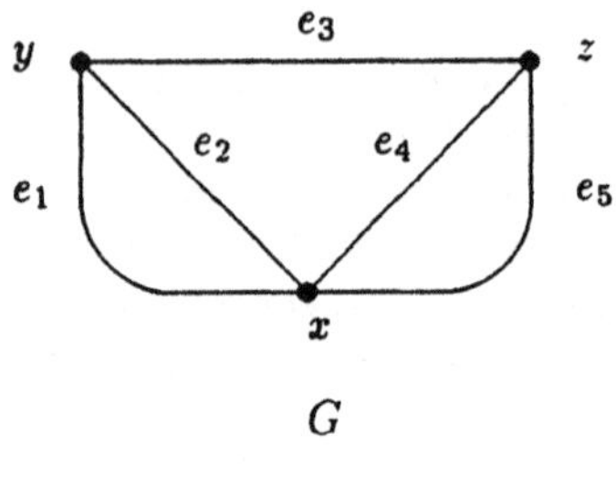

Fig.2. A 4-regular graph G minus one edge.

Then the sets of edges

$$A_x = \{e_2, e_5\}, A_y = \{e_1, e_3\}, A_z = \{e_4\}$$

give a decomposition of type (1,1) and the sets of edges

$$A'_x = \{e_1, e_5\}, A'_y = \{e_2, e_3\}, A'_z = \{e_3, e_4\}$$

give a decomposition of type (2,1). It is easy to find that for the given graph G, $N(1,1) = 14, N(2,1) = 16$.

The following theorem might be considered as a refinement of Theorem B.

Theorem 2.5. Let G be any 4-regular graph for which N, the number of distinct 2-decompositions is such that

$$N \equiv 1 (\text{mod } 3). \tag{1}$$

Let G_e be obtained from G by eliminating any edge $e = e(x, y)$, and let $N_e(1,1)$ and $N_e(2,1)$ be the corresponding numbers of partitions of G_e of type (1,1) and type (2,1). If these numbers are such that

$$N(1,1) - N(2,1) \not\equiv 1 (\text{mod } 3) \tag{2}$$

then the graph G has a 3-regular subgraph.

Proof. This proof is similar to the proof of Theorem B (see [7]). Let G be a 4-regular graph with n vertices and $m = 2n$ edges. Let $A = (a_{ij})$ be the $n \times m$ incidence matrix of $G : a_{ij} = 1$, if vertex v_i and edge j are incident and $a_{ij} = 0$, otherwise. Consider the system

$$A \begin{pmatrix} x_1^2 \\ x_2^2 \\ \vdots \\ x_m^2 \end{pmatrix} = \begin{pmatrix} 0 \\ 0 \\ \vdots \\ 0 \end{pmatrix}$$

of n quadratic equations in the m variables $x_1, x_2, \cdots, x_m$. This is actually a system of n multivariable polynomial equations:

$$F_1 = 0,\ F_2 = 0,\ \cdots,\ F_n = 0\ , \tag{3}$$

where F_i, $i = 1, \cdots, n$, has the form

$$F_i = x_{i_1}^2 + x_{i_2}^2 + x_{i_3}^2 + x_{i_4}^2. \tag{4}$$

where $1 \leq i_1, i_2, i_3, i_4 \leq m$. The sense of this equation is that the point v_i is incident to exactly four edges, which corresponds to four variables $x_{i_1}, x_{i_2}, x_{i_3}, x_{i_4}$.

Now consider the above system of equations over the Galois field $F = GF(3)$. According to the result of Alon, Friedland and Kalai [2] the existence of nontrivial solution of the system (3) (i.e. not all $x_i = 0$) implies the existence of a 3-regular subgraph of G.

We have that S, the number of solutions to the system above, is such that

$$S = \sum_{x_1, \cdots, x_m \in F} (1 - F_1^2)(1 - F_2^2) \cdots (1 - F_n^2). \tag{5}$$

In order to obtain only the trivial solution we must have $S \equiv 1 (\text{mod } 3)$ and therefore the above sum must also be 1, when carried out modulo 3. A direct calculation [7] shows that

$$S \equiv N (\text{mod } 3)\ ,$$

where N is the number of distinct 2-decompositions of G. The condition that $N \equiv 1 (\mathrm{mod}\ 3)$ gives that $S \equiv 1 (\mathrm{mod}\ 3)$. It means that we can not say anything about nontrivial solutions to the system (3).

Now we eliminate any edge (say $e = e(x, y)$) of G. Denote the resulting graph by G_e. Suppose that x_m is the variable which corresponds to the edge e. The incidence matrix A_e of G_e might be obtained from the matrix A by eliminating the last column, (which corresponds to the edge of e). Let

$$A_e \begin{pmatrix} x_1^2 \\ x_2^2 \\ \vdots \\ x_{m-1}^2 \end{pmatrix} = \begin{pmatrix} 0 \\ 0 \\ \vdots \\ 0 \end{pmatrix},$$

or

$$\overline{F_1} = 0,\ \overline{F_2} = 0,\ \cdots,\ \overline{F_n} = 0, \tag{6}$$

be the corresponding system of equations for the graph G_e. This system differs from the system (3) for the graph G only in two equations. Suppose that F_i and F_j correspond to the vertices $v_i = x$ and $v_j = y$. Now from (4) we have that

$$F_i = x_{i_1}^2 + x_{i_2}^2 + x_{i_3}^2 + x_m^2,$$

$$F_j = x_{j_1}^2 + x_{j_2}^2 + x_{j_3}^2 + x_m^2,$$

recall that variable x_m^2 corresponds to the edge $e = e(x, y)$, where $x = v_i$ and $y = v_j$. So

$$\overline{F_k} = F_k$$

for all $k = 1, \cdots, n$ except two values $k = i$ and $k = j$, for which

$$\overline{F_i} = x_{i_1}^2 + x_{i_2}^2 + x_{i_3}^2,$$

$$\overline{F_j} = x_{j_1}^2 + x_{j_2}^2 + x_{j_3}^2.$$

Now we have that the number S_e of solutions to the system (6) is such that

$$S_e = \sum_{x_1, \cdots, x_{m-1} \in F} (1 - \overline{F_1}^2)(1 - \overline{F_2}^2) \cdots (1 - \overline{F_n}^2). \tag{7}$$

Using the same arguments as in [7] we can write

$$(1-\overline{F_1}^2) \cdots (1-\overline{F_n}^2) = 1 + c(1, \cdots, m-1) x_1^2 \cdots x_{m-1}^2 + \sum_{i_1, \cdots, i_s} c(i_1, \cdots, i_s) x_{i_1}^2 \cdots x_{i_s}^2,$$

where $c(i_1, \cdots, i_s)$ are elements of $GF(3)$ and the sum is taken over all the possible monomials of any subset of the set of variables $\{x_1^2, \cdots, x_{m-1}^2\}$.

The result now follows from the following points:

(i) From the definitions of $N(1,1)$ and $N(2,1)$ it follows that

$$c(1,\cdots,m-1) \equiv (N(2,1)-N(1,1))\ (mod\ 3)$$

(we note that $x_i^4 = x_i^2$(modulo F))

(ii) For the case when $\{i_1,\cdots,i_s\} \neq \{1,\cdots,m-1\}$

$$\sum_{x_1,\cdots,x_{m-1}\in F} c(i_1,\cdots,i_s)x_{i_1}^2\cdots x_{i_s}^2 = 0\ .$$

(iii) For any $i = 1,\cdots,m-1$

$$\sum_{x_i\in F} x_i^2 \equiv -1\ (mod\ 3)\ ,$$

and, therefore,

$$c(1,\cdots,m-1) \sum_{x_1,\cdots,x_{m-1}\in F} x_1^2\cdots x_{m-1}^2 \equiv (N(1,1)-N(2,1))\ (mod\ 3)$$

(recall that $m-1$ is odd number).

Example 2.6. In the 4-regular graph G_4, as described on Fig.3, N, the number of distinct 2-decompositions is equal to 16, but it is very easy to see that it contains a 3-regular subgraph. If we eliminate the edge $e = e(x_1,x_3)$, then in the resulting graph G_4' the number of decompositions of type (1,1) and (2,1) correspondingly satisfies:

$$N(1,1) = 28,\ N(2,1) = 34\ ,$$

that is the sufficient conditions of Theorem 2.5 are satisfied, and therefore we also obtain that G_4 has a 3-regular subgraph.

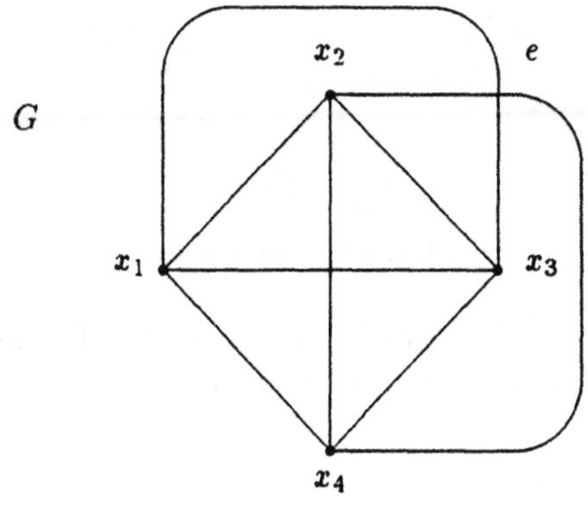

Fig. 3

Example 2.7. In both 4-regular graphs G_1 and G_2 as described on Fig.4, N, the number of distinct 2-decompositions is equal to $250 \equiv 1 (\mathrm{mod}\ 3)$. These graphs were given in [7] as graphs of the same type (built by pasting from double cycles with 3 vertices) and with the same value of N. The graph G_1 (Fig.4a) has a 3-regular subgraph but the graph G_2 (Fig.4b) has no 3-regular subgraph. Using our Theorem 2.5 we can now adequately explain the 3-regular behavior of these graphs. The graph G'_1, obtained by eliminating the edge e has the numbers

$$N(1,1) = 302, N(2,1) = 584 ,$$

hence

$$N(1,1) - N(2,1) = -282 \equiv 0 (\mathrm{mod}\ 3)$$

so we can conclude from Theorem 2.5 that G_1 has a 3-regular subgraph.

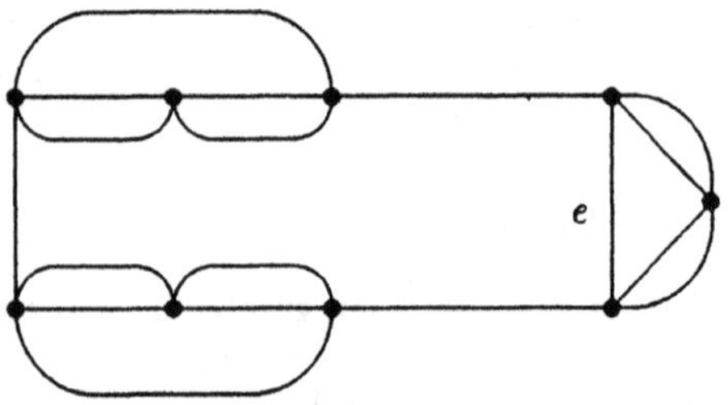

G_1, $N_{G_1} = 250$

Fig. 4.a. G_1 has a 3-regular subgraph

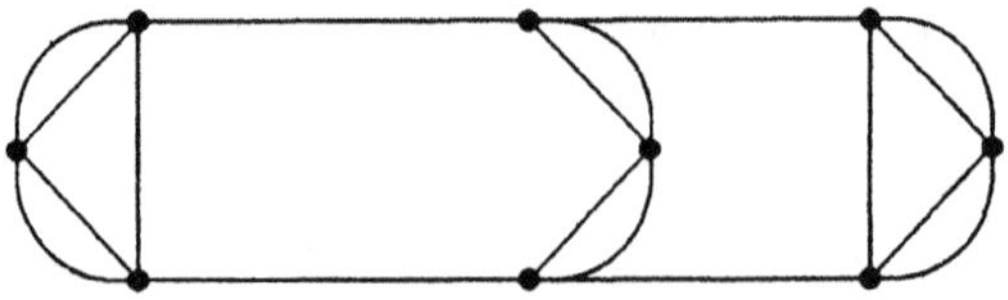

G_2, $N_{G_2} = 250$

Fig. 4.b. G_2 has no 3-regular subgraph

Acknowledgment. The authors would like to thank Carsten Thomassen and Noga Alon for their support and useful remarks.

References

[1] N. Alon, personal communication, 1993.

[2] N. Alon, S. Friedland and G. Kalai, "Every 4-regular graph plus an edge contains a 3-regular subgraph", *J. Combin. Theory Ser. B* 37 (1984), pp 92-93.

[3] N. Alon, S. Friedland and G. Kalai, "Regular subgraphs of almost regular graphs", *J. Combin. Theory Ser. B* 37 (1984), pp 79-91.

[4] J. A. Bondy and U. S. R. Murty, *Graph Theory with Applications*, p. 246, Macmillan and Co., London, 1976.

[5] V. Chvatal, H. Fleischner, J. Sheehan and C. Thomassen, "Three regular subgraphs of four regular graphs", *J. Graph Theory*, 3 (1979), pp 371-386.

[6] H. Fleischner, "Eulerian Graphs", *Selected Topics in Graph Theory 2*, pp.17-53, edited by L. W. Beineke and R. J. Wilson, Academic Press, London, 1983.

[7] O. Moreno V. A. Zinoviev, "3-Regular subgraphs of 4-regular graphs", submitted to *J. Combin. Theory Ser. B.*

[8] Charles Small, *Arithmetic of Finite Fields*, pp 50-51, Marcel Dekker Inc., New York, 1991.

[9] V. A. Tâskinov, "Regular subgraphs of regular graphs", *Soviet Math. Dokl.*, 26 (1982), pp 37-38.

Detection and Location of Given Sets of Errors by Nonbinary Linear Codes*

MARK G. KARPOVSKY SAEED M. CHAUDHRY
LEV B. LEVITIN
Research Laboratory of Design and Testing of Computer Hardware
Department of Electrical, Computer and Systems Engineering
Boston University
Boston, Massachusetts 02215, USA

CLAUDIO MORAGA
Department of Computer Science
University of Dortmund
4600 Dortmund 50, Federal Republic of Germany

Abstract—The problem of constructing codes capable of detection and location of a given set of errors is considered. Lower and upper bounds on a number of redundant symbols for an arbitrary set of errors are derived. These codes can be used for error detection and identification of faulty processing elements in multiprocessor systems. To this end, new classes of codes for several types of error sets such as stars, trees and FFTs meshes are presented. The concepts of strong and weak diagnostics (SD and WD, respectively) are introduced and discussed.

Index Terms—Detection and location of a given set of errors, diagnostic of multiprocessor systems or arrays, error detection, error location, linear codes.

1 Introduction

In order for processing elements to cooperate on solving a problem, an interconnection structure (network topology) is provided so that they can communicate with each other. Such multiprocessor systems are increasingly being used for high-performance computing [2], [3], [8], [9], [19]. In many cases, algorithms are developed to exploit specific multiprocessor interconnection structures like

*This work has been supported by the NSF under Grant MIP 9208487 and the NATO under Grant 910411.

meshes, rings, stars, trees, hyper cubes. Some well-known multiprocessor interconnection networks are described in [6], [10], [15].

Built-in self-test and self-diagnostic is becoming a viable alternative for a VLSI chip, board or system design. Techniques based on space and time compression are used to compress output test responses into signatures [1], [20]. Faulty processing elements can be identified by an analysis of distortions in the computed signatures [12]. These approaches aim at reducing the number of signatures required for diagnosis. Such approaches have been used for chip, board and system level testing and diagnosis (see e.g., [18]).

For the diagnosis (fault detection and location) of a system of processing elements the system is represented by a directed graph G whose vertices correspond to processing elements and directed edges correspond to communication links between processing elements [13]. Test data is applied to a set of input vertices of this graph and test responses flow through the graph to a set of output vertices. Due to failures in processing elements or communication links, a set of errors $E(G)$ is observed at the output vertices which corresponds to the structure of the underlying network topology.

In this paper we will consider the problem of constructing error detecting and locating codes for a network of processing elements modeled by a graph. The major problem in the design of error detecting and locating codes is the problem of minimization of the number of check symbols of a code for a given set of errors. In Section 3 we present upper and lower bounds for codes detecting and locating an arbitrary set of errors. Section 4 is devoted to analysis of some important communication topologies for multiprocessor systems and constructions of corresponding codes for detection and location of the underlying graph errors.

2 Definitions and Approach

Let V_q^n denote the n-dimensional vector space over $GF(q)$, $q = 2^b$ ($GF(q)$ is a Galois field of q elements). We denote by supp(v) the support of the n-tuple $v = (v_0, v_1, \ldots, v_{n-1})$, i.e. supp$(v) = \{i \in \{0, 1, \ldots, n-1\} | v_i \neq 0\}$.

Let X denote the set $\{X_0, X_1, \ldots, X_{N-1}\}$ of N processing elements. Consider a digraph G having X as a set of vertices and a set $U = \{U_0, U_1, \ldots, U_{M-1}\}$ of directed edges (b-bit communication links) between vertices of G. Let us denote by $I \subseteq X$ the set of input vertices and by $O \subseteq X$ the set of output vertices. We shall also assume that the graph has no cycles and all output vertices are reachable from at least one input vertex. Let $|O| = n$ ($|O|$ denotes the cardinality of O) and $Y = (y_0, y_1, \ldots, y_{n-1}) \in V_q^n$ be an output vector for the system represented by graph G where $y_i \in GF(q)$ is an output of the corresponding output vertex O_i, $i = 0, 1, \ldots, n-1$ and $q = 2^b$ (for simplicity we assume that all output vertices have b-bit outputs).

The problem to be considered is error detection and location under the assumption of single vertex failures in the graph G. A failure in the graph (system of processing elements) refers to a physical malfunction that cause the undesired event. The effect of a failure is the introduction of errors in the output vector Y. We consider a fault in the graph which alter its output value to $\tilde{Y} = (\tilde{y}_0, \tilde{y}_1, \ldots, \tilde{y}_{n-1})$ where $\tilde{y}_i \in GF(q)$. The error in the graph output Y can be characterized by the error vector $E = (e_0, e_1, \ldots, e_{n-1})$ where $e_i = \tilde{y}_i \oplus y_i$ for $i = 0, 1, \ldots, n-1$ ($\oplus$ is component-wise modulo 2 addition).

Let us first define an error set $E(G)$ characterized by the underlying graph G. In our definition of an error set we assume that at most one vertex or any number of incoming edges to this vertex may fail and a fault in the graph manifest itself by distorting all successor vertices outputs i.e. error propagates along a directed path. This assumption is reasonable for the case when the system of processing elements is tested by a large number of randomly chosen test patterns and, with probability close to one, a distortion will be propagated to all successor vertices [12].

Let $E_j = \{(e_0^{(j)}, e_1^{(j)}, \ldots, e_{n-1}^{(j)})\}$ denote a set of error patterns corresponding to a fault in vertex X_j where $e_i^{(j)} \in \{1, 2, \ldots, q-1\}$ if there exists a directed path from X_j to O_i and $e_i^{(j)} = 0$ otherwise ($e_i^{(j)} = \tilde{y}_i^{(j)} \oplus y_i^{(j)}$, $\tilde{y}_i^{(j)}, y_i^{(j)}$, are faulty and fault free outputs for the output vertex O_i, $i = 0, 1, \ldots, n-1$). The set $E(G) = \bigcup_{j=0}^{N-1} E_j$ of all possible error patterns corresponding to all single vertex failures in G is called the error set for G.

Let $E(G) \subset V_q^n$, $0 \notin E(G)$, be an error set for G. We shall call a linear (n, k), $k \leq n$, code defined by a $(n-k)$ by n parity check matrix H over $GF(q)$ a scheme that allows detection and/or location of error set $E(G)$ where the block length n is equal to the number of output vertices $|O| = n$ in G.

Let $E(G) \subset V_q^n$, $0 \notin E(G)$, be an error set for G. A linear (n, k) block code C over $GF(q)$ of length n defined by a $(n-k)$ by n parity check matrix H

1. detects $E(G)$ if and only if
   (a) for every $E_i \in E(G)$, $HE_i^{tr} \neq 0$ (E_i^{tr} is E_i transposed; all computations are over $GF(q)$).
2. locates $E(G)$ if and only if
   (a) for every $E_i \in E(G)$, $HE_i^{tr} \neq 0$ and
   (b) for every $E_i, E_j \in E(G)$ with $\mathrm{supp}(E_i) \neq \mathrm{supp}(E_j)$, $HE_i^{tr} \neq HE_j^{tr}$.
3. corrects $E(G)$ if and only if
   (a) for every $E_i \in E(G)$, $HE_i^{tr} \neq 0$ and
   (b) for every $E_i, E_j \in E(G)$ with distinct i, j, $HE_i^{tr} \neq HE_j^{tr}$.

A slightly different definition of error locating and correcting code has been given in [17].

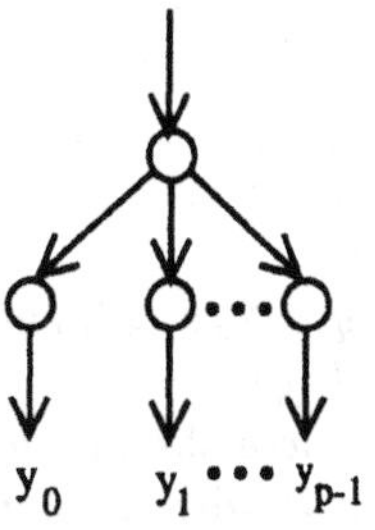

Figure 1: p-ary Star Network Topology.

To illustrate the above definitions let us now consider the problem of fault diagnosis for the p-ary star network topology (see Figure 1). For the p-ary star, the single central processing element (root) is connected to all others, $N = p + 1$, and $n = p$. Due to single vertex or processing element failures we have the following nonzero errors in the p-ary star:

$$E(G) = \left\{ \begin{array}{llll} (e_0, & e_1, & \ldots, & e_{p-1}), \\ (e_0, & 0, & \ldots, & 0), \\ (0, & e_1, & \ldots, & 0), \\ & & \vdots & \\ (0, & 0, & \ldots, & e_{p-1}) \end{array} \right\}, \tag{1}$$

where $e_i \in GF(q) - 0$. Thus, we have $(q-1)^p + p(q-1)$ nonzero error vectors for a p-ary star over $GF(q)$.

For error detection the problem is reduced to optimal construction of a parity check matrix H with n columns and a minimal number of rows $r \leq n$, such that for any error pattern $E_j \in E(G)$, the vector HE_j^{tr} has at least one nonzero component. Let h_i, $1 \leq i \leq r$, denote rows in a parity check matrix H, then to detect all single vertex failures in graph G it is sufficient for any $q > 2$ to have at least one row h_i in H such that $|\text{supp}(h_i) \cap \text{supp}(E_j)| = 1$ for any $E_j \in E(G)$. This condition ensures that any two or more nonzero components in error vectors corresponding to the same vertex failure, E_j, will not compensate and may not produce an all zero syndrome.

For the p-ary star, it is easy to see that to detect any single faulty output vertex, a row of all 1's in H is sufficient, since all error vectors corresponding to a single output vertex failure have only one nonzero component and for the central vertex fault one can take any row with one 1 and $p - 1$ 0s. Hence, the following parity check matrix H can be used for detection of p-ary star errors for any $p \geq 2$ and $q \geq 2$:

$$H = \begin{bmatrix} 1 & 1 & \cdots & 1 \\ 1 & 0 & \cdots & 0 \end{bmatrix}_{2\times p}. \tag{2}$$

Thus, for any p-ary star, $p \geq 2$, $q \geq 2$, $r = 2$ and we have a class of $(p, p-2)$ star error detecting codes over $GF(q)$.

For identification of faulty vertices the problem is reduced to optimal construction of a parity check matrix H with n columns and a minimal number of rows $r \leq n$, such that for any two error patterns $E_i, E_j \in E(G)$, with different support, $HE_i^{tr} \neq HE_j^{tr}$, where the number of errors with different support is equal to the number of vertices in the graph G.

Since, to locate all single vertex failures in graph G it is necessary that the number of error vectors with different support is equal to the number of vertices N in the graph G we have the following attainable lower bound on a number of outputs n (block length of an (n, k) error locating code) for a single vertex failure locatable graph:

$$n \geq \lceil \log_2(N+1) \rceil . \tag{3}$$

We now present a construction for star error locating codes over $GF(q)$. Let α be a primitive element in $GF(q)$ (α is a primitive element if and only if $\alpha^i \neq \alpha^j$ for $i \neq j$, $i, j = 0, 1, \ldots, q-2$ [16]). The code defined by the following H matrix locates errors in a p-ary star:

$$H = \begin{bmatrix} 1 & 1 & 1 & \cdots & 1 \\ 1 & \alpha & \alpha^2 & \cdots & \alpha^{p-1} \\ 1 & 0 & 0 & \cdots & 0 \\ 0 & 1 & 0 & \cdots & 0 \end{bmatrix}_{4\times p}, \tag{4}$$

where $4 \leq p < q$.

For an output vector $Y = (y_0, y_1, \ldots, y_{n-1})$, we define its syndrome $S(Y) = (S_1, S_2, \ldots, S_{r=n-k})$ as

$$S(Y) = HY^{tr}. \tag{5}$$

Diagnosis of a single vertex failure will consists of two steps: syndrome computation and association of the syndromes to a faulty vertex. A straightforward approach to the syndrome computation is via a combinational logic circuit that implements the parity check matrix and the association of the syndrome to a faulty vertex can be specified by a location algorithm.

The syndrome computation for the parity check matrix described in (4) can be implemented with $2(p-1)$ $GF(q)$ adders and $(p-1)$ multipliers (note that here we need a multiplier that multiplies a field element from $GF(q)$ by a fixed element from the field).

The location algorithm for the above $(p, p-4)$ p-ary star error locating codes is described as follows:

Let S_i, $i = 1, 2, 3, 4$ denote the syndromes obtained.

1. If $S_i = 0$, $i = 1,2,3,4$: no error, end.

2. If $S_3 \neq 0$ and $S_4 \neq 0$: error in the central vertex (root), end.

3. If $S_3 = 0$ or $S_4 = 0$: error in vertex (leaf) j, $0 \leq j \leq p-1$, where $\alpha^j = S_2/S_1$.

4. End. □

For the case of a p-ary star with $p = 2^i - 2$ and $q = 2$ one may choose as a check matrix for error location a matrix with $2^i - 2$ different nonzero i-tuples as its columns except for the all 1 vector. For example the following check matrix can locate all single vertex failures in the star with $p = 2^3 - 2 = 6$ and $q = 2$:

$$H = \begin{bmatrix} 000111 \\ 011001 \\ 101010 \end{bmatrix}. \tag{6}$$

Therefore, for the p-ary star with $p = 2^i - 2$:

$$r = i, \tag{7}$$

and we have a class of $(2^i - 2, 2^i - 2 - i)$ perfect star error locating codes over $GF(2)$ [11].

It is interesting to note that the parity check matrix for star error detecting codes (2) is over $\{0,1\} \subseteq GF(q)$. Such codes are particularly simple to implement since the check symbols are obtained using only additions in $GF(q)$, no multiplications are needed. Therefore, from the viewpoint of hardware implementation it is advantageous to have codes with parity check matrix over $\{0,1\}$ (see e.g., [12]).

In the next section, we present upper and lower bounds on the minimum number of check symbols for codes detecting and locating an arbitrary set of errors.

3 Error-Detection and Location Capabilities of Linear Codes

A lower bound on $r = n - k$, minimum number of redundant symbols, in any (n,k) block code capable of detecting an error set $E(G)$ can be proved as follows. Let $E(G) \subset V_q^n$, $0 \notin E(G)$, be the set of errors we wish to detect. A linear (n,k) code over $GF(q)$ has q^k code vectors. If this code is to detect the set $E(G)$ of errors, all error vectors must not be in the code. Thus the number of code vectors must be no greater than the total number of vectors in the space minus the number of error vectors in $E(G)$:

$$q^k \leq q^n - |E(G)|, \tag{8}$$

where $|E(G)|$ is the cardinality of $E(G)$. Thus

$$r \geq n - \lfloor \log_q(q^n - |E(G)|) \rfloor . \tag{9}$$

Another and more efficient lower bound on a number of redundant symbols for a code detecting a set of errors $E(G)$ can be proved as follows. Consider a graph $G(E) = (E(G), U)$ having the error set $E(G)$, $0 \notin E(G)$, as a set of vertices and U a set of edges $\{(E_i, E_j) | E_i \oplus E_j \in E(G)\}$. Let $E_i \oplus E_j = E_k \in E(G)$ and let H be a check matrix for a code detecting the set of errors $E(G)$. Since H is a linear code, $H(E_i^{tr} \oplus E_j^{tr}) = HE_i^{tr} \oplus HE_j^{tr} = HE_k^{tr} \neq 0$. Therefore, $HE_i^{tr} \neq HE_j^{tr}$, which means that the syndromes for two errors E_i and E_j must be different if E_i is connected to E_j in $G(E)$.

Let $\gamma(E)$ denote the chromatic number for $G(E)$ ($\gamma(E)$ is a minimal number of colors required to color vertices of $G(E)$ in such a way that no two neighboring vertices have the same color; techniques for graph coloring with lower and upper bounds for $\gamma(E)$ can be found e.g. in [4]. Then we have the following lower bound on a minimal number r of check symbols in a code detecting $E(G)$:

$$r \geq \lceil \log_q (\gamma(E) + 1) \rceil . \tag{10}$$

We note that the above lower bound is attainable. For example, for a p-ary star (see Figure 1), $p > 2$, and $q = 2$:

$$H = \begin{cases} [11\ldots1]_{1\times p} & p = \text{odd}, \\ \begin{bmatrix} 11\ldots1 \\ 10\ldots0 \end{bmatrix}_{2\times p} & p = \text{even}. \end{cases} \tag{11}$$

We also note that for the classical case when $E(G) = \{e | 0 < ||e|| \leq 2t\}$ ($||e||$ denote the number of nonzero components in e) we have $\gamma(E) = \sum_{i=1}^{t} \binom{n}{i}$ and (10) is the well known Hamming bound [16].

An upper bound on a number r of redundant symbols for a code detecting a set of errors $E(G)$ for any $n, k < n$, and any $E(G) \subset V_q^n$, $0 \notin E(G)$, is

$$r \leq \lceil \log_q(|E(G)|(q-1) + 1) \rceil . \tag{12}$$

This bound follows from the fact that a linear (n, k) code over $GF(q)$ with q^k code vectors detects error set $E(G)$ if and only if all code vectors are in the set $V_q^n - E(G)$ and in any set with $q^n - 1 - |E(G)| \geq q^n - \left(q^{n-k+1} - 1\right)(q-1)^{-1}$ nonzero vectors, there exists a linear subspace with q^k vectors [14].

Lower and upper bounds for a code locating an error set $E(G)$ can be obtained from the above bounds for error detection by replacing error set $E(G)$ with $E(G) \cup \{E_i \oplus E_j | \text{supp}(E_i) \neq \text{supp}(E_j), E_i, E_j \in E(G)\}$. This is due to the fact that if a code defined by the parity check matrix H detects a set $E(G) \cup \{E_i \oplus E_j | \text{supp}(E_i) \neq \text{supp}(E_j), E_i, E_j \in E(G)\}$, then for any $E_i, E_j \in E(G)$ with $\text{supp}(E_i) \neq \text{supp}(E_j)$, $HE_i^{tr} \neq HE_j^{tr}$ i.e., code locates error set $E(G)$.

Since all nonzero syndromes must be different for all single vertex failures the following attainable bounds on a minimum number of check symbols required for a (n,k) code over $GF(q)$ locating error set $E(G)$ hold:

$$\lceil \log_q(N+1) \rceil \leq r \leq n. \tag{13}$$

We note also that a linear (n,k) code with a parity check matrix H locating an error set $E(G)$, $0 \notin E(G)$ corrects $E(G)$ if and only if for every distinct $E_u, E_v \in E(G)$ with $\mathrm{supp}(E_u) = \mathrm{supp}(E_v)$ there exists a pair $E_i, E_j \in E(G) \cup 0$ with $\mathrm{supp}(E_i) \neq \mathrm{supp}(E_j)$ such that $E_u \oplus E_v = E_i \oplus E_j$. To show this assume that a code with check matrix H locates error set $E(G)$ but does not correct $E(G)$. Therefore, there exist at least one pair E_u, E_v such that $HE_u^{tr} = HE_v^{tr}$ where E_u, E_v have the same support. Since there exist at least one pair E_i, E_j such that $E_u \oplus E_v = E_i \oplus E_j$ and E_i, E_j have different support, we have $H(E_u^{tr} \oplus E_v^{tr}) = 0$ which implies $HE_i^{tr} = HE_j^{tr}$ which is a contradiction, because the code locates $E(G)$. We note that the above condition is a necessary and sufficient condition on the error set $E(G)$ such that if and only if this condition is satisfied any linear code locating $E(G)$ will also correct $E(G)$.

For example, for p-ary star with $p = 5$ and $q = 2^2$, $E(G) = \{(e_0, e_1, e_2, e_3, e_4),$ $(e_0, 0, 0, 0, 0), (0, e_1, 0, 0, 0), (0, 0, e_2, 0, 0), (0, 0, 0, e_3, 0), (0, 0, 0, 0, e_4\}$ where $e_i \in \{1 = 01, \alpha = 10, \alpha^2 = 11\}$. A code locating $E(G)$ does not guarantee error correction since for two errors $E_u = (1,1,1,1,1)$, $E_v = (1,1,\alpha,\alpha,\alpha)$ there is no pair E_i, E_j with different support such that $E_u \oplus E_v = E_i \oplus E_j = (0, 0, \alpha^2, \alpha^2, \alpha^2)$ (Note $\alpha^2 \oplus \alpha \oplus 1 = 0$).

Using the above arguments one can see that any code over $GF(q)$ locating up to t independent errors ($E(G) = \{e | 0 < ||e|| \leq 2t\}$) can also correct t errors. The same is also true for codes locating burst errors.

4 Codes for Diagnosis of Multiprocessor Systems

In Section 2 we have shown that the problem of hardware minimization for diagnosis of a system of processing elements modeled by a graph can be reduced to the design of a code with a minimal number of check symbols detecting and locating graph errors. In this section we will present several nearly optimal constructions for codes detecting and locating errors in tree and Fast Fourier Transform (FFT) interconnection networks. These interconnection networks have been widely used (see e.g. [5], [6] and [10]).

4.1 Detection and Location of Tree Errors

Let T_h be a p-ary full tree of height h ($p \geq 2, h \geq 2$) (see Figure 2). The height h is the length of a longest path from the root to any leaf. Here we assume that input vertex is the root and output vertices are $n = p^{h-1}$ leaves of the tree.

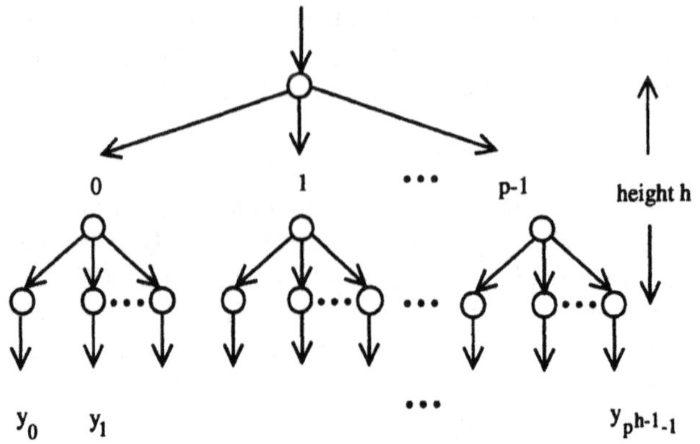

Figure 2: p-ary Full Tree Network Topology.

For a p-ary tree of height h, the set of errors $E(G)$ is:

$$\left\{\begin{array}{llllllllll}
(e_0, & e_1, & \ldots, & e_{p^{h-2}-1}, & e_{p^{h-2}}, & \ldots, & e_{2p^{h-2}-1}, & e_{(p-1)p^{h-2}}, & \ldots, & e_{p^{h-1}-1}), \\
(e_0, & e_1, & \ldots, & e_{p^{h-2}-1}, & 0, & \ldots, & 0, & 0, & \ldots, & 0), \\
(0, & 0, & \ldots, & 0, & e_{p^{h-2}}, & \ldots, & e_{2p^{h-2}-1}, & 0, & \ldots, & 0), \\
 & & & & & & & & \vdots & \\
(0, & 0, & \ldots, & 0, & 0, & \ldots, & 0, & e_{(p-1)p^{h-2}}, & \ldots, & e_{p^{h-1}-1}), \\
 & & & & & & & & \vdots & \\
(e_0, & 0, & \ldots, & 0, & 0, & \ldots, & 0, & 0, & \ldots, & 0), \\
(0, & e_1, & \ldots, & 0, & 0, & \ldots, & 0, & 0, & \ldots, & 0), \\
 & & & & & & & & \vdots & \\
(0, & 0, & \ldots, & 0, & 0, & \ldots, & 0, & 0, & \ldots, & e_{p^{h-1}-1})
\end{array}\right\}, \tag{14}$$

where $e_i \in GF(q) - 0$ and $|E(G)| = \sum_{i=0}^{h-1} p^i (q-1)^{p^{h-1-i}}$.

The recursive construction for check matrices of $(p^{h-1}, p^{h-1} - h)$, $p \geq 2$, $q > 2$, tree error detecting codes is given by:

$$H_h = \begin{bmatrix} \overbrace{H_{h-1} H_{h-1} \cdots H_{h-1}}^{p} \\ W \end{bmatrix}, \tag{15}$$

where W is a row vector of one 1 followed by $p^{h-1} - 1$ 0's and

$$H_2 = \begin{bmatrix} 11 \cdots 1 \\ 10 \cdots 0 \end{bmatrix}_{2 \times p}, \tag{16}$$

(H_h is a check matrix for the p-ary tree of height h).

It can be easily shown that all syndromes obtained by H_h for tree errors are not equal to zero and the number of rows r in H_h is equal to the height h of

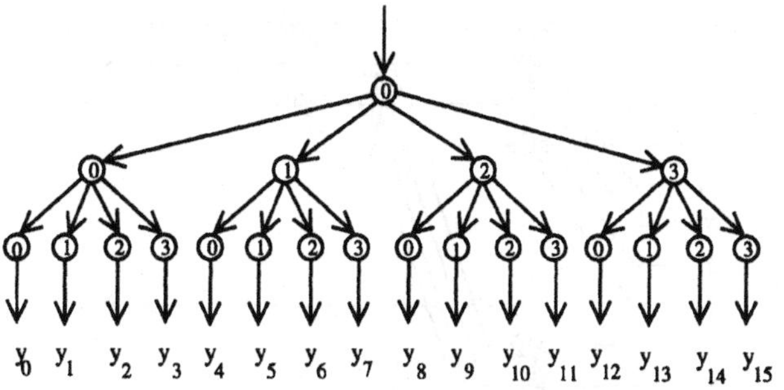

Figure 3: 4-ary Full Tree of Height $h = 3$ — Example.

the tree. Thus we have the class of $(p^{h-1}, p^{h-1} - h)$ p-ary tree error detecting codes.

The complexity L for the syndrome computing network in terms of a number of $GF(q)$ adders is

$$L = (p^{h-1} - 1)L_{\oplus}, \tag{17}$$

where $L_{\oplus}$ is the complexity of a $GF(q)$ adder and H_h given by (15) is optimal from the point of view of decoding complexity (note that elements of H_h are 0 or 1 for any q, therefore no multipliers are required).

Example. Consider the 4-ary full tree of height $h = 3$ over $GF(2^3)$ shown in Figure 3. This tree has $|E(G)| = (8-1)^{16} + 4(8-1)^4 + 16(8-1) \simeq 3.32 \times 10^{13}$ different error patterns due to 21 different single vertex faults. These errors are:

$$\left\{ \begin{array}{llllllllllllllll}
(e_0, & e_1, & e_2, & e_3, & e_4, & e_5, & e_6, & e_7, & e_8, & e_9, & e_{10}, & e_{11}, & e_{12}, & e_{13}, & e_{14}, & e_{15}), \\
(e_0, & e_1, & e_2, & e_3, & 0, & 0, & 0, & 0, & 0, & 0, & 0, & 0, & 0, & 0, & 0, & 0), \\
(0, & 0, & 0, & 0, & e_4, & e_5, & e_6, & e_7, & 0, & 0, & 0, & 0, & 0, & 0, & 0, & 0), \\
(0, & 0, & 0, & 0, & 0, & 0, & 0, & 0, & e_8, & e_9, & e_{10}, & e_{11}, & 0, & 0, & 0, & 0), \\
(0, & 0, & 0, & 0, & 0, & 0, & 0, & 0, & 0, & 0, & 0, & 0 & e_{12}, & e_{13}, & e_{14}, & e_{15}), \\
(e_0, & 0, & 0, & 0, & 0, & 0, & 0, & 0, & 0, & 0, & 0, & 0, & 0, & 0, & 0, & 0), \\
(0, & e_1, & 0, & 0, & 0, & 0, & 0, & 0, & 0, & 0, & 0, & 0, & 0, & 0, & 0, & 0), \\
 & \vdots & & & & & & & & & & & & & & \\
(0, & 0, & 0, & 0, & 0, & 0, & 0, & 0, & 0, & 0, & 0, & 0, & 0, & 0, & 0, & e_{15}),
\end{array} \right\}, \tag{18}$$

where $e_i \in GF(2^3) - 0$.

Based on the construction given in (15) we have the following parity check matrix for the $(16, 13)$ tree error detecting code over $GF(2^3)$

$$H_3 = \left[\begin{array}{cccc} 1111 & 1111 & 1111 & 1111 \\ 1000 & 1000 & 1000 & 1000 \\ 1000 & 0000 & 0000 & 0000 \end{array} \right], \tag{19}$$

and the combinational network for computing $S(Y) = (S_1, S_2, S_3) = H_3 Y^{tr}$ is shown in Figure 4. □

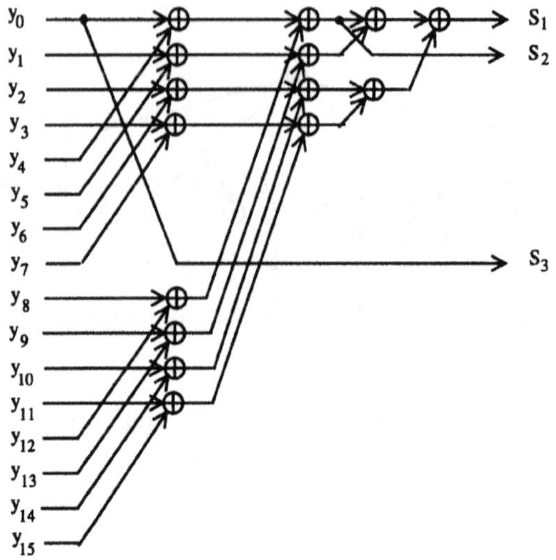

Figure 4: Syndrome Computing Network for the 4-ary (16, 13) Tree Error Detecting Code — Example.

For the case of a p-ary full tree over $GF(2)$ of any height h, $h \geq 2$, we have

$$r = \begin{cases} 1, & p = \text{odd}, \\ 2, & p = \text{even}. \end{cases} \tag{20}$$

A recursive construction for check matrices for the class of $(2^{h-1}, 2^{h-1} - 2)$ tree error detecting codes over $GF(2)$ can be obtained in the following way. We define the following mapping for columns in check matrix H_h:

$$\begin{bmatrix} 0 \\ 0 \end{bmatrix} \mapsto \begin{bmatrix} 00 \\ 00 \end{bmatrix}, \quad \begin{bmatrix} 0 \\ 1 \end{bmatrix} \mapsto \begin{bmatrix} 11 \\ 01 \end{bmatrix}, \quad \begin{bmatrix} 1 \\ 0 \end{bmatrix} \mapsto \begin{bmatrix} 10 \\ 11 \end{bmatrix}, \quad \begin{bmatrix} 1 \\ 1 \end{bmatrix} \mapsto \begin{bmatrix} 01 \\ 10 \end{bmatrix}. \tag{21}$$

The parity check matrix H_h can be obtained recursively from H_{h-1} by (21) where $H_2 = I_2$ (I_2 is an identity matrix of 2 by 2). For example, the matrix H_4 for the binary full tree of height four, $h = 4$, $n = 8$, is given by:

$$H_4 = \begin{bmatrix} 01 & 11 & 10 & 01 \\ 10 & 01 & 11 & 10 \end{bmatrix}. \tag{22}$$

We now present a construction for the class $(p^{h-1}, p^{h-1} - 3h + 2)$ tree error locating codes. The code defined by the following recursive definition of H_h

locates all tree errors:

$$H_h = \begin{bmatrix} H_{h-1} & H_{h-1} & \cdots & & H_{h-1} & & \\ 11\cdots1 & \alpha\alpha\cdots\alpha & \cdots & \alpha^{p-1} & \alpha^{p-1} & \cdots & \alpha^{p-1} \\ 10\cdots0 & 00\cdots0 & \cdots & 0 & 0 & \cdots & 0 \\ 00\cdots0 & 10\cdots0 & \cdots & 0 & 0 & \cdots & 0 \end{bmatrix}, \tag{23}$$

where

$$H_2 = \begin{bmatrix} 1 & 1 & 1 & \cdots & 1 \\ 1 & \alpha & \alpha^2 & \cdots & \alpha^{p-1} \\ 1 & 0 & 0 & \cdots & 0 \\ 0 & 1 & 0 & \cdots & 0 \end{bmatrix}_{4\times p}, \tag{24}$$

$4 \leq p < q$ and α is primitive in $GF(q)$. Thus for any p-ary, $4 \leq p < q$, tree $r = 3h-2$ and we have the class of $(p^{h-1}, p^{h-1}-3h+2)$ p-ary tree error locating codes over $GF(q)$. For the case of $p = 3$ one can also use the above construction (23) with $H_2 = I_3$ (I_3 is an identity matrix of 3 by 3). Therefore, for a p-ary tree with $p = 3$, $q > p$, $r = 3h - 3$.

For binary tree over $GF(q)$ the following recursive construction can be used for the class of $(2^{h-1}, 2^{h-1}-2h+2)$ binary tree error locating codes over $GF(q)$:

$$H_h = \begin{bmatrix} H_{h-1} & H_{h-1} \\ 10\cdots0 & 00\cdots0 \\ 00\cdots0 & 10\cdots0 \end{bmatrix}, \tag{25}$$

where $H_2 = I_2$. Thus, for a binary tree over $GF(q)$ of height h, $r = 2h - 2$ and we have the class of $(2^{h-1}, 2^{h-1} - 2h + 2)$ binary tree error locating codes over $GF(q)$.

The complexity L for the syndrome computing network in terms of numbers of $GF(q)$ adders and multipliers is:

$$L = (2(p^{h-1} - p) + h(p-1))L_{\oplus} + ((h-1)(p-1))L_{\otimes}, \tag{26}$$

where $L_{\otimes}$ is a complexity of a multiplier that multiplies a field element from $GF(q)$ by a fixed element from the same field.

The decoding procedure for tree codes is very simple. Let us denote the syndromes obtained as

$$S(Y) = H_h Y^{tr} = \begin{bmatrix} S^{h-1} \\ S_1^h \\ S_2^h \\ S_3^h \end{bmatrix}, \tag{27}$$

where S^{h-1} are syndromes due to the $[H_{h-1}H_{h-1}\cdots H_{h-1}]$ part of H_h (see (23)) and S_1^h, S_2^h, S_3^h are syndromes for the last three rows of the parity check matrix H_h. Let S^1 denotes the syndrome for the all 1 row. The location algorithm to find a faulty vertex is described as follows:

1. If $S_i = 0, i = 1, 2, \ldots, 3h-2$: no error, end.

2. Let $j = h$.

3. If both $S_2^j \neq 0$ and $S_3^j \neq 0$: error location is the root of the tree of height j, end.

4. For $j > 2$, if either $S_2^j = 0$ or $S_3^j = 0$: error location is in the subtree k, $0 \leq k \leq p-1$, where $\alpha^k = S_1^j / S_2^{j-1}$; for $j = 2$, if either $S_2^2 = 0$ or $S_3^2 = 0$: error location is in vertex (leaf) k, $0 \leq k \leq p-1$, where $\alpha^k = S_1^2/S^1$: end.

5. Repeat steps 3 and 4 for tree of height $j = j - 1$.

6. End.

Example continued: The parity check matrix H_3 for the 4-ary $(16, 9)$ tree error locating code is:

$$H_3 = \begin{bmatrix} 1 & 1 & 1 & 1 & 1 & 1 & 1 & 1 & 1 & 1 & 1 & 1 & 1 & 1 & 1 & 1 \\ 1 & \alpha & \alpha^2 & \alpha^3 & 1 & \alpha & \alpha^2 & \alpha^3 & 1 & \alpha & \alpha^2 & \alpha^3 & 1 & \alpha & \alpha^2 & \alpha^3 \\ 1 & 0 & 0 & 0 & 1 & 0 & 0 & 0 & 1 & 0 & 0 & 0 & 1 & 0 & 0 & 0 \\ 0 & 1 & 0 & 0 & 0 & 1 & 0 & 0 & 0 & 1 & 0 & 0 & 0 & 1 & 0 & 0 \\ 1 & 1 & 1 & 1 & \alpha & \alpha & \alpha & \alpha & \alpha^2 & \alpha^2 & \alpha^2 & \alpha^2 & \alpha^3 & \alpha^3 & \alpha^3 & \alpha^3 \\ 1 & 0 & 0 & 0 & 0 & 0 & 0 & 0 & 0 & 0 & 0 & 0 & 0 & 0 & 0 & 0 \\ 0 & 0 & 0 & 0 & 1 & 0 & 0 & 0 & 0 & 0 & 0 & 0 & 0 & 0 & 0 & 0 \end{bmatrix}. \tag{28}$$

Suppose that root of subtree 1 (see Figure 3) is faulty and the received message is:

$$E = \tilde{Y} \oplus Y = (0, 0, 0, 0, 1, \alpha, 1, \alpha^3, 0, 0, 0, 0, 0, 0, 0, 0). \tag{29}$$

Then the syndromes of this message are

$$S(E) = HE^{tr} = (\alpha, \alpha^5, \alpha, \alpha, \alpha^2, 0, \alpha)^{tr}. \tag{30}$$

This yields $S_2^3 = 0$, therefore the error is in the subtree 1 since, $S_1^3/S_2^2 = \alpha^2/\alpha = \alpha$, $i = 1$. Since $S_2^2 \neq 0$ and $S_3^2 \neq 0$, error is in the root of subtree 1 of height 2. The combinational network for computing $S(Y) = H_3 Y^{tr}$ is shown in Figure 5.
□

4.2 Detection of Errors in Fast Fourier Transform (FFT) Networks

The results presented above for detection of tree errors can be extended to other graphs of practical interest containing tree structures as subgraphs (i.e. any single vertex failure propagates through the graph in a tree-like manner). Below we consider an important application of tree-like codes for detection of errors in Fast Fourier Transform (FFT) network [5].

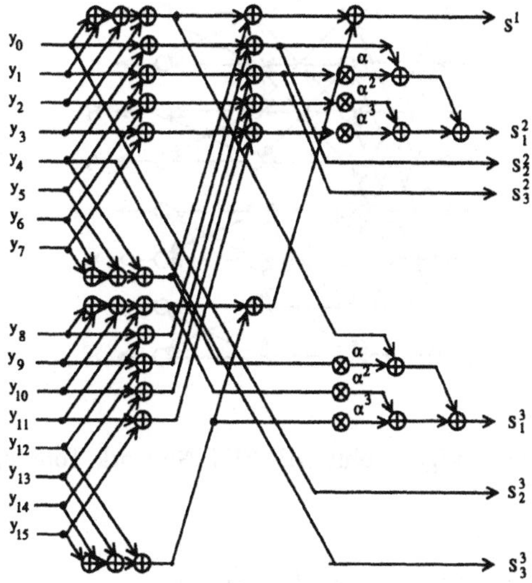

Figure 5: Syndrome Computing Network for the 4-ary (16, 9) Tree Error Locating Code — Example.

For n-point FFT, there are $N = n\log_2 n$ vertices interconnected with $\log_2 n$ levels of butterfly structures, e.g., the graph for the 8-point FFT (decimation-in-frequency (DIF)) is shown in Figure 6.

If we also consider input fanout branches as possible source of errors, there are $n(\log_2 n{+}1)$ single faults in the n-point FFT graph. Due to these single faults we have the following nonzero errors for the 8-point FFT graph of Figure 6:

$$E(G) = \left\{ \begin{array}{llllllll} (e_0, & e_1, & e_2, & e_3, & e_4, & e_5, & e_6, & e_7), \\ (e_0, & e_1, & e_2, & e_3, & 0, & 0, & 0, & 0), \\ (0, & 0, & 0, & 0, & e_4, & e_5, & e_6, & e_7), \\ (e_0, & e_1, & 0, & 0, & 0, & 0, & 0, & 0), \\ (0, & 0, & e_2, & e_3, & 0, & 0, & 0, & 0), \\ (0, & 0, & 0, & 0, & e_4, & e_5, & 0 & 0), \\ (0, & 0, & 0, & 0, & 0 & 0 & e_6, & e_7), \\ (e_0, & 0, & 0, & 0, & 0, & 0, & 0, & 0), \\ (0, & e_1, & 0, & 0, & 0, & 0, & 0, & 0), \\ & \vdots & & & & & & \\ (0, & 0, & 0, & 0, & 0, & 0, & 0, & e_7), \end{array} \right\}, \qquad (31)$$

where $e_i \in GF(q) - 0$ and $|E(G)| = \sum_{i=0}^{p} 2^i(q-1)^{2^{p-i}}$.

A recursive construction for the $(2^p, 2^p - p - 1)$ DIF FFT error detecting

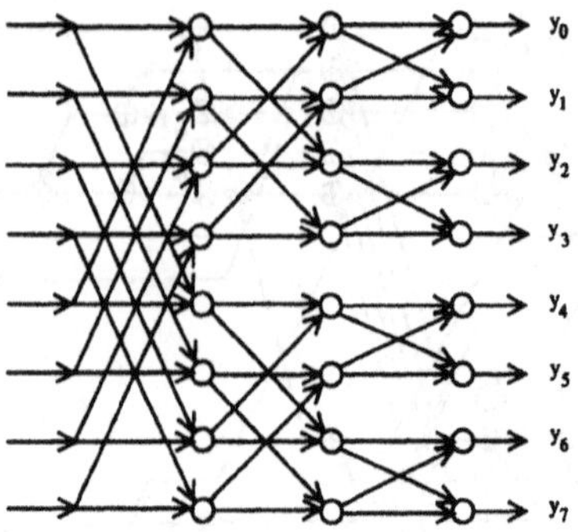

Figure 6: Eight-Point DIF FFT Network Topology.

codes is:

$$H_{2^p} = \begin{bmatrix} H_{2^{p-1}} H_{2^{p-1}} \\ W \end{bmatrix}, \tag{32}$$

where W be a row vector of one 1 followed by $2^p - 1$ 0's and

$$H_2 = \begin{bmatrix} 11 \\ 10 \end{bmatrix}. \tag{33}$$

We note that a similar construction for the class of $(2^p, 2^p - p - 1)$, decimation in time (DIT) FFT error detecting codes is also possible.

Based on the above construction we have the following parity check matrix H_{2^3} for the $(8,4)$ DIF FFT error detecting code over $GF(q)$:

$$H_{2^3} = \begin{bmatrix} 11 & 11 & 11 & 11 \\ 10 & 10 & 10 & 10 \\ 10 & 00 & 10 & 00 \\ 10 & 00 & 00 & 00 \end{bmatrix}. \tag{34}$$

The complexity L for the syndrome computing network is:

$$L = (2^p - 1)L_{\oplus}, \tag{35}$$

and H_{2^p} given by (32) is optimal from the point of view of decoding complexity (note that elements of H_{2^p} are 0 or 1 for any q, therefore no multipliers are required). The combinational network for computing $S(Y) = (S_1, \ldots, S_4) = H_{2^3}Y^{tr}$ is shown in Figure 7.

We note that for FFT graphs error location is not possible due the fact that different faults may have the same error patterns.

In Table 1 we summarize our results on linear error detecting and locating codes for different meshes (stars, binary and non binary full trees and FFTs).

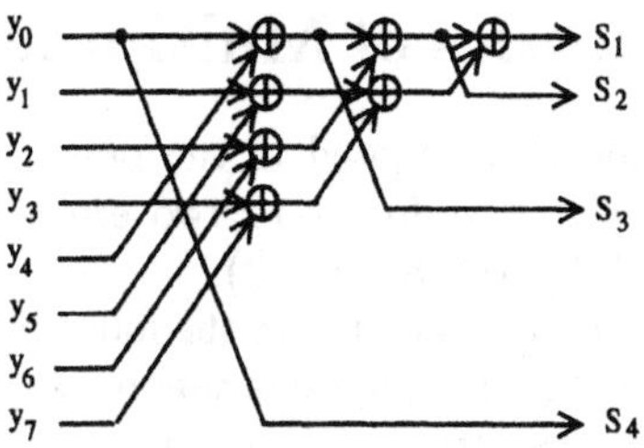

Figure 7: Syndrome Computing Network for the (8, 4) DIF FFT Error Detecting Code.

Table 1: Minimal Numbers of Check Symbols for Strong Diagnosis of Multiprocessor Interconnection Networks.

Graph	Parameters	Number of Check Symbols r	
	N, n	Error Detection	Error Location
p-ary star	$p+1, p$	2^*, $p \geq 2$, $q > 2$ 1, p odd, $q = 2$ 2, p even, $q = 2$	4, $4 \leq p < q$ i, $p = 2^i - 2$, $q = 2$
Binary full tree of height h	$2^h - 1, 2^{h-1}$	h^*, $h \geq 2$, $q > 2$ 2, $q = 2$	$2h - 2$, $h \geq 2$, $q > 2$
p-ary full tree of height h	$\frac{p^h - 1}{p-1}, p^{h-1}$	h^*, $p \geq 2$, $q > 2$ 1, p odd, $q = 2$ 2, p even, $q = 2$	$3h - 2$, $4 \leq p < q$ $3h - 3$, $p = 3$, $q > p$
2^p point FFT	$p2^p, 2^p$	$p + 1^*$	—

*Parity Check Matrix is over $\{0, 1\}$.

5 Weak Diagnostics of Multiprocessor Systems

The approach to error location adopted in the previous sections is based on rather a strong requirement that for every two errors, E_i, E_j if $\mathrm{supp}(E_i) \neq \mathrm{supp}(E_j)$, then $HE_i^{tr} \neq HE_j^{tr}$ (cf. Section 2).

This requirement can be reformulated in the following way. Denote by $U_i = \{E_{ik}\}$ the set of all errors E_{ik} with the same support $\mathrm{supp}(E_{ik}) = Y_i \subseteq Y$, such that $U_i \subseteq E(G)$. Then, for any two U_i, U_j, $i \neq j$,

$$H(U_i) \cap H(U_j) = \phi. \tag{36}$$

(Note that the check matrix H is a linear operator $H : V_q^n \rightarrow V_q^r$).

Condition (36) defines what can be called strong diagnostics (SD) of a given class of errors $E(G)$ in G. However this requirement may be too restrictive and, in some cases, may result in too large redundancy r. Moreover, it should be born in mind, that a fault that can manifests itself as an error with a largest possible support Y_i can, in fact, manifests itself as an error with a smaller support $Y_i' \subset Y_i$. If $|Y_i| = l$, then the fraction of such cases is of order l/q. Since in SD we assume that any fault manifest itself as an error with the maximum possible support (any output which can go wrong does so) we can not guarantee correct location of that fraction of faults.

It looks attractive, therefore, to consider a different approach which would relax the requirement (36), thereby considerably decreasing the value of r, and, on the other hand, would keep the fractions of faults denying location reasonably small for large q (of $\Theta(1/q)$).

Denote $V_i = \{E_{ik} | \mathrm{supp}(E_{ik}) \subseteq Y_i\}$ the set of all errors whose support are subsets of Y_i. Obviously, V_i is an m-dimensional coordinate subspace of V_q^n, where $m = |Y_i|$. Then the problem of error location can be formulated as follows: For a given set $\{V_i\}$ find a linear mapping $H : V_q^n \rightarrow V_q^r$ such that for any $i \neq j$,

$$H(V_i) \neq H(V_j). \tag{37}$$

We will call it weak diagnostics (WD).

The linear operator H can be viewed as a check matrix of a linear code of length n and dimension $(n - r)$ over $GF(q)$ which is a sub space of V_q^n and the kernel of the operator H.

Below we consider a few typical system topologies, give the explicit constructions for H and estimate the probability of correct fault location.

5.1 Faults in Stars

A star is a tree of height 2. Consider a p-ary star, where $p < q$. Here $n = p$ is the number of leaves. Let α be a primitive element in $GF(q)$ and consider a $(2 \times n)$ matrix

$$H_2 = \begin{bmatrix} 1 & 1 & 1 & \cdots & 1 \\ 1 & \alpha & \alpha^2 & \cdots & \alpha^{n-1} \end{bmatrix}. \tag{38}$$

Let (S_1, S_2) be the syndrome. Then the ratio S_2/S_1 reveals the faulty leaf:

$$S_2/S_1 = \alpha^i, \quad i = 0, 1, \ldots, n-1 \tag{39}$$

The misdiagnosis occurs if (39) holds, but in fact the root is faulty. For every i, (39) defines a subspace in V_q^n of dimension $(n-1)$. The intersection of all those subspaces corresponds to the case $S_1 = 0$, $S_2 = 0$, which defines a subspace of dimension $(n-2)$. Thus the fraction of misdiagnosis cases is

$$\omega_e = \frac{nq^{n-1} - q^{n-2}}{q^n} = \frac{nq-1}{q^2} \approx \frac{n}{q}. \tag{40}$$

In a practical situation, $n \leq 100$, $q = 2^{32}$, and $\omega_e \approx 2.5 \times 10^{-8}$. This negligible probability of mislocation allows us to reduce the redundancy by a factor of 2: $r = 2$, instead of $r = 4$ for SD (cf. Section 4.1).

5.2 Faults in Trees

Consider now a more general case of a p-ary tree of height h (Fig. 2). Here $n = p^{h-1}$ is the number of leaves. The check matrix H_h has the following form:

$$H_h = \begin{bmatrix} H_{h-1} & H_{h-1} & \cdots & & H_{h-1} & & \\ 11\cdots 1 & \alpha\alpha\cdots\alpha & \cdots & \alpha^{p-1} & \alpha^{p-1} & \cdots & \alpha^{p-1} \end{bmatrix}, \tag{41}$$

where H_2 is given by (38) (with p substituted for n). The number of rows in this matrix $r = h$, whereas $r = 3h - 2$ for SD (cf. (25)). The location procedure is simple. Let $(S_1, S_2, \ldots, S_h)$ be the syndromes. Then, similar to the procedure in Section 5.1, the ratio S_h/S_1 reveals the faulty subtree of height $h-1$: if $S_h/S_1 = \alpha^i$, $i = 0, 1, \ldots, p-1$ we decide the fault is in the subtree whose root is the node i at the second level from the top (Fig. 2), otherwise the tree root is faulty. Similarly, if $S_{h-1}/S_1 = \alpha^i$, $i = 0, 1, \ldots, p-1$, it reduces the faulty part to a subtree of height $h-2$, etc.

However, there is a possibility of making a wrong decision at any step of the procedure, due to the fact that it may happen $S_{h-k}/S_1 = \alpha^i$, $i = 0, 1, \ldots, p-1$; $k = 0, 1, \ldots, h-2$, in spite of the fact that the fault is at level $k+1$ from the top. Using the expression (40) and omitting terms of higher order in q^{-1}, we obtain the fraction of misdiagnosis cases:

$$\begin{aligned} \omega_e &= pq^{-1} + (1-pq^{-1})pq^{-1} + (1-pq^{-1})^2 pq^{-1} + \cdots + (1-pq^{-1})^{h-2} pq^{-1} \\ &= 1 - (1-pq^{-1})^{h-1} \end{aligned} \tag{42}$$

For large q, $\omega_e \approx (h-1)pq^{-1}$, which is, again, very small for practical values of the parameters.

5.3 Faults in Disconnected Processors

Consider now the situation when we have a set of n processors disconnected in the testing mode. Then a fault in a single processor manifests itself as a single error. Our goal is to locate faulty processors, i.e. to locate errors up to multiplicity l. Hence the set of errors is a ball B_l of radius l in V_q^n centered at the origin.

Obviously, the minimum number of check digits that would allow us to locate at least some errors is $r = l+1$. We choose matrix H to be the check matrix of a q-ary Reed-Solomon code:

$$H = \begin{bmatrix} 1 & 1 & 1 & \cdots & 1 \\ 1 & \alpha & \alpha^2 & \cdots & \alpha^{n-1} \\ 1 & \alpha^2 & \alpha^4 & \cdots & \alpha^{2(n-1)} \\ \vdots & \vdots & \vdots & \ddots & \vdots \\ 1 & \alpha^l & \alpha^{2l} & \cdots & \alpha^{l(n-1)} \end{bmatrix}. \tag{43}$$

The remarkable property of this matrix is that any $(l+1)$ its columns are linearly independent. Since Reed-Solomon is an MDS code, it attains the Singleton bound, and its distance $d = r+1 = l+2$.

It should be pointed out the difference between our problem of error location and the more usual problem of correct decoding [7], [21]. In spite of the fact that, in the situation considered, any code that locates all errors up to a given multiplicity corrects all such errors as well, the problems are quite different.

In decoding the goal is to find a coset leader of the minimum weight, thereby minimizing the probability of error. In our problem we do not consider probabilities; we are interested to find out how many errors from B_l can be uniquely identified by their syndromes, in other words, which cosets intersects with B_l in exactly one point.

Consider the set $\{Y_i\}$, $i = 1, 2, \ldots, \binom{n}{l}$, of all possible l-tuples from n. Then $V_i = \{E_{ik} | \text{supp}(E_{ik}) \subseteq Y_i\}$ is an l-dimensional coordinate subspace in V_q^n, $|V_i| = q^l$. Obviously, $\bigcup_i V_i = B_l$ It is easy to see that the mapping $H : V_q^n \to V_q^{l+1}$ maps any V_i to V_q^{l+1} injectively. Indeed, the opposite would mean a linear dependence of l columns of H. Also we observe that

$$\sum_i |V_i| = \binom{n}{l} q^l \approx |B_l| = \sum_{m=0}^{l} \binom{n}{m} (q-1)^m. \tag{44}$$

The ratio of these numbers differ from 1 in terms of order $nl^{-1}q^{-1}$. Henceforth we neglect these terms assuming that $l \ll n \ll q$.

Denote the image $H(V_i) = W_i$. All W_i are l-dimensional subspaces in the $(l+1)$-dimensional space of syndromes $H(V_q^n) = \text{W}$. Let us estimate how many members of W_i do not belong to any other subspace $W_j = H(V_j)$. Note that any two subspaces W_i, W_j intersect over a subspace of dimension $(l-1)$. The

Table 2: Minimal Numbers of Check Symbols for Weak Diagnosis of Multiprocessor Interconnection Networks.

Graph	Parameters N, n	Check Symbols r	Fractions of misdiagnosed errors ω_e
p-ary star	$p+1, p$	2	$(nq-1)q^{-2}$
p-ary full tree of height h	$\frac{p^h-1}{p-1}, p^{h-1}$	h	$1-(1-pq^{-1})^{h-1}$
n disconnected processors with up to l faults	n, n	$l+1$	$\leq 1-(1-q^{-1})^{\binom{n}{l}-1}$

minimum dimension of intersection of three subspace is $(l-2)$, etc. (The probability that randomly chosen t subspaces intersect over a subspace of dimension higher than $l-t+1$ is of order $q^{-(l-t+2)}$, $t \leq l+1$). Alternatively subtracting and adding terms corresponding to intersections of two, three, etc. subspaces we obtain the following estimate for the fraction of errors in any W_i that can be located:

$$\begin{aligned} \omega_c &\approx q^{-l}\left[q^l - \left[\binom{n}{l}-1\right]q^{l-1} + \binom{\binom{n}{l}-1}{2}q^{l-2} - \cdots\right] \\ &= \sum_{t=0}^{l+1}(-1)^t\binom{\binom{n}{l}-1}{t}q^{-t}, \end{aligned} \tag{45}$$

where $\omega_e = 1-\omega_c$.

We are interested in the case when ω_c is close enough to 1. It requires that $\binom{n}{l}q^{-1} < 1$. Then the terms in (45) decrease monotonically and we can extend the summation up to $\binom{n}{l}$ terms. Moreover, since the largest intersection term (the second) is negative, one can believe that the expression will give us a lower bound for the fraction of localizable errors. Therefore,

$$\omega_c \geq \sum_{t=0}^{\binom{n}{l}-1}(-1)^t\binom{\binom{n}{l}-1}{t}q^{-t} = (1-q^{-1})^{\binom{n}{l}-1} \approx e^{-\frac{\binom{n}{l}}{q}} \tag{46}$$

For example, if $l=5$, $n=100$, $q=2^{32}$, then $\omega_c = 0.979$. Thus, by allowing a small fraction of errors not to be located we reduce substantially the redundancy from $r=2l+1$ to $r=l+1$.

In Table 2 we summarize our results on linear error locating codes for different meshes (stars, trees and disconnected) with weak diagnostic.

6 Conclusions

In this paper we have presented bounds on numbers of check symbols required for codes detecting and locating arbitrary set of errors. These codes can be used for identification of faulty processing elements in multiprocessor systems or array processors. We presented several nearly optimal error detecting and locating codes for tree and FFT errors. Hardware implementation based on the proposed codes results in considerable savings of redundant overhead. The concepts of strong and weak diagnostics are introduced and estimates of probabilities of correct fault location for weak diagnostics are presented.

Acknowledgment

The authors would like to thank Professor Tatyana D. Roziner of Boston University, Boston MA, for valuable discussions on the results presented in this paper.

References

[1] P. H. Bardell, W. H. McAnney, and J. Savir. *Built-in Self Test for VLSI: Pseudorandom Techniques.* Wiley Interscience, New York, NY, 1987.

[2] K. E. Batcher. Design of a Massively Parallel Processor. *IEEE Transaction on Computers*, C-29:836–840, Sept. 1980.

[3] J. Bently and H. T. Kung. A Tree Machine for Searching Problems. In *International Conference on Parallel Processing*, pages 257–266, 1979.

[4] C. Berge. *Graphs and Hypergraphs.* North-Holland, New York, NY, 1973.

[5] D. F. Elliott and K. R. Rao. *Fast Transforms: Algorithms, Analyses, Applications.* Academic Press, New York, NY, 1982.

[6] T. Y. Feng. A Survey of Interconnection Networks. *IEEE Computer*, 14:960–965, 1981.

[7] C. R. P. Hartmann. Decoding Beyond the BCH Bound. *IEEE Transaction on Information Theory*, pages 441–444, May 72.

[8] J. P. Hayes et al. A Microprocessor-Based Hypercube Supercomputer. *IEEE Micro*, 6:6–17, Oct. 1986.

[9] W. D. Hillis. *The Connection Machine.* MIT Press, Cambridge, MA, 1985.

[10] K. Hwang and F. A. Briggs. *Computer Architecture and Parallel Processing.* Academic Press, New York, NY, 1982.

[11] M. G. Karpovsky. Weight Distributions of Translates, Covering Radius, and Perfect Codes Correcting Errors of Given Weights. *IEEE Transaction on Information Theory*, IT-27:462–472, July 1981.

[12] M. G. Karpovsky and S. M. Chaudhry. Built-in Self Diagnostic by Space-Time Compression of Test Responses. In *IEEE VLSI Test Symposium*, pages 149–154, 1992.

[13] M. G. Karpovsky, L. B. Levitin, and F. S. Vainstein. Identification of Faulty Processing Elements by Space-Time Compression of Test Responses. In *International Test Conference*, pages 638–647, 1990.

[14] M. G. Karpovsky and V. D. Milman. On Subspace Contained In Subsets Of Finite Homogeneous Space. *Discrete Mathematics*, 22:273–280, 1978.

[15] S. Y. Kung. *VLSI Array Processors*. Prentice-Hall, Englewood Cliffs, NJ, 1988.

[16] F. J. MacWilliams and N. J. A. Sloane. *The Theory of Error-Correcting Codes*. North-Holland, New York, NY, 1977.

[17] B. Masnick and J. Wolf. On Linear Unequal Error Protection Codes. *IEEE Transaction on Information Theory*, IT-3:600–607, Oct. 1967.

[18] E. J. McCluskey. Built-in Self Test Techniques. *IEEE Design and Test of Computers*, pages 21–28, Apr. 1985.

[19] F. P. Preparata and J. Vuillemin. The Cube-Connected Cycles: A Versatile Network for Parallel Computation. *Communication of the ACM*, 24:568–572, May 1981.

[20] S. R. Reddy, K. K. Saluja, and M. G. Karpovsky. A Data Compression Technique for Test Responses. *IEEE Transaction on Computers*, C-38:1151–1156, Sept. 1988.

[21] U. K. Sorger. A New Reed-Solomon Code Decoding Algorithm Based on Newton's Interpolation. *IEEE Transaction on Information Theory*, IT-39:358–365, Mar. 1993.

Quaternary constructions of formally self-dual binary codes and unimodular lattices

Alexis Bonnecaze and Patrick Solé

CNRS, I3S,
250, rue A. Einstein,
06 560 Valbonne, France
Email: sole@mimosa.unice.fr
bonnecaz@mimosa.unice.fr

Abstract. Quaternary codes have been studied recently in connection with the construction of sequences with low correlation, lattices and good non linear codes (Kerdock, Preparata). In this paper, we construct formally self-dual binary codes and unimodular lattices using quaternary codes. Two different processes are studied: constructions using Hensel lifting and $(u|u+v)$ construction. We give a number of examples of formally self-dual binary codes of length $n \leq 64$. We obtain a new construction of the Leech lattice, and two new constructions of the Gosset lattice.

Key words: Codes, Weight Enumerators, Self-Dual Quaternary Codes, Formally Self-Dual Binary Codes, Unimodular Lattices

1 Introduction

We call formally self-dual code (f.s.d.) a binary code C whose weight enumerator is invariant under the Mac-Williams transform. Recently, several authors [1] have found a new construction of the Nordstrom-Robinson code which is a binary non linear f.s.d. code. They have shown that this code (which is optimal) is linear over $\mathbb{Z}_4$, the set of integers modulo 4, and that as $\mathbb{Z}_4$-code it is self-dual (In fact, the $\mathbb{Z}_4$ version of the Nordstrom-Robinson code turns out to be the 'octacode'). We know that non linear codes can be better than any other linear codes of the same minimum distance and that at certain lengths, f.s.d. codes often have a higher minimum weight than self-dual codes of that length (see [2] for more details). Our idea is to construct binary codes which are the image under the Gray-map of self-dual quaternary codes. The codes obtained are f.s.d. codes [1] and are often non linear in spite of their $\mathbb{Z}_4$ linearity.

Another application of self-dual quaternary codes is to construct unimodular lattices by construction A (mod 4). This secondary application turned out to yield more interesting results.

In Section 2 we give some definitions and properties about the Gray map and quaternary codes.

In section 3 we construct quaternary self dual codes generated by a "lifted" binary polynomial.

The binary $(u|u+v)$ construction is studied in section 4. In section 5 we give two new constructions of E_8, a new construction of the Leech lattice, and a new construction of the shorter Leech lattice O_{23}. In section 6 we give quaternary constructions for the extended Hamming code of length 8 and F_{16}, a code introduced by Vera Pless [3]. In section 7 a numerical table summarizes our results.

2 Notations and Definitions

2.1 Quaternary codes

We define a quaternary code as an additive subgroup of $\mathbb{Z}_4^n$. Then it is a linear code over $\mathbb{Z}_4$. To be able to define the notions of dual code ($C^\perp$) and then of self-dual code ($C = C^\perp$), we define an inner product on $\mathbb{Z}_4^n$ by $a.b = a_1b_1 + \ldots + a_nb_n$ (mod 4).
The complete weight enumerator (c.w.e.) of C is

$$cwe_C(W,X,Y,Z) = \sum_{a\in C} W^{n_0(a)}X^{n_1(a)}Y^{n_2(a)}Z^{n_3(a)}, \quad (1)$$

where $n_j(a)$ is the number of components of a that are congruent to j (mod 4).
We say that two codes are equivalent if one can be obtained from the other by a permutation of the n coordinates, and changing the signs of certain coordinates. A code which is equivalent to its dual is called **isodual**.
Equivalent codes may have distinct c.w.e.'s, so we define the **symmetrized weight enumerator** (s.w.e.), which is obtained by identifying X and Z in (1):

$$swe_C(W,X,Y) = cwe_C(W,X,Y,X). \quad (2)$$

The form of the canonical generator matrix of a quaternary code is :

$$G = \begin{bmatrix} I_{k_1} & M & N \\ 0 & 2I_{k_2} & 2P \end{bmatrix},$$

where M and P are $\mathbb{Z}_2$-matrices and N is a $\mathbb{Z}_4$-matrix. We have

$$|C| = 4^{k_1}2^{k_2}.$$

We call "dim" the dimension over $\mathbb{Z}_4$ (dim(C)=$log_4|C|$); therefore we have

$$dim(C) = k_1 + \frac{k_2}{2}.$$

The **Lee weights** of the elements of $\mathbb{Z}_4$ 0,1,2,3 are respectively 0,1,2,1. The Lee weight $wt_L(a)$ of a codeword $a \in \mathbb{Z}_4^n$ is the rational sum of the Lee weights of its components. So the Lee distance $d_{\mathcal{L}}(a,b)$ between two codewords a, b, is the smallest absolute value of any integer congruent, modulo 4, to the difference $a - b$.
Over $\mathbb{Z}_4$, the MacWilliams identity gives the symmetric weight enumerator for the dual $C^\perp$ of a quaternary code C:

$$swe_{C^\perp}(W,X,Y) = \frac{1}{|C|}swe_C(W+2X+Y, W-Y, W-2X+Y).$$

In order to obtain binary codes, we use the **Gray map** ϕ which provides a 1-1 corrrespondance between $\mathbb{Z}_4^n$ and $\mathbb{Z}_2^{2n}$. First, we define three maps from $\mathbb{Z}_4$ to $\mathbb{Z}_2$ by

a	$\alpha(a)$	$\beta(a)$	$\gamma(a)$
0	0	0	0
1	1	0	1
2	0	1	1
3	1	1	0

then, we can extend α, β, γ componentwise to maps from $\mathbb{Z}_4^n$ to $\mathbb{Z}_2^n$.
ϕ is then defined as follows:

$$\phi : \mathbb{Z}_4^n \to \mathbb{Z}_2^{2n}$$

$$\phi = (\beta(a), \gamma(a)),\ a \in \mathbb{Z}_4^n.$$

The map ϕ is an isometry from ($\mathbb{Z}_4^n$, Lee distance) to ($\mathbb{Z}_2^{2n}$, Hamming distance).

A quaternary code C_4 is linear over $\mathbb{Z}_4$ but in general, it is not linear over $\mathbb{Z}_2$. Then we define the $\mathbb{Z}_4$-dual of $C = \phi(C_4)$ to be $C_\perp = \phi(C_4^\perp)$ and we have the following non commuting diagram:

$$\begin{array}{rccc} & C_4 & \xrightarrow{\phi} & C = \phi(C_4) \\ dual & \downarrow & & \\ & C_4^\perp & \xrightarrow{\phi} & C_\perp = \phi(C_4^\perp). \end{array}$$

In the following, we denote by $\otimes$ the componentwise product of two vectors.

2.2 Some properties of quaternary codes

Quaternary codes and their image under the Gray-map have some interesting properties (see [1] for details):

- If C_4 is quaternary linear code, then its binary Gray representation $C = \phi(C_4)$ is distance invariant (if $z \in C$, the distribution of distances

$$d_H(z, z'),\ z' \in C$$

  does not depend on the choice of z).
- If C_4 and $C_4^\perp$ are dual quaternary codes, then $C = \phi(C_4)$ and $C_\perp = \phi(C_4^\perp)$ are related by the binary MacWilliams transform.
- Let $a,\ b \in C$, we define $a**b$ as follows:

$$a**b := 2\alpha(a) \otimes \alpha(b).$$

  The binary image $C = \phi(C_4)$ of a quaternary linear code C_4 is linear if and only if

$$a,\ b \in C \Rightarrow\ a**b \in C_4. \tag{3}$$

3 Quaternary Quadratic-Residue Codes

We construct codes by "lifting" over $\mathbb{Z}_4$ a generator poynomial g of a binary Quadratic Residue code of length p (p has to be a prime of the form $8m \pm 1$). We give two constructions of codes: The "augmented" codes and the "extended" codes.

3.1 Augmented codes: The E_p^+ construction

This construction is studied for $p = 7$ in [4], page 41. Let $\underline{a}$ be the all a word (aaa...aaa). Let p ($p = \pm 1 \, mod \, 8$) be a prime number. Let Q_1 denote the set of quadratic residues modulo p and Q_2 the set of nonresidues. Q_1 is then a disjoint union of cyclotomic cosets modulo p. We have, modulo 2

$$x^p - 1 = (x-1)m_1(x)m_2(x).$$

where $m_i = \prod_{r \in Q_i}(x - \alpha^r)$ for $i \in 1,2$ and α is a primitive p^{th} root of unity in some field containing $GF(2)$. We consider the cyclic code E_{p2} which is generated by

$$h_2(x) = (x-1)m_1(x)$$

of degree m. We will "lift" $h_2(x)$ over $\mathbb{Z}_4$ in order to generate a quaternary cyclic code. There is a unique monic polynomial $h(x) \in \mathbb{Z}_4[x]$ of degree m such that $h(x) \equiv h_2(x)$ (mod 2) and $h(x)$ divides $X^p - 1$ (mod 4). This polynomial $h(x)$ may be found ([1], page 14) as follows.
We can write that $h_2(x) = e(x) - d(x)$, where $e(x)$ contains only even powers and $d(x)$ only odd powers. Then $h(x^2) = \pm(e^2(x) - d^2(x))$ and $h(x)$ is obtained by replacing x^2 with x.
For example, for $p = 23$, we have

$$X^{23} - 1 = (x-1)(x^{11} + x^{10} + x^6 + x^5 + x^4 + x^2 + 1)(x^{11} + x^9 + x^7 + x^6 + x^5 + x + 1).$$

If E_{p2} is generated by

$$(x-1)(x^{11} + x^{10} + x^6 + x^5 + x^4 + x^2 + 1),$$

then E_p is generated by

$$(x-1)(x^{11} + 3x^{10} + 2x^7 + x^6 + x^5 + x^4 + x^2 + 2x + 3)$$

and is a cyclic code. We now define E_p^+, the code generated by E_p and $\underline{2}$.

Lemma 1. *If* $p \equiv$ *-1 mod 8,* E_p^+ *is a quaternary self-dual code.*
If $p \equiv$ *1 mod 8,* E_p^+ *is a quaternary isodual code.*

Proof. $p \equiv$ -1 mod 8
We know that E_p is a cyclic code and $\underline{1}$ is not a codeword of E_p. Moreover, we have $\underline{2} = 2(\frac{x^p-1}{x-1}) = 2m_1(x)m_2(x) = 2(\sum_{i=0}^{p-1} x^i)$. If we take $x = 1$ (1 is a zero of the code) we have $2p \not\equiv 0 \; mod\, 4$. Then $\underline{2}$ is not a codeword of E_p.
It is easy to show that $dim(E_p^+) = dim(E_p^+)^\perp = p/2$ using the fact that $(E_p^+)^\perp = E_p^\perp \cap (\underline{2})^\perp$ and $dim(E_p^{+\perp}) = p/2$. We have:

$$\begin{aligned} dim(E_p^\perp \cap (\underline{2})^\perp) &= dim(E_p^\perp) + dim(\underline{2}^\perp) - dim(E_p^\perp + (\underline{2})^\perp) \\ &= (3p)/2 - dim(E_p^\perp + (\underline{2})^\perp) \\ &= p/2. \end{aligned}$$

because $dim(E_p^\perp + (\underline{2})^\perp) = p$.
Moreover, we have $E_p^+ \subset (E_p^+)^\perp$ because $(E_p^+)^\perp = E_p^\perp \cap (\underline{2})^\perp$, $E_p \subset E_p{}^\perp$ and $\underline{2} \subset (\underline{2})^\perp$. So, E_p^+ is a self-dual code over $\mathbb{Z}_4$.

$p \equiv 1$ mod 8
The proof is similar to that of the first case. Let N_p be the code generated by $m_2(x)$ and $\overline{N_p}$ its expurgated code. We have $dim(E_p^+)^\perp = dim(\overline{N_p} + \underline{2})$, and $(\overline{N_p} + \underline{2}) \subset (E_p^+)^\perp$ because $\overline{N_p} \subset N_p$ and $N_p = E_p{}^\perp$. □

Example 1.

- E_7^+ is a self-dual code and $\phi(E_7^+)$ is $(14, 2^7, 4)$ formally self-dual code.
- E_{17}^+ is an isodual code with s.w.e.

$$\begin{aligned}
&z^{17} + 34\, z^{12}x^5 + 68\, z^{11}x^6 + 136\, z^{10}y^6x + \\
&68\, z^{10}x^7 + 170\, z^9y^8 + 1224\, z^9y^6x^2 + \\
&85\, z^9x^8 + 442\, z^8y^8x + 2720\, z^8y^6x^3 + \\
&85\, z^8x^9 + 272\, z^7y^{10} + 3400\, z^7y^8x^2 + \\
&4896\, z^7y^6x^4 + 68\, z^7x^{10} + 1904\, z^6y^{10}x + \\
&7208\, z^6y^8x^3 + 8432\, z^6y^6x^5 + 72\, z^6x^{11} + \\
&544\, z^5y^{12} + 5712\, z^5y^{10}x^2 + 10540\, z^5y^8x^4 + \\
&8432\, z^5y^6x^6 + 34\, z^5x^{12} + 2720\, z^4y^{12}x + \\
&9520\, z^4y^{10}x^3 + 10540\, z^4y^8x^5 + 4896\, z^4y^6x^7 + \\
&5440\, z^3y^{12}x^2 + 9520\, z^3y^{10}x^4 + 7208\, z^3y^8x^6 + \\
&2720\, z^3y^6x^8 + 5440\, z^2y^{12}x^3 + 5712\, z^2y^{10}x^5 + \\
&3400\, z^2y^8x^7 + 1224\, z^2y^6x^9 + 2720\, zy^{12}x^4 + \\
&1904\, zy^{10}x^6 + 442\, zy^8x^8 + 136\, zy^6x^{10} + 544\, y^{12}x^5 + \\
&272\, y^{10}x^7 + 170\, y^8x^9 + x^{17}.
\end{aligned}$$

3.2 Extended codes: The QR_p Construction

We construct QR_p by "lifting" over $\mathbb{Z}_4$ a generator poynomial g of a binary Quadratic Residue code of length p. Let $\Phi(h)$ be the "lifted" polynomial. Then QR_p is the extended cyclic code generated by $\Phi(h)$ (and extended by adding an overall parity symbol).

Lemma 2. *QR_p is a quaternary self-dual code if $p \equiv -1 \bmod 8$, and is isodual if $p \equiv 1 \bmod 8$.*

Proof. For the first case, the proof is similar to that of theorem 7, chapter 16 of [5] and in the second case we have $QR_p = EN_p$ where EN_p denote the extended code with generator polynomial $m_2(x)$. □

Example 2.

- $\phi(QR_{17})$ is the $(36, 2^{18}, 8)$ code with w.e. $= 71708x^{18} + 54621x^{16} + 28152x^{14} + 10608x^{12} + 1530x^{10} + 306x^8 + 1$.
- The binary image of the "lifted Golay" (QR_{23}) is a $(48, 2^{24}, 12)$ formally self-dual code, with a weight enumerator symmetric with respect to 24.

Number Of Words	*Weights*
1	0, 48
12144	12, 36
61824	14, 34
195063	16, 32
1133440	18, 30
1445136	20, 28
4080384	22, 26
2921232	24

4 The $(u|u+v)$ construction

The $(u|u+v)$ construction is the following. Let $C_i (i = 1, 2)$ be an (n, M_i, d_i) binary code. Then $C = \phi(C_1 + 2C_2) = \{(u|u+v) \ : \ u \in C_1, \ v \in C_2\}$ is a $(2n, 2M_1M_2, min(2d_1, d_2))$ binary code.
C is always $\mathbb{Z}_2$-linear (by construction).

Lemma 3. *If C_1 and C_2 are $\mathbb{Z}_2$-linear, then $C_1 + 2C_2$ is $\mathbb{Z}_4$-linear if and only if $\forall u, v \in C_1$, $u \otimes v \in C_2$, where $\otimes$ is the Hadamard product.*

Proof. It is clear that this condition is necessary. We must show that it is sufficient. Let $u, v \in C_1 + 2C_2$ such that $u = u_1 + 2u_2$ and $v = v_1 + 2v_2$. We have

$$u + v = (u_1 \oplus v_1) + 2(u_2 \oplus v_2) \oplus (u_1 \otimes v_1),$$

where $\oplus$ denote the addition modulo 2 in $\mathbb{Z}_2^n$.
Then, using the condition, $u + v \in C$ and C is $\mathbb{Z}_4$-linear. □

Example 3. The $\mathcal{K}_8$ code is a self-dual code introduced by Klemm [6].

$$\phi(\mathcal{K}_8) = \phi(R_8 + 2P_8) = (P_8|P_8 + R_8).$$

The code $\mathcal{K}_8$ is [16,8,4] $\mathbb{Z}_2$-linear and self-dual.

5 Unimodular lattices

There are many parallels between self-dual codes and certain unimodular lattices. In particular between self-dual codes in which the weight of every codeword is a multiple of four and even unimodular lattices (also called Type II lattices) like E_8 and Λ_{24}. We first establish some terminology and definitions:

5.1 Definitions

We call n-dimensional lattice Λ a discrete abelian subgroup of $\mathbb{R}^n$. This subgroup is spanned by n linearly independant vectors $v_1, \ldots, v_n \in \mathbb{R}^n$. The **determinant** of Λ is the volume of the parallelepiped spanned by $v_1, \ldots, v_n$. The dual of a n-dimensional lattice Λ is noted $\Lambda^\star$ and is given by

$$\Lambda^\star = \{x \in \mathbb{R}^n : x.a \in \mathbb{Z} \ \forall a \in \Lambda\},$$

and $det(\Lambda^\star) = (det\Lambda)^{-1}$. A lattice Λ is **integral** if and only if $\Lambda \subset \Lambda^\star$. An integral lattice Λ with $det(\Lambda)$=1, or equivalently with $\Lambda = \Lambda^\star$ is called **unimodular**. The **norm** of Λ, $\mathcal{N}(\Lambda)$, is the subgroup generated by $\{N(x) : x \in \Lambda\}$, where $N(x) = x.x$. If $\mathcal{N}(\Lambda) = 2\mathbb{Z}$, Λ is called **even**.
When the minimal euclidean distance between two points of the lattice Λ is $2r$, then, some spheres of radius r centered at the points of Λ can only touch each other without intersecting. This set of spheres forms an **n-dimensional sphere packing**. The radius r is then the packing radius of Λ.The proportion of the space that is occupied by the spheres is called the **density** and is given by the formula

$$\Delta = \frac{V_n r^n}{det\Lambda},$$

where V_n is the volume of a sphere of radius 1. Obviously, when the radius increase, the set of spheres covers more space. The smaller r such that $\mathbb{R}^n$ is totaly covered is called the covering radius of Λ. A **theta series** of lattice gives the number of points which are at equal distance from the origin and the **kissing** number the number of spheres touching one sphere in the packing.

5.2 Constructions of lattices from quaternary codes : Construction A

We consider lattices obtained from quaternary codes by using the construction A (generalized over $\mathbb{Z}_4$):
Let $\Lambda = C + 4\mathbb{Z}^n$, $C \subset \mathbb{Z}_4^n$. .

We first establish two properties

Theorem 4. *If C is a quaternary self-dual code then $\frac{\Lambda}{2}$ is unimodular.*

Proof. First, we show that $\frac{\Lambda}{2}$ is integral (i.e $\frac{\Lambda}{2} \subset (\frac{\Lambda}{2})^*$). Let λ_1, λ_2 be two elements of the lattice Λ such that $\lambda_1 = \frac{c_1+4n_1}{2}$ and $\lambda_2 = \frac{c_2+4n_2}{2}$ (with $c_1, c_2 \in C$ $n_1, n_2 \in \mathbb{Z}^n$.) We have $(c_1 . c_2) \in 4\mathbb{Z}$ because $C = C^\perp$,
so,

$$2\lambda_1 . 2\lambda_2 = (c_1 . c_2) + 4(n_2 . c_1) + 4(n_1 . c_2) + 16(n_1 . n_2) \in 4\mathbb{Z}.$$

and

$$\lambda_1 . \lambda_2 \in \mathbb{Z}.$$

Moreover, $\det(\Lambda) = \frac{4^n}{|C|} = 2^n = |C|$ so $\det(\frac{\Lambda}{2}) = 1$ and we have the equality

$$\frac{\Lambda}{2} = (\frac{\Lambda}{2})^*,$$

and $\frac{\Lambda}{2}$ is unimodular. □

Theorem 5. *A lattice* (Λ) *is even if and only if there exists* $P \in \mathbb{Z}[x]$ *such that*

$$s.w.e.(1, y, y^4) = P(y^8).$$

Proof. By the same calculation $\frac{\Lambda}{2}$ is even if every codeword of C has euclidean norm divisible by 8. But $s.w.e.(1, y, y^4) = \sum_{c \in C} y^{\|c\|^2}$. □

We are now in measure to construct our lattices.

Lemma 6. *The lattices* $\mathbb{Z}^6$ *and* $\mathbb{Z}^7$ *are obtained by* $2\mathbb{Z}^6 = \mathcal{D}_6^{\oplus} + 4\mathbb{Z}^6$ *and* $2\mathbb{Z}^7 = E_7^+ + 4\mathbb{Z}^7$.

Proof. The two codes $\mathcal{D}_6^{\oplus}$ and E_7^+ are defined in [4]. They are both quaternary self-dual codes so the lattices constructed are unimodular with dimensions 6 and 7 and we know that $\mathbb{Z}^6$ and $\mathbb{Z}^7$ are the unique unimodular lattices with these dimensions. □

Constructions of the E_8 lattice E_8 is an even unimodular 8-dimensional lattice. It is the unique lattice with density $\Delta = \pi^4/384 = 0.2537\ldots$, and minimal norm 2. Its theta series is $1 + 240q^2 + 2160q^4 + \ldots$.

Theorem 7. *The four following equations give four constructions of the lattice* E_8.

$$\begin{aligned} &I. \ \ 2E_8 = \mathcal{K}_8' + 4\mathbb{Z}^8 \\ &II. \ \ 2E_8 = \mathcal{O}_8 + 4\mathbb{Z}^8 \\ &III. \ 2E_8 = \mathcal{Q}_8 + 4\mathbb{Z}^8 \\ &IV. \ \ 2E_8 = \mathcal{K}_8 + 4\mathbb{Z}^8. \end{aligned}$$

Here $\mathcal{K}_8'$ *is the code introduced by Conway and Sloane [4]. Its generator matrix is given in the next section.* $\mathcal{O}_8$ *denote the "octacode",* $\mathcal{Q}_8$ *is the self dual code (see [4]) with generator matrix*

$$G_{Q_8} = \begin{bmatrix} 0\,0\,1\,1\,0\,2\,1\,3 \\ 0\,0\,0\,2\,1\,3\,1\,1 \\ 1\,1\,0\,2\,0\,0\,1\,3 \\ 0\,2\,0\,2\,0\,2\,0\,2 \\ 0\,0\,0\,0\,0\,0\,2\,2 \end{bmatrix},$$

(note that this code is $\mathbb{Z}_2$-linear but not self-dual over $\mathbb{Z}_2$) and $\mathcal{K}_8$ is the self dual code introduced in the preceding section with generator matrix

$$G_{\mathcal{K}_8} = \begin{bmatrix} 1\,1\,1\,1\,1\,1\,1\,1 \\ 0\,2\,0\,0\,0\,0\,0\,2 \\ 0\,0\,2\,0\,0\,0\,0\,2 \\ 0\,0\,0\,2\,0\,0\,0\,2 \\ 0\,0\,0\,0\,2\,0\,0\,2 \\ 0\,0\,0\,0\,0\,2\,0\,2 \\ 0\,0\,0\,0\,0\,0\,2\,2 \end{bmatrix}.$$

Proof. We know that E_8 is the only unimodular lattice of minimal norm 2 in dimension 8. It is easy to see that all these self-dual codes have euclidean norm 8, so the norm of the lattice is 2 and (See [7] page 121) it is E_8.
Moreover, we can find the theta series of E_8 by replacing the three variables x, y, z in the s.w.e.(x, y, z) of each code by

$$\begin{aligned} &1 + 2q^{16} + 2q^{64} \\ &q + q^9 + q^{25} \\ &2q^4 + 2q^{36} + 2q^{100} \end{aligned}$$

respectively. So, the kissing number is 240 and we know ([7] chapter 14, page 342) that E_8 is the unique 8-dimensional lattice with 240 as kissing number. □

Construction of the lattice O_{23} The shorter Leech lattice O_{23} is the unique 23-dimensional unimodular latice of norm 3. Its theta series is $1+4600q^3+93150q^4+\ldots$. (See [7] page 179 for more details).

Theorem 8. $2O_{23} = E_{23}^+ + 4\mathbb{Z}^{23}$

Proof. The euclidean norm of the self-dual code E_{23}^+ is 12, so the norm of the lattice is 3 and its dimension is 23. □

Construction of the BW_{32} This is an even unimodular lattice of norm 4. We have (See [8] page 1169):

$$\begin{aligned} BW_{32} &= \Lambda(0,4) \\ &= 4\mathbb{Z}^{32} + 2(32,26,4) + (32,6,16) \\ &= 4\mathbb{Z}^{32} + 2RM(3,5) + RM(1,5), \end{aligned}$$

with the notations of [8].
This shows that BW_{32} is obtained from a quaternary self dual code by construction A.

Construction of the Leech lattice The Leech lattice is an exceptionally dense sphere-packing in 24 dimensional space. It contains all good lattices of lower dimensions and plays a pivotal role in mathematical theory of lattices and finite simple groups. The following construction appears to be new.

Theorem 9. $2\Lambda_{24} = QR_{24} + 4\mathbb{Z}^{24}$

Proof. QR_{24} is the "lifted Golay" of length 24 over $\mathbb{Z}_4$ with minimal Lee distance 12 and symmetrized weight enumerator

$$
\begin{aligned}
&z^{24} + 24288\, y^{16}z^8 + 4096\, y^{24} + 61824\, x^{11}y^{12}z + 61824\, xy^{12}z^{11} + 2576\, x^{12}z^{12} \\
&+12144\, x^2y^8z^{14} + 680064\, x^2y^{16}z^6 + 1133440\, x^3y^{12}z^9 + 170016\, x^4y^8z^{12} \\
&+1700160\, x^4y^{16}z^4 + 4080384\, x^5y^{12}z^7 + 765072\, x^6y^8z^{10} + 680064\, x^6y^{16}z^2 \\
&+4080384\, x^7y^{12}z^5 + 759\, x^8z^{16} + 1214400\, x^8y^8z^8 + 24288\, x^8y^{16} \qquad\qquad (4)\\
&+1133440\, x^9y^{12}z^3 + 765072\, x^{10}y^8z^6 + 170016\, x^{12}y^8z^4 + 12144\, x^{14}y^8z^2 \\
&+759\, x^{16}z^8 + x^{24}.
\end{aligned}
$$

The theta series of Λ is obtained from the s.w.e. of the QR_{24} (Eq. 4) by replacing x, y, z by

$$
\begin{aligned}
&1 + 2q^{16} + 2q^{64} + \dots \\
&q + q^9 + q^{25} + \dots \\
&2q^4 + 2q^{36} + 2q^{100} + \dots
\end{aligned}
$$

respectively. Then the theta series of Λ is

$$
\begin{aligned}
&1 + 196560\, q^{16} + 16773120\, q^{24} + 398034000\, q^{32} + 4629381120\, q^{40} \\
&+34417656000\, q^{48} + 187489935360\, q^{56} + 814877513808\, q^{64} \\
&+2975352422400\, q^{72} + \dots
\end{aligned}
$$

and we know that the Leech lattice is the only 24-dimensional lattice with kissing number 196560 (See [7], chapter 14). $\qquad\square$

Another proof using theorem 4 & 5 shows that $\frac{\Lambda}{2}$ is an even unimodular lattice of norm 4. The result follows by Conway's characterization of the Leech lattice. QR_{23} is a self-dual code and the lattice is even because $s.w.e._{QR_{23}}(1, y, y^4) = y^{96} + 12903\, y^{64} + 231840\, y^{56} + 1925376\, y^{48} + 5974848\, y^{40} + 6546375\, y^{32} + 1987616\, y^{24} + 98256\, y^{16} + 1 = P(y^8)$ with $P \in \mathbb{Z}[x]$.

Forney gives a similar construction with a different quaternary code, see [8] for more details. The code used is $\mathcal{K}_8 \nabla O_8'$ where ∇ denote the cubing construction: let $C1$, $C2$ be two quaternary codes, $C1\nabla C2$ consists of all the vectors

$$
|a + x|b + x|a + b + x|, \ a, b \in C1, x \in C2.
$$

Let x be a vector of type $2^2 0^6$ and $a, b = 0^8$, then we notice that the minimum distance of $\mathcal{K}_8 \nabla O_8'$ is at most 8 and its euclidean norm is 16 [8].

6 Sporadic isomorphisms

$\mathcal{K}_8'$ is a self dual code introduced by Conway and Sloane [4] having generator matrix

$$G_{\mathcal{K}_8'} = \begin{bmatrix} 1\,1\,1\,1\,0\,0\,0\,2 \\ 0\,0\,0\,2\,1\,1\,1\,1 \\ 0\,2\,0\,2\,0\,0\,0\,0 \\ 0\,0\,2\,2\,0\,0\,0\,0 \\ 0\,0\,0\,0\,0\,2\,0\,2 \\ 0\,0\,0\,0\,0\,0\,2\,2 \end{bmatrix}$$

F_{16} is introduced by Pless [3]. It has generator matrix

$$G_{F_{16}} = \begin{bmatrix} 1\,1\,1\,1\,0\,0\,0\,0\,0\,0\,0\,0\,0\,0\,0\,0 \\ 0\,0\,1\,1\,1\,1\,0\,0\,0\,0\,0\,0\,0\,0\,0\,0 \\ 0\,0\,0\,0\,1\,1\,1\,1\,0\,0\,0\,0\,0\,0\,0\,0 \\ 0\,0\,0\,0\,0\,0\,0\,0\,1\,1\,1\,1\,0\,0\,0\,0 \\ 0\,0\,0\,0\,0\,0\,0\,0\,0\,0\,1\,1\,1\,1\,0\,0 \\ 0\,0\,0\,0\,0\,0\,0\,0\,0\,0\,0\,0\,1\,1\,1\,1 \\ 1\,0\,1\,0\,1\,0\,1\,0\,0\,0\,0\,0\,0\,0\,1\,1 \\ 1\,1\,0\,0\,0\,0\,0\,0\,1\,0\,1\,0\,1\,0\,1\,0 \end{bmatrix}$$

Lemma 10. $\phi(\mathcal{K}_8') = F_{16}$

Proof. There exist only three self-dual codes of length 16 with $d = 4$. Moreover, only one of these codes has its weight enumerator identical to that of F_{16} and it is easy to see that $\phi(\mathcal{K}_8')$ is linear. □

Lemma 11. $\phi(R_4 + 2P_4) = \mathcal{H}_8$, *where* R_4, P_4, $\mathcal{H}_8$ *denote respectively the repetition code, the parity check of length* 4 *and the extended Haming code of length* 8.

Proof. $\mathcal{H}_8$ is the only one linear code with parameters [8, 4, 4]. $\phi(R_4+2P_4)$ is doubly even f.s.d. hence linear. In fact there is no non-linear extremal Type II (i.e doubly even) code (see [7] page 195). □

7 Table

The following table gives our principal results. The first two columns give the length of the code over $\mathbb{Z}_4$ and $\mathbb{Z}_2$ respectively. The minimum distance is denoted by D (the minimum distance of a code and its binary image by the Gray-map is the same). The fourth column gives the minimum distance of the best binary linear code of length N. The fifth and sixth columns give respectively the denomination of the code over $\mathbb{Z}_4$ and its parameters over $\mathbb{Z}_2$ (brackets are used when the code is linear). The last column gives the lattice which is obtained by the construction A.

$N(\mathbb{Z}_4)$	$N(\mathbb{Z}_2)$	D	$BestKnown$ $AndTypeI$ $(linear)$	C $(Over\mathbb{Z}_4)$	$\phi(C)$ $(Over\mathbb{Z}_2)$	$\frac{C+4\mathbb{Z}^n}{2}$
4	8	4	4	$R_4 + 2P_4$	$[8,4,4]$	D_4^+
6	12	4	4	$\mathcal{D}_6^{\oplus}$ [4]	$(12, 2^6, 4)$	$\approxeq \mathbb{Z}^6$
7	14	4	4	E_7^+ [4]	$(12, 2^7, 4)$	$\approxeq \mathbb{Z}^7$
8	16	4	4	$\mathcal{K}_8 = R_8 + 2P_8$	$[16,8,4]$	E_8
8	16	6	4	O_8	$(16, 2^8, 6)$	E_8
8	16	4	4	$\mathcal{K}_8'$	$[16,8,4]$	E_8
8	16	4	4	$\mathcal{Q}_8$	$[16,8,4]$	E_8
16	32	8	8	$RM(1,4) + 2RM(2,4)$	$[32,16,8]$	${E_8}^2$
17	34	8	8	E_{17}^+	$(34, 2^{17}, 6)$	
18	36	8	8	QR_{17}	$(36, 2^{18}, 8)$	
23	46	10	10	E_{23}^+	$(46, 2^{23}, 10)$	O_{23}
24	48	8	10	$\mathcal{K}_8 \nabla O_8'$	$(48, 2^{24}, 8)$	Λ_{24}
24	48	12	10	QR_{23}	$(48, 2^{24}, 12)$	Λ_{24}
32	64	8	12	$RM(1,5) + 2RM(3,5)$	$[64,32,12]$	BW_{32}

8 Acknowledgement

We thank Dave Forney for helpful discussions concerning his construction of the Leech lattice, Vera Pless for sending us [2] and Neil Sloane for sending us [4].

References

1. R. Hammons, P.V.Kumar, A.R. Calderbank, N.J.A Sloane, and P. Solé. Kerdock, Preparata, Goethals and Other Codes are linear over Z4. *submitted to IEEE Transactions on information theory*, 1992.
2. G. T. Kennedy and V. Pless. On Designs and Formally Self-Dual Codes. *preprint*, 1992.
3. V. Pless. A Classification of Self-Orthogonal Codes over GF(2). *Discrete Math*, pages 209–246, 1972.
4. J.H. Conway and N.J.A Sloane. Self-dual codes over the integers modulo 4. *JCT A 62,30-45*, 1993.
5. F.J. MacWilliams and N.J.A. Sloane. *The theory of Error-Correcting Codes*. North-Holland, 1977.
6. M.Klemm. Selbstduale Codes uber dem Ring der ganzen Zalen modulo 4. *Arch. Math.*, 53:201–207, 1989.
7. J.H.Conway and N.J.A. Sloane. *Sphere Packings, Lattices and Groups*. Springer-Verlag, 1988.
8. JR D. Forney. Coset Codes - Part II: Binary lattices and related codes. *IEEE, Transactions on information theory*, 34:1152–1187, 1988.

New lower bounds for some spherical designs

Peter Boyvalenkov, Svetla Nikova
Institute of Mathematics, Bulgarian Acad. of Sciences,
8 G.Bonchev str, 1113 Sofia, Bulgaria
e-mail sectmoi@bgearn.bitnet

Abstract

A new method for obtaining lower bounds for spherical t-designs is proposed. Applications and new bounds are given for $t = 9, 10$. Some restrictions on the distribution of the inner products of the points of putative designs are obtained.

1. Introduction.

A non-empty finite set $W \subset \mathbf{S}^{n-1}$ is called a spherical $\tau-$design on $\mathbf{S}^{n-1}$ if and only if

$$\sum_{x \in W} f(x) = 0$$

for all homogeneous harmonic polynomials $f(x) = f(x_1, x_2, ..., x_n)$ of degree $1, 2, ..., \tau$. There is also the following equivalent definition of spherical designs: W is a spherical $\tau-$design on $\mathbf{S}^{n-1}$ if and only if

$$\int_{\mathbf{S}^{n-1}} f(x) d\mu(x) = \frac{1}{|W|} \sum_{x \in W} f(x)$$

holds for all polynomials $f(x)$ of degree at most τ. Here μ is the normalized Lebesgue measure, i.e. $\mu(\mathbf{S}^{n-1}) = 1$. Thus, the spherical designs can be used for numerical integration.

Delsarte, Goethals and Seidel [1] gave the following necessary lower bound for the cardinality of a $\tau-$design on $\mathbf{S}^{n-1}$.

$$| W | \geq R(n, \tau) = \begin{cases} \binom{n+e-1}{n-1} + \binom{n+e-2}{n-1}, & \text{if } \tau = 2e; \\ 2\binom{n+e-1}{n-1}, & \text{if } \tau = 2e+1. \end{cases} \quad (1)$$

A spherical design is called tight [1] if it attains the bound (1). Bannai and Damerell [2,3] proved that if $n \geq 3$ then tight spherical $\tau-$designs on $\mathbf{S}^{n-1}$ do not exist if $\tau = 2e$ and $e \geq 3$ or $\tau = 2e+1$ and $e \geq 4$ except the case $\tau = 11$, $n = 24$ (about this case see [4]). There are only few examples of tight $\tau-$designs for $\tau = 4, 5, 7$ and $n \geq 3$ (the only open cases after [2,3]).

Seymour and Zaslavsky [5] proved the existence of spherical $\tau-$designs on $\mathbf{S}^{n-1}$ for all values of n and τ, provided $|W|$ is sufficiently large. Let us remark that two disjoint $\tau-$designs with r and s points can be combined to give a $\tau-$design with $r+s$ points. Thus, together with the explicit constructions, the most interesting problem in this topic is to find the mimimum number of points of (non-tight) spherical $\tau-$design on $\mathbf{S}^{n-1}$.

All known explicit constructions, except Hardin-Sloane's [6] construction of 4-designs, give designs with large number of points (see [7,8,9]).

In this paper we consider a new method for obtaining lower bounds improving (1) in some cases. In fact, we obtain the first improvements of the bound (1) by more than 1. Our method gives good polynomials for applying in Theorem 1 ([1,Th.5.10.]) below. Some properties of these polynomials are considered in Section 2. Indeed, we obtain linear programming bounds using polynomials of higher degree than these of Delsarte, Goethals and Seidel.

New lower bounds for spherical $\tau-$designs are presented in [10] for $\tau = 6, 7$ and 8. Here we give the new bounds we have obtained for $\tau = 9$ and 10. In Section 4 some restrictions on the structure of putative designs, attaining our bounds, are given.

2. Method for obtaining lower bounds.

Delsarte, Goethals and Seidel [1] obtain the bound (1) using the following theorem [1,Th.5.10.].

Theorem 1. *Let $f(t)$ be a real polynomial such that*
(A1) $f(t) \geq 0$ for $-1 \leq t \leq 1$;
(A2) The coefficients in the expansion of $f(t)$ in terms of Gegenbauer polynomials [11]

$$f(t) = \sum_{i=0}^{k} f_i P_i^{(n)}(t)$$

satisfy $f_{\tau+1} \leq 0, \ldots, f_k \leq 0$.
Then the cardinality of a spherical $\tau-$design $W \subset \mathbf{S}^{n-1}$ is bounded below by

$$|W| \geq f(1)/f_0.$$

The polynomials, used in [1] to estimate the size of spherical $\tau-$designs, have degree τ. Here we search for good polynomials of higher degree ($\tau+3$) using some restrictions on the form of the best polynomials (i.e. the polynomials giving the best lower bounds).

Definition 1. Let us denote

$$B_{n,\tau} = \{f(t) : \ f(t) \text{ satisfies the conditions of Theorem 1}\},$$

and $L(f) = f(1)/f_0$ for $f(t) \in B_{n,\tau}$.

Since for every $a > 0$ one has $L(f(t)) = L(af(t))$, we can normalize the considered polynomials or their divisors.

Definition 2. A polynomial $f(t) \in B_{n,\tau}$ of degree k is called extremal for $B_{n,\tau}$ if

$$L(f) = \max\{L(g) : \ g(t) \in B_{n,\tau}, \ \deg(g) \le k\}, \text{ and}$$
$$L(f) > \max\{L(g) : \ g(t) \in B_{n,\tau}, \ \deg(g) < k\}.$$

The next theorem we have proved in [10]. It concerns the number of the double zeros of the extremal polynomials.

Theorem 2. *Let $f(t)$ be an extremal for $B_{n,\tau}$ ($n \ge 3$, $\tau \ge 4$) polynomial of degree $k \ge \tau + 3$. If τ is odd or if τ is even and the number -1 is an even zero of $f(t)$ (or $f(-1) > 0$), then $f(t)$ has at least $[\frac{\tau}{2}] + 1$ double zeros in $[-1, 1]$. If τ is even and the number -1 is an odd zero of $f(t)$, then $f(t)$ has at least $\frac{\tau}{2}$ double zeros in $[-1, 1]$.*

We restrict ourselves to search for extremal polynomials of degree $\tau + 3$. These polynomials will have the following form

$$f(t) = \begin{cases} A^2(t)[q(t+1) + 1 - t] & \text{if } \tau = 2e \\ A^2(t)[q(t-p)^2 + 1 - t^2] & \text{if } \tau = 2e+1 \end{cases} \tag{2}$$

where $\deg(A) = e + 1$ and $0 < q < 1$. The polynomial

$$A(t) = t^{e+1} + a_1 t^e + \cdots + a_{e-1} t + a_e$$

has $e+1$ zeros in $[-1, 1]$. Indeed, it follows from Theorem 2 that the polynomial $f(t)$ for $\tau = 2e$ must have at least e double zeros. However, we search for polynomials of degree $2e + 3$ with $e + 1$ double zeros.

In order to reach the condition (A2), we have to require $f_i \le 0$ for $i = \tau+1, \ \tau+2, \ \tau+3$. The condition $0 < q < 1$ implies immediately $f_{\tau+3} < 0$. The next theorem shows that we can take $f_{\tau+1} = 0$ without loss of generality.

Theorem 3. *Let $f(t)$ be an extremal for $B_{n,\tau}$ ($n \ge 3$, $\tau \ge 4$) polynomial of degree $k \ge \tau + 1$. If $L(f) < R(n, \tau + 1)$ then $f_{\tau+1} = 0$ in the expansion $f(t) = \sum_{i=0}^{k} f_i P_i^{(n)}(t)$.*

Proof. Let us suppose that $f_{\tau+1} < 0$ under the assumptions of the theorem. The polynomial $C_{n,\tau+1}(t)$, used in [1] for obtaining the bound

$$|W| \ge R(n, \tau + 1) > R(n, \tau),$$

has degree $\tau+1$ and $c_{\tau+1} > 0$ in the expansion $C_{n,\tau+1}(t) = \sum_{i=0}^{\tau+1} c_i P_i^{(n)}(t)$.

Since $f_{\tau+1} < 0$, there exist positive numbers α and β, such that $\alpha f_{\tau+1} + \beta c_{\tau+1} \leq 0$. We consider the polynomial $H(t) = \alpha f(t) + \beta C_{n,\tau+1}(t)$. It is easy to see that $H(t)$ belongs to the set $B_{n,\tau}$, it has degree k or lower and the number $L(H) = H(1)/f_0(H)$ lies between the numbers $L(f)$ and $R(n,\tau{+}1) = L(C_{n,\tau+1})$. Hence we have

$$L(f) < L(H) < R(n, \tau+1) ,$$

a contradiction with the extremality of the polynomial $f(t)$. The proof is completed.

The numbers $R(n, \tau+1)$ is very large compared with $R(n,\tau)+1$ (the best known lower bound for the size of a non-tight $\tau-$design on $\mathbf{S}^{n-1}$). Therefore, without loss of generality, one can search for improvements over (1), assuming $f_{\tau+1} = 0$. In particular, extremal polynomials of degree $\tau+1$ do not exist.

One can use some polynomials of degree $\tau+3$ in order to obtain assertions that are similar to Theorem 3 and concern the coefficient $f_{\tau+2}$. We do not apply here such polynomials. We require $f_{\tau+2} = 0$. Thus, one can express the first two coefficients a_1 and a_2 of $A(t)$ as functions of q and n (see Eq.(2)).

We have to find q and the remaining coefficients of $A(t)$ which minimize the rational function

$$F(q, a_3, ..., a_e) = L(f) = 2qA^2(1)/f_0$$

The denominator f_0 is expressed by q, n and $a_1, a_2, ..., a_e$ by the next trivial Lemma.

Lemma 1. *If $f(t) = \sum_{i=0}^{k} a_i t^i = \sum_{i=0}^{k} f_i P_i^{(n)}(t)$ is a real polynomial, then*

$$f_0 = a_0 + \frac{a_2}{n} + \frac{3a_4}{n(n+2)} + \cdots = a_0 + \sum_{i=1}^{[k/2]} \frac{(2i-1)!!a_{2i}}{n(n+2)\cdots(n+2i-2)} \tag{3}$$

(where $(2i-1)!! = 1.3.5.\cdots.(2i-1)$.)

The partial derivatives of $F(q, a_3, ..., a_e)$ give a system of linear (with respect to $a_3, a_4, ..., a_e$) equations:

$$\begin{aligned} f'_{0\,a_3} - f'_{0\,a_4} &= 0 \\ f'_{0\,a_3} - f'_{0\,a_5} &= 0 \\ &\cdots \\ f'_{0\,a_3} - f'_{0\,a_e} &= 0 \\ 2f_0 - A(1)f'_{0\,a_3} &= 0. \end{aligned} \tag{4}$$

In fact, the last equation is not linear, however it becomes after the setting the parameters a_4, a_5, ..., a_e as functions of a_3 by the first $e-3$ equations.

Thus, one can resolve this system with respect to the parameters $a_3, a_4, ..., a_e$. These become functions of q and the dimension n. It does not seem possible to apply further analytical methods in order to find the optimal values of $q \in (0,1)$. So, we use a PC and a method Monte Carlo to find good aproximations of the extremal polynomials.

The lower bounds, we have obtained, are better than (1) in some cases, according to the existence of extremal polynomials of degree $\tau+3$ satisfying (2) and the requirements $f_{\tau+1} = f_{\tau+2} = 0$.

The above method is similar to the method (proposed by the first author [12,13,14]) for finding good polynomials estimating the maximum cardinality of spherical codes. Indeed, Odlyzko and Sloane [18] were first who showed that polynomials of higher degrees can be used for obtaining new upper bounds for spherical codes by linear programming (see also [15,16], where Levenshtein gave polynomials of lower degrees without making use of computer).

In our previous work [10] we gave the new bounds for spherical $\tau-$designs we have obtained for $\tau = 6, 7$, and 8. Below we continue our list of numerical examples of improvements of the Delsarte-Goethals-Seidel bound (1).

3. New lower bounds for spherical 9- and 10-designs.

3.1. 9-designs.

According to Section 2 we consider polynomials of degree 12 having the following form:

$$f(t) = (t^5 + at^4 + bt^3 + ct^2 + dt + e)^2(t+1)[q(t+1)+1-t] = \sum_{i=0}^{12} f_i P_i^{(n)}(t),$$

where $f_{10} = f_{11} = 0$ and $0 < q < 1$.

Using the equations $f_{10} = f_{11} = 0$, we obtain

$$a = \frac{q}{1-q} = a(q),$$

$$b = \frac{3a^2+a}{2} - \frac{33}{n+20} - \frac{1}{2(q-1)} = b(q,n),$$

respectively.

Next, we find c, d and e as functions of q and n by the system (4). After the minimizing the ratio $f(1)/f_0$ one obtains improvements of the bound $R(n,9)+1$ (tight spherical 9-designs do not exist for $n \geq 3$ [3,Th.1.]) in dimensions $4 \leq n \leq 14$. For example, the polynomial we have obtained for $n = 4$ has parameters $a = 0.18516$, $b = -0.63841$, $c = -0.11109$, $d = 0.0519$, $e = 0.00694$, $q = 0.15623$.

The new lower bounds are presented in Table 1. The old bounds in the second column are equal to the quantity

$$R(n,9)+1=\frac{n(n+1)(n+2)(n+3)}{12}+1$$

as follows by Theorem 1 from [3].

Table 1. New lower bounds on the size of the spherical 9-designs on $\mathbf{S}^{n-1}$, $4 \le n \le 14$.

n	[3,Th.1]	New bounds
4	71	73
5	141	149
6	253	272
7	421	458
8	661	724
9	991	1087
10	1431	1565
11	2003	2173
12	2731	2924
13	3641	3828
14	4761	4892

3.2. 10-designs.

We consider polynomials of degree 13

$$f(t)=(t^6+at^5+bt^4+ct^3+dt^2+et+f)^2[q(t+1)+1-t]=\sum_{i=0}^{13} f_iP_i^{(n)}(t),$$

where $f_{11}=f_{12}=0$ and $0<q<1$.

As in the previous case we express a and b by the equations $f_{11}=f_{12}=0$. Then we obtain the remaining coefficients c, d, e, and f by the system (4). Finally, we find by a computer the last parameter q. Several new bounds (in dimensions $4 \le n \le 15$) are presented in Table 2 below. In some cases the improvements are almost 20 percents.

Table 2. New lower bounds on the size of the spherical 10-designs on $\mathbf{S}^{n-1}$, $4 \le n \le 15$.

n	[2,Th.1.]	New bounds
4	92	97
5	197	215
6	379	424
7	673	770
8	1123	1305
9	1783	2097
10	2718	3222
11	4005	4771
12	5734	6845
13	8009	9556
14	10949	13028
15	14689	17394

4. Restrictions on the structure of putative spherical designs.

In this section we obtain some restrictions on the distribution of the scalar products of the points of putative spherical designs meeting the bounds obtained by the method from Section 2. Probably such restrictions can be used for obtaining further improvements of the lower bounds for the cardinality of spherical designs.

We make use of the following well known equality [1,Cor.3.8], [15,16,Eq.(1.7)]:

$$| W | f(1) + \sum_{t_{x,y} \in M_W} f(t_{x,y}) = | W |^2 f_0 + \sum_{i=1}^{k} \frac{f_i}{r_i} \sum_{j=1}^{r_i} (\sum_{x \in W} v_{ij}(x))^2, \qquad (5)$$

where

$f(t) = \sum_{i=0}^{k} f_i P_i^{(n)}(t)$ is an arbitrary real polynomial;

$W \subset \mathbf{S}^{n-1}$ is any non-empty finite subset (spherical code);

$M_W = \{t_{x,y} \mid t_{x,y} = (x,y), x, y \in W, x \neq y\}$;

$\{v_{ij}(x) | j = 1, 2, ..., r_i\}$ is an orthonormal basis of the space V_i of all homogeneous harmonic polynomials of total degree i;

and $r_i = \dim V_i = \binom{n+i-1}{n-1} - \binom{n+i-3}{n-1}$.

The equality (5) has been proved fundamental for the investigation of cardinality of spherical codes and designs.

If W is a spherical τ-design and $f(t) \in B_{n,\tau}$, then we have by (5)

$$\sum_{t \in M_W} f(t) \le |W|(f_0|W| - f(1)) \qquad (6)$$

There is an equality in (6) for $k = \deg(f) \leq \tau$. We are interested in applications with polynomials of higher degree. Indeed, we use polynomials we have found in the way from Section 2. In this case the number $|W|(f_0|W| - f(1))$ (see (6)) takes a minimal value. We remark that we count in M_W at least twice all the points (because $t_{x,y} = (x,y) = (y,x) = t_{y,x}$).

Using the graphics of the polynomials from Section 2 one can obtain some restrictions on the set M_W.

Let us denote

$$I_k = \{t \in [-1,1] \mid f(t) > \frac{|W|}{2k}(f_0|W| - f(1)), t \notin I_{k-1}\}$$

for $k = 1, 2, \ldots$.

Theorem 4. *The set $M_W \bigcup I_k$ does not contain more than $2k-2$ points (counting with the multiciplities).*

Proof. If $| M_W \bigcup I_k | \geq 2k$, then we immediately obtain a contradiction with (6).

In particular, we obtain an upper bound for the quantity

$$s(W) = \max\{(x,y) \mid x, y \in W, x \neq y\}.$$

Corollary 1. *If* $[s_1, 1] \subset I_1$, *then* $s(W) < s_1$.

Example 1. Applying the method from Section 2 we obtained in [10] that tight 4-designs on $\mathbf{S}^{n-1}$ do not exist for $n = 3, 4, 5$ (see the remark at the end of Example 8.3 in [1]). The smallest possible designs would have sizes 10, 15 and 21, respectively. Using the corresponding polynomials one can see that if such 4-designs exist, then the sets M_W do not contain points in the intervals $[0.7805, 1]$, $[-1, -0.977] \bigcap [0.688, 1]$, and $[-1, -0.954] \bigcap [0.678, 1]$ respectively.

Example 2. It follows by [10] that any 6-design $W \subset \mathbf{S}^3$ has at least 32 points ($R(4,6) = 30$). Applying Corollary 1 with the polynomial from [10], we obtain that if $|W| = 32$, then M_W does not contain points in the set $[-1, -0.96] \bigcap [0.746, 1]$.

Lower bounds for $s(W)$ can be found in [16,p.69] and [17]. One can obtain slightly better bounds using the following theorem.

Theorem 5. *Let $s_0 \in (0,1)$, $f(t)$ be a real polynomial, and $W \subset \mathbf{S}^{n-1}$ be a spherical $\tau-$design such that:*
(B1) $f(t) \leq 0$ for $-1 \leq t \leq s_0$ and $f(t) \geq 0$ for $s_0 \leq t \leq 1$;

(B2) The coefficients in the expansion of $f(t)$ in terms of Gegenbauer polynomials [11]

$$f(t) = \sum_{i=0}^{k} f_i P_i^{(n)}(t)$$

satisfy $f_0 > 0$, $f_{\tau+1} \geq 0, \ldots, f_k \geq 0$;
(B3) $|W| > f(1)/f_0$.
Then $s(W) > s_0$.

Proof: Since W is a $\tau-$design, we have by (5)

$$f(1)|W| + \sum_{t \in M_W} f(t) = f_0 |W|^2 + \sum_{i=\tau+1}^{k} \frac{f_i}{r_i} \sum_{j=1}^{r_i} (\sum_{x \in W} v_{ij}(x))^2.$$

The second term in the right side is nonnegative because of (B2). If $s(W) \leq s_0$, then we have $\sum_{t \in M_W} f(t) \leq 0$ by (B1) and we obtain a contradiction with (B3). Thus, $s(W) > s_0$.

Example 3. We consider a putative 6-design on $\mathbf{S}^3$ from Example 2. One can apply Theorem 1 with the polynomial

$$f(t) = (t^3 + at^2 + bt + c)^2 (t^2 + et + f)(t - s_0)$$

where $a = 1.38504$, $b = 0.508877$, $c = 0.032728$, $e = 0.094752$, $f = 0.745672$, $s_0 = 0.559$. Therefore, we have $s(W) > 0.559$. The last polynomial was found by the method for obtaining upper bounds for spherical codes [12,13,14].

Acknowledgements. This research was partially supported by the Bulgarian NSF Contract I-35/1993. The first author was partially supported by the Foundation "Eureka". The authors would like to thank Professor Stefan Dodunekov for his helpfull comments and suggestions throughout their work on this paper.

REFERENCES

[1] Delsarte,P., Goethals,J.-M., Seidel,J.J.: Spherical codes and designs. Geom. Dedicata 6, 363-388 (1977).

[2] Bannai,E., Damerell,R.M.: Tight spherical designs I, J. Math. Soc. Japan 31, 199-207 (1979).

[3] Bannai,E., Damerell,R.M.: Tight spherical designs II, J. London Math. Soc. 21, 13-30 (1980).

[4] Bannai,E., Sloane,N.J.A.: Uniqueness of certain spherical codes, Canad. J. Math. 33, 437-449 (1981).

[5] Seymour,P.D., Zaslavsky,T.: Averaging sets: a generalization of mean values and spherical designs, Adv. in Math. 52, 213-240 (1984).

[6] Hardin,R.H., Sloane,N.J.A.: New spherical 4-designs, Discrete Math. 106/107, 255-264 (1992).

[7] Bajnok,B.: Construction of spherical 4- and 5-designs, Graphs Combin. 7, 219-233 (1991).

[8] Bajnok,B.: Construction of designs on the 2-sphere, European J. Combin. 12, 377-382 (1991).

[9] Neutsch,W.: Optimal spherical designs and numerical integration on the sphere, J. Comput. Phys. 51, 313-325 (1984).

[10] Boyvalenkov,P.G., Nikova,S.: Improvements of the lower bounds for some spherical designs, submitted.

[11] Abramowitz,M., Stegun,I.A.: Handbook of Mathematical Functions. National Bureau of Standarts Appl. Math. Series 55: US Dept. Commerce: Washington DC 1972.

[12] Boyvalenkov,P.G.: PhD Thesis: Sofia University 1993.

[13] Boyvalenkov,P.G.: On the upper bounds for the kissing numbers, Serdica 18, 278-285 (1992).

[14] Boyvalenkov,P.G.: On the extremality of the polynomials used for obtaining the best known upper bounds for the kissing numbers, to appear in Journal of Geometry.

[15] Levenshtein,V.I.: On bounds for packings in n-dimensional Euclidean space, DAN SSSR 245, 1299-1306 (1979) (in Russian).

[16] Levenshtein,V.I.: Bounds for packings in metric spaces and certain applications, Problemy Kibernetiki 40, 44-110 (1983) (in Russian).

[17] Fazekas,G., Levenshtein,V.I.: On upper bounds for code distance and covering radius of designs in polynomial metric spaces, Fifth Joint Soviet-Swedish Int. Workshop on Inf. Theory, Moscow 65-68 (1990).

[18] Odlyzko,A.M., Sloane,N.J.A.: New bounds on the number of unit spheres that can touch a unit sphere in n dimensions, J. Comb. Theory, A 26, 210-214 (1979).

LATTICES BASED ON LINEAR CODES

Gregory Poltyrev *
Department of Electrical Engineering - Systems,
Tel Aviv University,
Ramat Aviv 69978, Israel

Abstract. The Conway and Sloane construction A method is applied to evaluation of a class of lattices. The points of the lattices belong to a scaled set of n-dimensional vectors with integer components and the lattices constructed with the aide of linear over $GF(p)$ codes, p is prime. It is known that the Minkowski-Hlawka bound on the normalized logarithm of packing density is attained asymptotically by such construction of lattices. We show that the capacity of an AWGN channel without restriction can be also attained by these class of lattices.

1 Introduction

The generalized capacity C_∞ of a channel with additive white Gaussian noise (AWGN) without restriction was introduced in a previous paper of the author [1]. C_∞ is the maximal value of the normalized logarithmic density (NLD) δ of an infinite constellation (IC) such that for any $\delta < C_\infty$ it is possible to find, for a sufficiently large n, an IC with arbitrarily small decoding error probability. It was shown in [1] that this generalized capacity can be attained by linear IC's, i.e., lattices. Moreover, it was shown that there are lattices such that their packing density satisfies the Minkowski-Hlawka bound and, simultaneously, the exponent of upper and lower bounds on their decoding error probability coincides asymptotically over some region of NLD close to C_∞.

Previously, Rush and Sloane [2], and Rush [3] have considered linear packings (lattices) of n-dimensional "superballs" $|x_1|^\sigma + |x_2|^\sigma + \cdots + |x_n|^\sigma \leq 1$ ($\sigma = 1, 2, \cdots$), and some other bodies, which were obtained by construction A using linear codes over $GF(p)$, where p is an odd prime ($p \to \infty$, as $n \to \infty$). In particularly, it was shown that there exist packings of "superballs" for $\sigma \leq 2$ with the density that asymptotically satisfies the Minkowski-Hlawka bound [3], and considerably exceeds this bound for $\sigma > 2$ [2], [3]. In the present work we shall show that C_∞ can also be attained by the same class of lattices.

2 Ensemble of linear, over $GF(p)$, codes

Analogously to [2] and [3] we identify the elements of $GF(p)$ with the subset of integers $\{\frac{-p+1}{2}, \cdots, 0, \cdots \frac{p-1}{2}\}$. Consequently, the Euclidean norm of $\mathbf{x} \in GF(p)^n$

* This research was supported by the Ministry of Science and Technology.

can be defind as follows $|\mathbf{x}| = \sqrt{x_1^2 + \cdots + x_n^2}$. We shall assign a linear over $GF(p)$ code by means of its check matrix H. Let H be some matrix with r rows and n columns, $r \leq n$, such that any row belongs to $GF(p)^n$. Let $\mathbf{y} = \mathbf{x}H^t$, where $\mathbf{x} \in GF(p)^n$, the multiplication is over $GF(p)$ and the superscript t denotes the transposition. Clearly, $G(p,n,r) \equiv \{\mathbf{x} : \mathbf{x}H^t = \mathbf{0},\ \mathbf{x} \in GF(p)^n\}$ is a linear over $GF(p)$ code.

For given r and n, we consider all $r \times n$ matrixes H with elements belonging to $GF(p)^n$. Let the probability distribution on this set of matrixes is defined as follows: the elements of H are chosen independently of each other according to the uniform distribution on $GF(p)$. The ensemble of linear codes over $GF(p)$ defined by this ensemble of check matrixes will be denoted by $\mathcal{G}(p,n,r)$. The following lemma is a direct generalization of Theorem 2.1 in [4].

Lemma 1. *Let B be some subset of $GF(p)^n$, $B \subset GF(p)^n, \mathbf{0} \notin B$. Then the average M_B, over the ensemble $\mathcal{G}(p,n,r)$, of the number of nonzero codewords belonging to B is given by*

$$M_B = p^{-r}|B|, \tag{1}$$

where $|B|$ denotes the cardinality of B.

Proof. Let some $\mathbf{x} \in GF(p)^n, \mathbf{x} \neq \mathbf{0}$ be fixed. The probability that $\mathbf{x} \neq \mathbf{0}$ satisfies some check, i.e., $\mathbf{x}h_i^t = 0$, where h_i is a row of the matrix H, is equal to p^{-1}. Because the rows of the matrix H are choosed independently of each other the probability that $\mathbf{x}$ satisfies all checks, i.e., $\mathbf{x}h_i^t = 0$ for all $i = 1, \cdots, r$, is equal p^{-r}. Hence the expected value, over the ensemble $\mathcal{G}(p,n,r)$, of the number of nonzero codewords belonging to B is calculated according to (1). @

In particularly, it follows from Lemma 1 that the average, over the ensemble $\mathcal{G}(p,n,r)$, of the number of codewords in a code $G(p,n,r)$ is equal to $p^{n-r} - p^{-r} + 1$.

We define $B_{n,w,p}(\mathbf{0})$ to be a discrete ball of radius w centered at $\mathbf{0}$, i.e.,

$$B_{n,w,p}(\mathbf{0}) = \{\mathbf{x} \in GF(p)^n : \ |\mathbf{x}| \leq w\} \subset GF(p)^n, \tag{2}$$

where $|\mathbf{x}|$ is the Euclidean norm of $\mathbf{x}$. Let $G(p,n,r,w_0)$ be a code such that $|\mathbf{x}| \geq w_0$ for any nonzero codeword $\mathbf{x}$.

Corollary 2. *(*The Gilbert-Varshamov bound*). For any r such that*

$$r > \log_p |B_{n,w_0-1,p}(\mathbf{0})| \tag{3}$$

there is a code $G(p,n,r,w_0)$ in the ensemble $\mathcal{G}(p,n,r)$.

Proof. If (3) holds then Lemma 1 implies that the average M_B, over $\mathcal{G}(p,n,r)$, of the number of nonzero codewords belonging to $B_{n,w_0-1,p}(\mathbf{0})$ is less than 1. There exists in $\mathcal{G}(p,n,r)$ a code possessing not more than M_B such words. The corollary now follows by the fact that the number of nonzero codewords belonging to any set must be an integer. @

Now we proceed to consider the transmission of information through an AWGN channel by means of a code $G(p,n,r)$. More precisely, we consider the transmission of words of the code $C_\alpha \equiv \alpha G(p,n,r) \equiv \{\mathbf{x} : \alpha^{-1}\mathbf{x} \in G(p,n,r)\}, \alpha > 0$. Let $\phi(\cdot)$ be the Euclidean spectrum of the code $G(p,n,r)$, i.e., $\phi(w)$ is the number of words of $G(p,n,r)$ with Euclidean norm w. The decoding error probability $\lambda_0(p,r,\alpha)$ of the code C_α, given the zero codeword was transmitted, can be bounded [1] as follows

$$\lambda_0(p,r,\alpha) \le \sum_{w \le 2\alpha^{-1}t} \phi(w)\mathbf{Pr}(z_1 \ge 0.5\alpha w, \mathbf{z} \in \mathrm{C}(t,\alpha w)) + \mathbf{Pr}(|\mathbf{z}| > t), \text{ for any } t > 0, \tag{4}$$

where $\mathrm{C}(t,\alpha w)$ is the part of the n-dimensional Euclidean sphere $V(t) = \{\mathbf{x} : |\mathbf{x}| \le t\}$ which is cut off by the hyperplane that is orthogonal to a radius of $V(t)$ and intersects it at a distance $\alpha w/2$ from the center and $\mathbf{z} = (z_1, z_2, \cdots, z_n)$ is a noise vector.

A bound on the average, over $\mathcal{G}(p,n,r)$, of $\lambda_0(p,r,)$ can be obtained by (4). First we shall prove the following lemma.

Lemma 3. *Let $\overline{\phi}(w)$ be the average, over $\mathcal{G}(p,n,r)$, of $\phi(w)$. Then*

$$\overline{\phi}(w) \le p^{-r}|V(1)|(w + 0.5\sqrt{n})^n = e^{n(0.5\ln 2\pi e(\omega+0.5)^2 - \rho\ln p + o(1))} \equiv e^{n(\eta(\omega)+o(1))}, \tag{5}$$

where $|V(1)|$ is the volume of n-dimensional Euclidean sphere of radius 1, $\omega = w/\sqrt{n}$, and $\rho = r/n$.

Proof. Let $Q = \{\mathbf{x} : |x_i| < 0.5, i = 1, \cdots, n\}$ be an n-dimensional cube centered at the point $\mathbf{0}$. Let us associate with any point $\mathbf{x} \in B_{n,w,p}(\mathbf{0})$ the cube $Q(\mathbf{x}) = Q + \mathbf{x}$. Clearly, the cubes corresponding to different points are disjoint. Thus

$$|B_{n,w,p}(\mathbf{0})| \le |V(w + 0.5\sqrt{n})| = |V(1)|(w + 0.5\sqrt{n})^n. \tag{6}$$

Now (6) implies (5) by Lemma 1 and by Stirling formula for the Gamma function. @

We have by (4) and (5)

$$\overline{\lambda}_0(p,r,\alpha) \le \sum_{w \le 2\alpha^{-1}t} e^{n(\eta(\omega)+o(1))}\mathbf{Pr}(z_1 \ge 0.5\alpha w, \mathbf{z} \in C(t,\alpha w)) + \mathbf{Pr}(|\mathbf{z}| > t), \text{for any } t > 0. \tag{7}$$

It was shown in [5] that the right side of (7) is minimized by t_r which satisfies the following equation

$$\sum_{w \leq 2\alpha^{-1}t} e^{n(\eta(\omega)+o(1))} \int_0^{\theta(w)} \sin^{n-2}\theta d\theta = \frac{\sqrt{\pi}\Gamma(\frac{n-1}{2})}{\Gamma(\frac{n}{2})}, \tag{8}$$

where $\theta(w) = \arccos(\frac{w}{2\alpha^{-1}t})$ and $\Gamma(\cdot)$ is the Gamma function.

Lemma 4. *Let $\tau_r = \frac{t_r^2}{n}$, where t_r is the solution to Equation 8. Then*

$$\tau_r = 0.5\alpha^2(\omega_r^2 + 0.25\omega_r)(1+o(1)), \tag{9}$$

where ω_r is the real positive solution to the following equation

$$e^{2\eta(\omega)} - 1 = \frac{\omega}{\omega + 0.5} \tag{10}$$

and $\eta(\omega) = 0.5 \ln 2\pi e(\omega+0.5)^2 - \rho \ln p$ (see (5)).

Proof. Let

$$\nu(\omega) = \frac{\Gamma(\frac{n}{2})}{\sqrt{\pi}\Gamma(\frac{n-1}{2})} e^{n(\eta(\omega)+o(1))} \int_0^{\theta(\omega)} \sin^{n-2}\theta d\theta. \tag{11}$$

Then (8) can be written as follows

$$\sum_{w \leq 2\alpha^{-1}t} \nu(\omega) = 1. \tag{12}$$

The integral in (11) can be approximated by the following expression [6]

$$\int_0^{\theta(\omega)} \sin^{n-2}\theta d\theta = \frac{\sin^{n-1}\theta}{(n-1)\cos\theta}(1+o(1)) = \frac{(1-\frac{\alpha^2\omega^2}{4t^2})^{\frac{n-1}{2}}}{(n-1)\frac{\alpha\omega}{2t}}(1+o(1)). \tag{13}$$

We have by (11) and (13)

$$\ln \nu(\omega) = n(0.5 \ln(1 - \frac{\alpha^2\omega 2}{4\tau^2}) + \eta(\omega) + o(1)) \equiv n(\mu(\omega) + o(1)). \tag{14}$$

Let $\tau(\omega)$ be the solution to $\mu(\omega) = 0$, i.e.,

$$\tau(\omega) = \frac{\alpha^2\omega^2 e^{2\eta(\omega)}}{4(e^{2\eta(\omega)} - 1)}. \tag{15}$$

Since $\nu(\omega)$ increases with t, (12) implies

$$\tau_r = \min_{\omega} \tau(\omega)(1+o(1)). \tag{16}$$

It is easy to check that the minimum in (16) is attained by ω_r which is the positive solution to (10). Eq.(9) now follows by (15) and (10). @.

The following theorem presents an upper bound on the average, over $\mathcal{G}(p,n,r)$, of $\lambda_0(p,r,\alpha)$.

Theorem 5. *For an AWGN channel with noise variance σ^2, the average $\overline{\lambda}_0(p,r,\alpha)$, over $\mathcal{G}(p,n,r)$, of the probability $\lambda_0(p,r,\alpha)$ satisfies the following inequality*

$$\overline{\lambda}_0(p,r,\alpha) \le e^{-n(E_{rc}(\rho,\alpha,\sigma^2)-o(1))}, \tag{17}$$

where

$$E_{rc}(\rho,\alpha,\sigma^2) \equiv \begin{cases} \frac{\alpha^2\omega_\sigma^2}{8\sigma^2} - \eta(\omega_\sigma), & \sigma^2 < \tau_r - 0.25\alpha^2\omega_r^2, \\ \frac{\tau_r-\sigma^2}{2\sigma^2} - 0.5\ln\frac{\tau_r}{\sigma^2}, & \tau_r - 0.25\alpha^2\omega_r^2 \le \sigma^2 \le \tau_r, \\ 0, & \sigma^2 > \tau_r, \end{cases} \tag{18}$$

and

$$\omega_\sigma = 0.25(\sqrt{1+64\alpha^{-2}\sigma^2} - 1). \tag{19}$$

Proof. It follows by (7) that

$$\overline{\lambda}_0(p,r,\alpha) \le 2\alpha^{-1}t_r \max_{w\le 2\alpha^{-1}t_r} \left(e^{n\eta(w)}\mathbf{Pr}(z_1 \ge 0.5\alpha w, \mathbf{z} \in \overline{V}(t_r,\alpha w)) + \mathbf{Pr}(|\mathbf{z}| > t_r) \right). \tag{20}$$

By applying Lemma 4 and the following bounds (see e.g. [1]) for the probabilities in (20)

$$\mathbf{Pr}(\sum_{i=1}^{n} z_i^2 \ge n\tau) \le \begin{cases} \exp\left(-n\left(\frac{\tau}{2\sigma^2} - 0.5\ln\frac{\tau e}{\sigma^2}\right)\right), & \tau \ge \sigma^2 \\ 1, & \tau \le \sigma^2, \end{cases} \tag{21}$$

and

$$\mathbf{Pr}\left(z_1 \ge 0.5\alpha\omega\sqrt{n}, \sum_{i=1}^{n} z_i^2 \le n\tau\right) \le$$

$$\begin{cases} \exp\left(-n\left(\frac{\tau}{2\sigma^2} - 0.5\ln\frac{(\tau-0.25\alpha^2\omega^2)e}{\sigma^2}\right)\right), & \tau - 0.25\alpha^2\omega^2 \le \sigma^2 \\ exp\left(-n\frac{\alpha^2\omega^2}{8\sigma^2}\right), & \tau - 0.25\alpha^2\omega^2 \ge \sigma^2, \end{cases} \tag{22}$$

we obtain (17). @

3 Lattices based on linear, over $GF(p)$, codes

Let L be some lattice in the n-dimensional Euclidean space R^n. For any point $\mathbf{y} \in L$ the Voronoi region $W(\mathbf{y}) \subset R^n$ is defined as follows: for any $\mathbf{v} \in W(\mathbf{y})$ we have $|\mathbf{v} - \mathbf{y}| < |\mathbf{v} - \mathbf{y}'|$ for all $\mathbf{y}' \in L, \mathbf{y}' \neq \mathbf{y}$, and for any $\mathbf{v} \notin W(\mathbf{y})$ there is an $\mathbf{y}' \neq \mathbf{y}$ such that $|\mathbf{v} - \mathbf{y}| \geq |\mathbf{v} - \mathbf{y}'|$. The Voronoi regions are congruent, i.e., $W(\mathbf{y}) = W(\mathbf{0}) + \mathbf{y}$. The value $\gamma = |W(\mathbf{0})|^{-1}$ is the lattice density. The largest number w_p such that $V(w_p) \subset W(\mathbf{0})$ is called the packing radius of the lattice L. The value $\Delta = \gamma |V(w_p)|$ is the packing density of L.

Let $G(p, n, r, w_0)$ be some linear code over $GF(p)$. The infinite constellation L_G defined by $L_G = \{\mathbf{y} : \mathbf{x} = p\mathbf{u} + \mathbf{x},\ \mathbf{u} \in Z^n, \mathbf{x} \in G(p, n, r, w_0)\} \subset Z^n$ is a lattice of construction A based on $G(p, n, r, w_0)$. (We shall call L_G "the lattice constructed on G".) For the lattice L_G we have: $|W(\mathbf{0})| \leq p^r$ ($|W(\mathbf{0})| = p^r$ iff all r rows of the check matrix H are linear independent); $w_p = 0.5 \min\{p, w_0\}$; $\Delta = p^{-r}|V(w_p)|$. Therefore, if $p \geq w_0$, then we conclude by Corollary 2 and (6) that it is possible to choose $G(p, n, r, w_0) \in \mathcal{G}(p, n, r)$ and to construct lattice L_G such that

$$\Delta \geq \frac{|V(0.5w_0)|}{|B(n, w_0, p)|} \geq 2^{-n} \left(\frac{w_0}{w_0 + 0.5\sqrt{n}} \right)^n . \tag{23}$$

If we select now $p = w_0 \geq n$ then we obtain the following bound

$$n^{-1} \log_2 \Delta \geq -1 - \log_2(1 + 0.5n^{-0.5}) = -1 - o(1), \tag{24}$$

which coincides asymptotically with the Minkowski-Hlawka bound. (This result was obtained earlier in [3]).

Let now the lattice αL_G, $\alpha > 0$, be used for transmitting information through an AWGN channel with noise variance σ^2. The Voronoi region $W(\mathbf{y}), \mathbf{y} \in \alpha L_G$, is the decision region for maximum likelihood decoding, hence the value of the decoding error probability $\lambda(\mathbf{y})$ for any point $\mathbf{y} \in \alpha L_G$ does not depend on $\mathbf{y}$ and is equal to $\lambda(\mathbf{0})$. If $p \geq 2\alpha^{-1} t_r$, where t_r is the solution to (8), then we have by (7) and Theorem 1 that there is a code $G(p, n, r) \in \mathcal{G}(p, n, r)$ such that the decoding error probability $\lambda(\mathbf{0})$ of αL_G satisfies the bound (17). The density of the lattice αL_G is equal to $\gamma = \alpha^{-n} p^{-r}$, hence $\delta \equiv n^{-1} \ln \gamma = -\ln \alpha - \rho \ln p$. It was shown in [1] that the generalized capacity of an AWGN channel is equal to $C_\infty = -0.5 \ln 2\pi e \sigma^2$. We proceed to show now that C_∞ can be attained by lattices constructed on G. Moreover, we shall show the the exponent of the upper bound on the decoding error probability of the best lattices αL_G coincides asymptotically with the exponent of the upper bound obtained in [1] for the best general lattice with the same NLD. We start with the following lemma.

Lemma 6. *The exponent $E_{rc}(\rho, \alpha, \sigma^2)$ of the bound (16) satisfies the following inequality*

$$E_{rc}(\rho, \alpha, \sigma^2) \leq E^*_{rc}(\rho, \alpha, \sigma^2) \equiv$$

$$\begin{cases} 0.5 - \eta_0^*(\omega_\sigma), & \sigma^2 < 0.5\alpha^2\omega 0^{*2}u^2 \\ \frac{\alpha^2\omega_0^{*2}-\sigma 2}{2\sigma^2} - 0.5\ln\frac{\alpha^2\omega_0^{*2}}{\sigma^2}, & 0.5\alpha^2\omega_0^{*2}u^2 \le \sigma^2 \le \alpha^2\omega_0^{*2}u^2, \end{cases} \tag{25}$$

where $\omega_0^* = \frac{p^\rho}{\sqrt{2\pi e}}$ *and* $u = 1 - 0.5\omega_0^{*-1}$.

Proof. It follows by (10) that $e^{2\eta(\omega)} > 2$, and therefore

$$\omega_r > \sqrt{2}(\omega_0^* - 0.5) = \sqrt{2}\omega_0^* u. \tag{26}$$

Let the function $\eta^*(\omega)$ be defined as follows

$$\eta^*(\omega) = \begin{cases} \eta(\omega), & \omega < \omega_0^* u, \\ 0.5\ln 2\pi e u^2\omega_0^{*2} - \rho\ln p, & \omega \ge \omega_0^* u, \end{cases} \tag{27}$$

and let

$$\tau^*(\omega) = \frac{\alpha^2\omega^2 e^{2\eta^*(\omega)}}{4(e^{2\eta^*(\omega)} - 1}. \tag{28}$$

It follows by $\eta^*(\omega) \ge \eta(\omega)$ and (15) that $\tau^*(\omega) \le \tau(\omega)$. Hence (see (16))

$$\tau_r'(\omega) \equiv \min_\omega \tau(\omega) \ge \min_\omega \tau^*(\omega) = \alpha^2\omega_0^{*2}u^2, \tag{29}$$

where the last equality is obtained by means of direct calculations of the minimum of $\tau^*(\omega)$. Inequality (25) now follows by (29) and $\omega_\sigma \ge \frac{2\sigma}{\alpha}$ (see (19)).@

Theorem 7. *Let $p \ge n$. Then there is a code $G(p,n,r)$ such that the decoding error probability λ of the lattice αL_G in an AWGN channel with noise variance σ^2 satisfies the following inequality*

$$\lambda \le e^{(-n(E(\delta,\sigma^2)-o(1))}, \tag{30}$$

where $\delta = -\ln\alpha - \rho\ln p$ is the NLD of L_G, and

$$E(\delta,\sigma^2) \equiv \begin{cases} \frac{1}{16\pi e^{2\delta+1}\sigma^2}, & \delta \le \delta_{ex} = -0.5\ln 8\pi e\sigma^2, \\ -0.5\ln 8\pi\sigma^2 - \delta, & \delta_{ex} < \delta \le \delta_{cr} = -0.5\ln 4\pi e\sigma^2, \\ \frac{e^{-2}}{4\pi e\sigma^2} + \delta + 0.5\ln 2\pi\sigma^2, & \delta_{cr} < \delta \le C_\infty = -0.5\ln 2\pi e\sigma^2. \end{cases} \tag{31}$$

Proof. Let the code $G(p,n,r)$ be fixed and let $\mathcal{A}_{w_0}$ and $\overline{\mathcal{A}}_{w_0}$ be the events of decoding into a codeword $\mathbf{x}$ with norm $|\mathbf{x}| < w_0$ and $|\mathbf{x}| \geq w_0$, respectively, given that the zero codeword was transmitted. Let $\lambda_1 = \mathbf{Pr}(\mathcal{A}_{w_0})$ and $\lambda_2 = \mathbf{Pr}(\overline{\mathcal{A}}_{w_0})$. Then we have, analogously to (7),

$$\overline{\lambda}_0(p,r,\alpha) \leq \overline{\lambda}_1 + \sum_{w_0 \leq w \leq 2\alpha^{-1}t_r} e^{n(\eta(\omega)+o(1))}\mathbf{Pr}(z_1 \geq 0.5\alpha w, \mathbf{z} \in \mathbf{V}(t_r, \alpha w)) + \mathbf{Pr}(|\mathbf{z}| > t_r). \tag{32}$$

It follows by (6) that

$$\overline{N}_0 \leq 0.25, \quad \text{for } \omega_0 = \omega_0^* u - o(1), \tag{33}$$

where $\overline{N}_0$ is the average, over $\mathcal{G}(p,n,r)$, of the number of nonzero codewords of norm less than $n\omega_0$. Therefore we obtain, by the Chebyshev inequality, that there is a code $G(p,n,r)$ in the ensemble $\mathcal{G}(p,n,r)$ such that $N_0 \leq 0.5$ and $\lambda_2 \leq 4\overline{\lambda}_2$. Clearly, $\lambda_1 = 0$ for this code.

Now, if $p \geq n$ then $u = 1 - o(1)$ and the decoding error probability λ of the lattice αL_G satisfies $\lambda \leq 4\overline{\lambda}_2$. Taking into account that, for a fixed value σ^2 of the noise variance, the restrictions on σ^2 in (25) must be converted into restrictions on δ, we obtain the statement of the theorem by Theorem 5, by Lemma 6 and by the inequality

$$-n^{-1}\ln\overline{\lambda}_2 \geq \frac{1}{16\pi e^{2\delta+1}\sigma^2} - o(1), \qquad \delta^2 \leq \delta_{ex} = -0.5\ln 8\pi e\sigma^2. \tag{34}$$

@

The exponent $E(\delta,\sigma^2)$ of (30) coincides with the exponent of the upper bound obtained in [1] for the best general lattice with the same NLD.

References

1. G. Poltyrev, "On Coding without Restrictions for the AWGN Channel", to be appear in *IEEE Trans. Information Theory.*
2. J.A. Rush and N.J. Sloane, "An Improvement to the Minkowski Hlawka Bound for Packing Superballs," *Mathematika,* vol. 34, pp. 8-18, 1987.
3. J.A. Rush, " A lower bound on packing density," *Inventiones mathematicae,* vol.98, pp. 499-509, 1989.
4. R. Gallager "Low-density parity-check codes", *M.I.T.Press, Cambridge, Massachusets,* 1963.
5. H. Herzberg and G. Poltyrev , "Techniques for Bounding the Probability of Decoding Error in Block Coded Modulation Structures", to be appear in *IEEE Trans. Information Theory.*
6. C.E. Shannon, "Probability of error for optimal codes in a Gaussian channel," *Bell Syst. Tech. J.,* vol.38, pp. 611-656, May 1959.

Quantizing and decoding for usual lattices in the L_p-metric

P. Loyer and P. Solé

I3S, 250 avenue Albert Einstein, 06560 Valbonne, France
loyer@mimosa.unice.fr, sole@mimosa.unice.fr

Abstract. We generalize the Conway-Sloane algorithm for the decoding of usual lattices to the L_p norm. In particular, the Voronoi diagrams and covering radii of some of these lattices, namely $\mathbb{Z}^n, D_n, D_n^*, A_n$, are computed. Some of the results hold only for the L_1-metric.

1 Introduction

In the context of lattice vector quantization, the encoding problem consists in finding the closest lattice point of any given point of the ambiant space. The problem occurs twice :

- when encoding the data by a vector of the codebook
- when recovering a vector of the codebook from its index.

Such an algorithm was proposed by Conway and Sloane for the euclidean distance and for certain lattices (see [4], [2, chapter 20]). Here, we generalize it to another class of metrics : the so-called L_p-norms.

The engineering motivations of our work are well developped in [8] and our paper is a partial response to the questions pointed out in the paper cited. In the first application, the motivation is to have a more general distorsion criterion than the mean square error. In the second application it is to use a voronoi codebook better suited to the law of the source. For example, the euclidean norm is relevant for gaussian sources while the L_1-norm is relevant for laplacian sources. Some authors even use the $L_{0.7}$-norm (see [6]) !

A more general application is to obtain a characterization of geometric parameters of lattices under any L_p-norm like the voronoi diagram, the covering radius and so on.

The lattices we will consider are some of the (irreducible) root lattices. Root lattices are of great practical interest because of their regularity. The voronoi cell consists of a fundamental polytope and of its reflexions through its own roofs. All information about that is summarized in an efficient manner by Coxeter diagrams. Then, the parameters of the lattice can be easily computed (see [5], [1], [3]).

2 Case of $\mathbb{Z}^n, D_n, A_n$

2.1 Background on L_p-norms

Assume that E is a real vector space of finite dimension n, with a given basis. $(x_1 \cdots x_n)$ denotes the coordinates of a vector x on this basis. We can define the so-called L_p-*norm* on E : if p is a nonnegative real, we set

$$\|x\|_p = (\sum_{i=1}^{n} |x_i|^p)^{1/p}$$

where $\|x\|_p$ denotes the L_p norm of x.

Strictly, L_p is a norm, in topological sense, iff $p \geq 1$ (see [7]).

For $p = 2$, we recognize the classical euclidean norm. This is sometimes defined without the square root. But here, we need it, among other reasons, to generalize the definition to $p = \infty$.

When p tends to infinity, $\|x\|_p$ converges to an other norm, denoted $\|x\|_\infty$, which is equal to

$$\sup_{i=1\ldots n} |x_i|.$$

2.2 Case of $\mathbb{Z}^n$: the fundamental result

Each of these norms gives rise to a particular sense of the notion of nearest neighbour and we shall investigate it in a first step before computing the parameters we are interested in. The basic lattice for our investigation is $\mathbb{Z}^n$. For any real x, let $f(x)$ equal the closest integer and for any vector $x \in \mathbb{R}^n$ let

$$f(x) = (f(x_1), \ldots, f(x_n)).$$

By an integer point, we mean a point in $\mathbb{Z}^n$.

Proposition 1. *$f(x)$ is the closest integer point to x for every L_p-norm, including L_∞. Then the voronoi diagram of $\mathbb{Z}^n$ is independent of p, the voronoi cell being an hypercube of side 1.*

Proof. let us compare $f(x)$ with any other integer point, namely $k = (k_1, \ldots, k_n)$. By definition, the i-th coordinate of $f(x)$ is $f(x_i)$, chosen so that $|f(x_i) - x_i| \leq |k_i - x_i|$. Then,when p is finite, we deduce successively

$$\begin{aligned} |f(x_i) - x_i| \leq |k_i - x_i| \Rightarrow & \quad |f(x_i) - x_i|^p \leq |k_i - x_i|^p \\ & \Rightarrow \textstyle\sum_{i=1}^{n} |f(x_i) - x_i|^p \leq \sum_{i=1}^{n} |k_i - x_i|^p \\ & \Rightarrow \quad \|f(x) - x\|_p \leq \|k - x\|_p. \end{aligned}$$

When p is infinity, we have

$$\begin{aligned} (\forall i\ |f(x_i) - x_i| \leq |k_i - x_i|) \Rightarrow & \sup |f(x_i) - x_i| \leq \sup |k_i - x_i| \\ \Rightarrow & \quad \|f(x) - x\|_\infty \leq \|k - x\|_\infty. \end{aligned}$$

□

Let us give an example : let x be a vector of $\mathbb{R}^3$

$$x = (0.12, 1.87, -5.3)$$

Then

$$f(x) = (0, 2, -5)$$

is the closest point of x, *independently of* p (for the case of $p = 2$, see [4]).

The distance from the center of a cell to any vertex gives the covering radius because the vertices maximize each of the coordinates. The argument holds for every L_p-norm, but the value is not the same. Considering V(O), the voronoi-cell centered in O, and the vertex $(\frac{1}{2}, \ldots, \frac{1}{2})$, we obtain, when p is finite

$$R_p = \frac{n^{\frac{1}{p}}}{2}$$

and when p is infinity

$$R_\infty = \lim_{p \to \infty} R_p = \frac{1}{2}$$

which can also be computed as $\sup |x_i|$.

2.3 Case of D_n : quantizing method, voronoi-cell and covering radius

D_n consists of integer points in $\mathbb{R}^n$ whose coordinates add up to an even number.

Given a vector x in $\mathbb{R}^n$ we will show that its closest point in D_n can be choosen among its two closest points in $\mathbb{Z}^n$. Let us define $g(x)$ by rounding all x_i to the next integer (namely $f(x_i)$) except of the one for wich $|f(x_i) - x_i|$ is the greatest (in case of a tie any one can be choosen), wich we round the wrong way. For the vector x given in the previous example, we have

$$g(x) = (0, 2, -6).$$

Proposition 2. *$g(x)$ is the second closest integer point to x for every L_p-norm, including L_∞.*

Proof. Consider a point

$$k = (k_1, \ldots, k_n) \in \mathbb{Z}^n - \{f(x)\}.$$

Denoting $g_i(x)$ the i-th coordinate of $g(x)$, we have by definition, $|g_i(x) - x_i| \leq |k_i - x_i|$. From there, like for $f(x)$ in $\mathbb{Z}^n$, we can deduce that $g(x)$ is the closest point to x in $\mathbb{Z}^n - \{f(x)\}$ or the second closest point in $\mathbb{Z}^n$. That holds for every p, possibly infinite. □

Note that the two coordinates sums of $f(x)$ and $g(x)$ differ of exactly one unit, so that one is even. In other words, it belongs to D_n lattice (see [2, chapter 4]). That leads to a simple algorithm to determine the nearest neighbour in D_n of a real point (we mean a point in $\mathbb{R}^n$) :

Algorithm 1 *:*

1. *compute the two candidates $f(x)$ and $g(x)$;*
2. *choose the one whose coordinates add up to an even number.*

The output point is clearly independent of p. This allows to state the

Proposition 3. *the closest point to x in D_n does not depend on the L_p-norm, p possibly being infinity. Thus the voronoi-diagram as well does not depend on the L_p-norm. It is the same as for the L_2-norm.*

The voronoi cell is described by coxeter diagram of figure 1 (see [2, chapter 21], [5]).

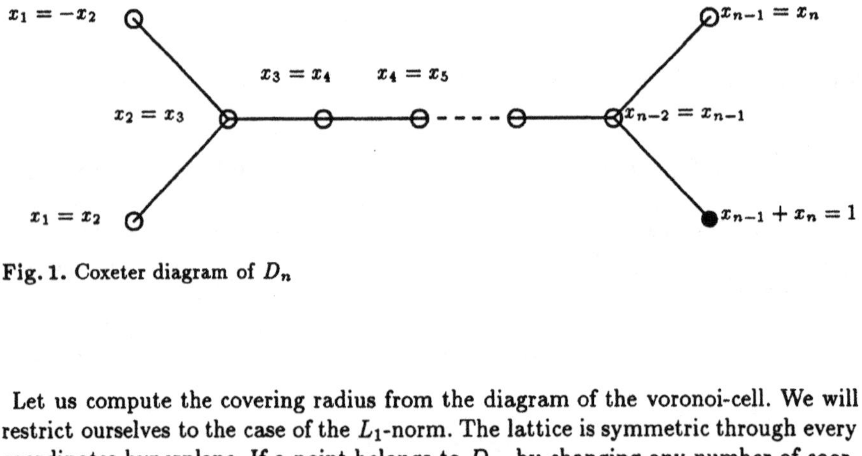

Fig. 1. Coxeter diagram of D_n

Let us compute the covering radius from the diagram of the voronoi-cell. We will restrict ourselves to the case of the L_1-norm. The lattice is symmetric through every coordinates hyperplane. If a point belongs to D_n, by changing any number of coordinates into their opposites, we will find a point in D_n again. Thus we can look for the longest vector in the voronoi-cell only in the positive orthant, defined by $x_i \geq 0$ for all i. The idea is that, in this orthant, the L_1-norm of a vector is nothing else than the sum of its coordinates. As shown on the coxeter diagram, the hyperplanes bounding the voronoi-cell in this region are the ones of equation $x_i + x_j = 1$. In particular, we have

$$\begin{aligned} x_1 + x_2 &\leq 1 \\ x_2 + x_3 &\leq 1 \\ &\dots \\ x_n + x_1 &\leq 1. \end{aligned}$$

Adding up, we see that

$$\sum_{i=1}^{n} x_i \leq \frac{n}{2}.$$

There is one point in the cell which reaches this bound, namely $(\frac{1}{2}, \ldots, \frac{1}{2})$. Thus it is a longest vector in it. Then the covering radius is

$$R_1 = \frac{n}{2}.$$

2.4 Case of A_n : quantizing method and voronoi-cell

A_n consists of integer points of $\mathbb{R}^{n+1}$ whose coordinates add to 0 (see [2, chapter 4]). Thus A_n is contained in the hyperplane

$$H : \sum_{i=1}^{n+1} x_i = 0$$

Since A_n is defined by the mean of a space of upper dimension, the algorithm of quantization is quite more complicated.

Let x be a point of $\mathbb{R}^{n+1}$ and assume $x \in H$. Looking for its nearest neighbour in A_n we proceed as follows :

Algorithm 2 :

1. *compute the closest integer point $f(x)$ and the deficiency*

$$\Delta = \sum_{i=1}^{n+1} f(x_i)$$

2. *sort the coordinates of $f(x)$ in increasing order of the difference $x_i - f(x_i)$ (we still denote the result $f(x)$);*
3. *discuss on Δ :*
   - *if $\Delta = 0$: end ;*
   - *if $\Delta < 0$ we add one to the $(-\Delta)$ last coordinates of $f(x)$ for the new deficiency to be null ;*
   - *if $\Delta > 0$ we substract one to the Δ first coordinates of $f(x)$ for the new deficiency to be null.*

(At the end, it is necessary to permute back the coordinates of the resulting vector).

Δ can be interpreted as a gap between $f(x)$ and A_n.

We have chosen the first or the last of the coordinates of the reordered $f(x)$ so that $|f(x_i) - x_i|$ is maximal and then, the distance from $f(x_i)$ to the second nearest integer minimal. Thus we have made the smallest change to $\|f(x)\|_p$ that brings the vector in A_n. From this remark, we deduce the

Proposition 4. *when x is in H, the closest point to x in A_n does not depend on the L_p-norm, p possibly being infinity. Thus the voronoi diagram as well does not depend on p. It is the same as for the L_2-norm.*

The case of x not being in H is not of interest for practical engeeniring.

3 Lattices defined as a union of cosets

3.1 Quantization method

General case : we become able to investigate more complicated lattices, those defined as union of cosets of the previous ones. Suppose the general case of a lattice Λ defined as a union of cosets of one of its sublattices, say Λ' :

$$\Lambda = \bigcup_{i=1}^{k} (\Lambda' + \boldsymbol{r}_i).$$

One of the $\boldsymbol{r}_i$ is generally the 0-vector what means that one of the cosets is nothing else than Λ' itself. Let $\boldsymbol{x}$ be a vector of $\mathbb{R}^n$, $\psi(\boldsymbol{x})$ be its nearest neighbour in Λ' and $\psi_i(\boldsymbol{x})$ be its nearest neighbour in $\Lambda' + \boldsymbol{r}_i$, we obtain the nearest neighbour of $\boldsymbol{x}$ in Λ by the following

Algorithm 3 *:*

1. *for all i compute*

$$\psi_i(\boldsymbol{x}) = \psi(\boldsymbol{x} - \boldsymbol{r}_i) + \boldsymbol{r}_i.$$

We thus obtain n candidates ;

2. *for each candidate compute the error*

$$\varepsilon_i = ||\psi_i(\boldsymbol{x}) - \boldsymbol{x}||_p$$

and choose the candidate which minimizes it.

The first step traduces the fact that $\boldsymbol{x}$ is to $\Lambda' + \boldsymbol{r}_i$ what $\boldsymbol{x} - \boldsymbol{r}_i$ is to Λ'.

The procedure can be applied for every L_p-norm but the result is generally different (for the case of $p = 2$, see [4]). We shall detail this remark in the following section.

Applications : the A_n^*, D_n^*, D_n^+ and E_8 lattices correspond to this description (for more details about this paragraph, see [2, chapter 4], [2, chapter 21], [4]).

A_n^* is the dual of A_n. Since A_n is an integral lattice (the dot product of any two vectors is integer) (see [2, chapter 2], [3]), it is a sublattice of its dual and the latter can be described as a union of cosets of the the former :

$$A_n^* = \bigcup_{i=0}^{n} (A_n + [i])$$

where

$$[i] = ((\frac{i}{n+1})^j, (\frac{j}{n+1})^i), \; i + j = n + 1.$$

D_n^* is the dual of D_n. Like for A_n, D_n^* can be described as a union of cosets of D_n. But there is a more pleasant description by cosets of $\mathbb{Z}^n$, namely :

$$D_n^* = \mathbb{Z}^n \cup (\mathbb{Z}^n + (\tfrac{1}{2} \cdots \tfrac{1}{2}\,))$$

This writing uses only two cosets while the one with D_n uses four. Furthermore, the repartition of the points appears more clearly.

D_n^+ is defined as

$$D_n^+ = D_n \cup (D_n + (\tfrac{1}{2} \cdots \tfrac{1}{2}))$$

It is a lattice iff n is even because we then clearly have an additive subgroup of $\mathbb{R}^n$. E_8 is nothing else than D_8^+. It is of special interest, among other reasons, because it gives the best sphere packing in dimension 8.

3.2 Limits of the method

The chosen candidate and then the notion of nearest neighbour and the voronoi diagram generally depend on p. The reason why is that the mediator of two points is no longer the same when we take an other norm than the euclidean norm. We shall not only give a counterexample in dimension 2 for the L_1-norm but also describe the mediator of two points in sense of this norm.

Given two points A and B, we compute their mediator by starting from their middle (which belongs to the mediator), say O. We let this point move of a certain vector $\boldsymbol{\Delta}$ and study whether we stay on the mediator. We compare $\|\mathbf{AO}+\boldsymbol{\Delta}\|_1$ and $\|\mathbf{BO}+\boldsymbol{\Delta}\|_1$ and thus obtain a condition on $\boldsymbol{\Delta}$ for the two norms to be equal.

The L_p-norms do not depend on the origin of the space we are working in. Then we can take O as the origin. Then

$$\|\mathbf{AO}+\boldsymbol{\Delta}\|_1 = \|B+\boldsymbol{\Delta}\|_1$$

and

$$\|\mathbf{BO}+\boldsymbol{\Delta}\|_1 = \|A+\boldsymbol{\Delta}\|_1.$$

We compute

$$\|B+\boldsymbol{\Delta}\|_1 - \|A+\boldsymbol{\Delta}\|_1$$

according to the coordinates of A and B. We set

$$A(-x,-y),\ B(x,y),\ \boldsymbol{\Delta} = \Delta_x \imath + \Delta_y \jmath$$

We can suppose $0 \leq x \leq y$ without loss of generality (of course, x and y are not both 0). The other cases are easily deduced from this one. We have :

$$\begin{aligned}\|A+\boldsymbol{\Delta}\|_1 &= |-x+\Delta_x| + |-y+\Delta_y| \\ \|B+\boldsymbol{\Delta}\|_1 &= |x+\Delta_x| + |y+\Delta_y| \\ \|B+\boldsymbol{\Delta}\|_1 - \|A+\boldsymbol{\Delta}\|_1 &= (|x+\Delta_x| - |-x+\Delta_x|) \\ &\quad +(|y+\Delta_y| - |-y+\Delta_y|)\end{aligned}$$

Let us summarize the variations of the x-term in the following table :

Δ_x	$-\infty$ … $-x$	$-x$ … x	x … ∞
$\|x+\Delta_x\|$	$-x-\Delta_x$	$x+\Delta_x$	$x+\Delta_x$
$\|-x+\Delta_x\|$	$x-\Delta_x$	$x-\Delta_x$	$-x+\Delta_x$
$\|x+\Delta_x\|-\|-x+\Delta_x\|$	$-2x$	$2\Delta_x$	$2x$

We have similar variations for the y-term. We compute the difference and summarize all the results in figure 2:

The following cases are to be excluded :

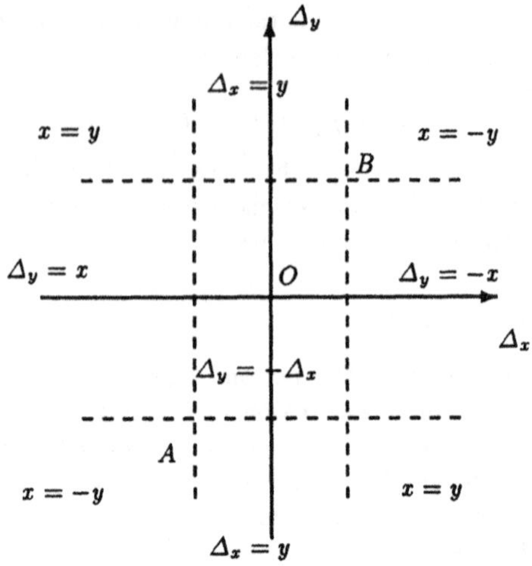

Fig. 2. value of $\|B + \Delta\|_1 - \|A + \Delta\|_1$

- $x = -y$ because x and y are both positive and not both null ;
- $\Delta_x = y$ because it occurs in the interval $-x \leq \Delta_x \leq x$. Combined with $x \leq y$ it gives rise to an absurdity

Whether $x = y$ or not, we obtain two different kinds of mediator, represented on figures 4 and 3 respectively.

It is a remarkable fact that in the first case, the fronter of $V(A) \cap V(B)$ is of no-null area.

It is now clear that the notion of nearest neighbour is not the same for the L_1-norm as for the L_2 one and then that the Voronoi diagram is generally not the same. We shall however show in the following paragraph that in the particular case of D_n^* it is the same again.

3.3 Particular case of D_n^*

Proposition 5. *the Voronoi diagram of D_n^* is the same for the L_1-norm as for the L_2-norm.*

Proof. we use the description of D_n thanks to cosets of $\mathbb{Z}^n$. The only case to be considered for the candidates is O and $(\frac{1}{2} \cdots \frac{1}{2})$ and the only region in which the choice raises a problem is the intersection of the Voronois i.e. an hypercube whose two opposites vertices are the candidates.

By moving the origin to the middle of those points, we obtain two points whose coordinates are all equal. It is the generalization in dimension n for the case $x = y$ in the previous study. Then the mediator of the two points, restricted to the region of interest, in sense of the L_1-norm is the hyperplane $\sum x_i = 0$, i.e. it is nothing else

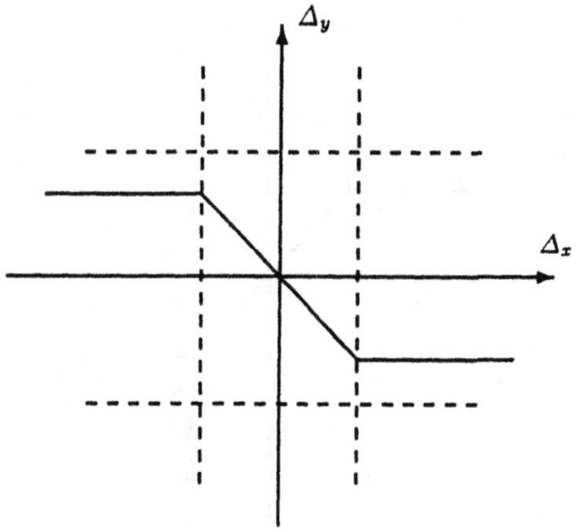

Fig. 3. $x \neq y$

than the mediator for the L_2-norm. Thus the choice between the two candidates will be the same either for one norm or for the other and the notion of nearest neighbour is the same. □

3.4 Why such a method does not hold for D_n^+

One could think that such a result would hold for D_n^+. But it does not. Let us give a counterexample in dimension 2. We denote C the $(\frac{1}{2}, \frac{1}{2})$ point. Consider the point $P(-\frac{1}{8}, \frac{3}{4})$ (see figure 5). Applying algorithm 1 we find that the closest point of P in D_2 is O and applying algorithm 3 we also find that the closest point of P in $D_2 + C$ is C.

But

$$\|\mathbf{OP}\|_1 = \frac{7}{8} = \|\mathbf{CP}\|_1$$

and

$$\|\mathbf{OP}\|_2^2 = \frac{37}{64}$$
$$\|\mathbf{CP}\|_2^2 = \frac{29}{64}$$
$$\|\mathbf{OP}\|_2 > \|\mathbf{CP}\|_2$$

C is the closest neighbour to P in the sense of L_2 but not of L_1. The result can easily be generalized to dimension n by adding $n-2$ coordinates to $\frac{1}{4}$. Thus we can conclude that the Voronoi diagram of D_n^+ is never the same for L_1 and L_2.

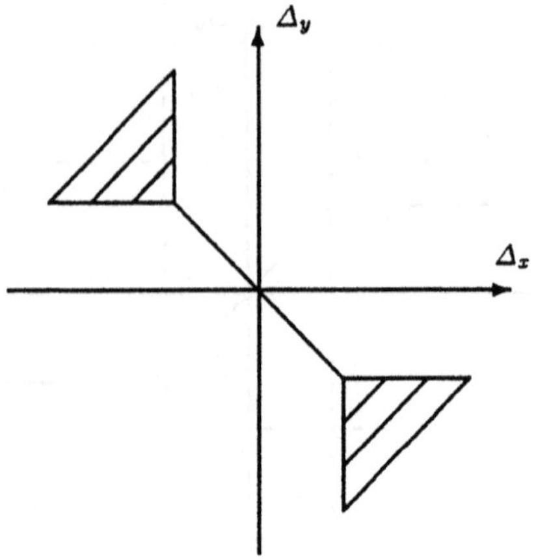

Fig. 4. $x = y$

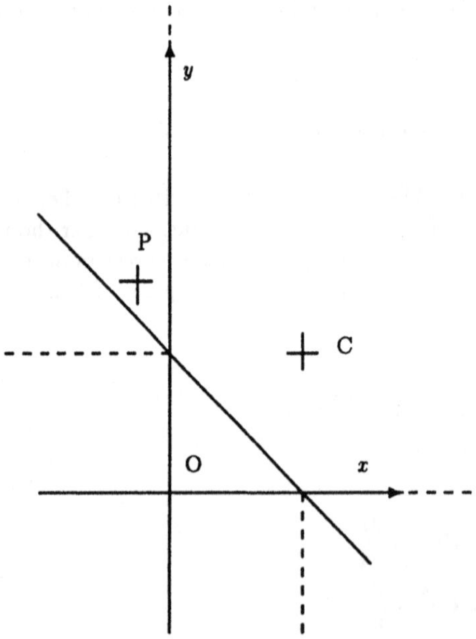

Fig. 5. counterexample for D_n^+

4 Conclusion

Making a simple remark, we have deduced that Conway's and Sloane's algorithm for decoding could be applied unchanged for any L_p-norm to each of the lattices they have studied in [4]. Furthermore, the output point in the algorithms is often the same as for the L_2-norm. Then the Voronoi diagram is the same and the covering radius of the lattice can be computed by using similar methods. In a forthcoming paper, we will give supplementary results concerning this parameter.

References

1. N. Bourbaki. *Groupes et algèbres de Lie, Chapitres 4, 5 et 6.* Hermann, 1968.
2. J.H. Conway and N.J.A. Sloane. *Spheres packings, lattices and groups.* Springer-Verlag, 1st edition, 1988.
3. J.H. Conway and N.J.A. Sloane. The cell structures of certain lattices. In P. Hilton, F. Hirzebruch, and R. Remmert, editors, *Miscellanea mathematica*, pages 71–107. Springer-Verlag, 1991.
4. J.H. Conway and N.J.A. Sloane. Fast quantizing and decoding algorithms for lattices quantizers and codes. *IEEE Trans. Inform. Theory*, 28(2):227–232, March 92.
5. H.S.M. Coxeter. *Regular Polytopes.* Dover, 3rd edition, 1973.
6. N. Farvardin and J.W. Modestino. Optimum quantizer performance for a class of non-gaussian memoryless sources. *IEEE Trans. Inform. Theory*, 30:485–497, May 84.
7. Robert M. Gray. *Source coding theory.* Kluwer academic publishers, 1990.
8. M. Vedat Eyuboğlu and G. David Forney. Lattice and trellis quantization with lattice-and trellis-bounded codebooks–high-rate theory for memoryless sources. *IEEE Trans. Inform. Theory*, 39(1):46–59, January 93.

Bounded-Distance Decoding of the Leech Lattice and the Golay Code

Ofer Amrani[1] Yair Be'ery[1] and Alexander Vardy[2] **

[1] Tel-Aviv University
Department of Electrical Engineering - Systems
Ramat-Aviv 69978
Tel-Aviv, Israel

[2] IBM Research Division
Almaden Research Center
650 Harry Road
San Jose, CA 95120

Abstract. We present an efficient algorithm for bounded-distance decoding of the Leech lattice. The new bounded-distance algorithm employs the partition of the Leech lattice into four cosets of Q_{24}, beyond the conventional partition into two H_{24} cosets. The complexity of the resulting decoder is only 1007 real operations in the worst case, as compared to about 3600 operations for the best known maximum-likelihood decoder and about 2000 operations for the original bounded-distance decoder of Forney. Restricting the proposed Leech lattice decoder to $GF(2)^{24}$ yields a bounded-distance decoder for the binary Golay code which requires at most 431 operations as compared to 651 operations for the best known maximum-likelihood decoder. Moreover, it is shown that our algorithm decodes correctly at least up to the guaranteed error-correction radius of the Leech lattice. Performance of the algorithm on the AWGN channel is evaluated analytically by explicitly calculating the effective error-coefficient, and experimentally by means of a comprehensive computer simulation. The results show a loss in coding-gain of less than 0.1 dB relative to the maximum-likelihood decoder for BER ranging from 10^{-1} to 10^{-7}.

* This work together with an independent work by Feng-wen Sun and Henk van-Tilborg will appear in the Information Theory Trans. as a joint paper entitled : "The Leech lattice and the Golay code: Bounded distance decoding and multilevel constructions."

**Research supported in part by the Eshkol Fellowship administered by the Israel Ministry of Science and in part by the Rothschild Fellowship administered by the Rothschild Yad Hanadiv Foundation.

1. Introduction

The Leech lattice Λ_{24} is certainly one of the most interesting and well studied lattices (cf. [4]). Maximum-likelihood decoding of the Leech lattice has been intensively investigated: Conway and Sloane [3], Forney [5], Lang and Longstaff [7], Be'ery, Shahar, and Snyders [2], and Vardy and Be'ery [12] have devised various decoding algorithms with complexities ranging from 55968 down to 3595 operations. While the problem of maximum-likelihood decoding of Λ_{24} is interesting in its own right, in practice it may be rewarding to use a slightly suboptimal but more efficient bounded-distance decoding algorithm. One such algorithm was developed by Forney [6]. The computational complexity of Forney's original algorithm is somewhat less than 2000 operations. Since Forney's decoder is based on soft-decision decoding of the binary Golay code, this complexity may be reduced by incorporating the latest Golay decoder of [11] in the algorithm of [6]. This results in a bounded-distance Leech lattice decoder which requires about 1500 operations. In this correspondence we show that a more efficient bounded-distance decoder for the Leech lattice is possible, by introducing an algorithm with an average complexity of only 953 operations and worst case complexity of at most 1007 operations. Furthermore, it is shown that this algorithm achieves the error-correction radius of the Leech lattice.

Our decoding algorithm is based on the construction of Λ_{24} from the quaternary (6,3,4) hexacode H_6, described in [12]. We now briefly review the relevant notation of [12], as so far as it's essential to enable the reader follow the algorithm. Let D_2 be the two-dimensional checkerboard lattice. Partition D_2 into 16 subsets and label each subset with either A_{ijk} or B_{ijk}, where $i, j, k \in GF(2)$. The correspondence between the points of D_2 and the labels of the 16 subsets is established by tiling the plane with scaled and rotated version of the following constellation:

$$\begin{array}{cccc} A_{000} & B_{000} & A_{110} & B_{110} \\ B_{100} & A_{010} & B_{010} & A_{100} \\ A_{111} & B_{111} & A_{001} & B_{001} \\ B_{011} & A_{101} & B_{101} & A_{011} \end{array} \tag{1}$$

We shall represent the points of Λ_{24} by 2×6 arrays whose entries are the points of D_2, such as

$$\begin{bmatrix} A_{i_1j_1k_1} & A_{i_3j_3k_3} & \cdots & A_{i_{11}j_{11}k_{11}} \\ A_{i_2j_2k_2} & A_{i_4j_4k_4} & \cdots & A_{i_{12}j_{12}k_{12}} \end{bmatrix} \tag{2}$$

The array in (2) is called type–A since it contains only A_{ijk} points. Similarly, type–B array will consist of only B_{ijk} points. Let $[A_{i_1j_1k_1}, A_{i_2j_2k_2}]^t$ be a column of a type–A array and let $u = (i_1, j_1, i_2, j_2)$ be the corresponding binary 4-tuple.

If u contains an even number of nonzeros then the column is said to be even, otherwise the column is said to be odd. The index i_1 is called the h-parity of the column. The overall h-parity of the array is defined as the modulo-2 sum of the h-parities of the six columns. The overall k-parity of the array is the modulo-2 sum of the k subscripts of the 12 points. As in [4, 11], any binary 4-tuple u is regarded as an *interpretation* of a character $x \in \{0,1,\omega,\bar{\omega}\} = GF(4)$. Conversely any $x \in \{0,1,\omega,\bar{\omega}\}$ may be regarded as the *projection* of four different binary 4-tuples u, such that $x = (0,1,\omega,\bar{\omega}) \cdot u$. The projection of a 2×6 array, such as (2), is a vector $\underline{x} \in GF(4)^6$ consisting of the projections of the six columns. Conversely, any such 2×6 array which projects on $\underline{x} \in GF(4)^6$ is called an *image* of $\underline{x}$.

Using the above notation the Leech lattice may be defined as follows (cf. [12]). The Leech lattice is the set of all the 2×6 arrays whose entries are points of D_2, such that each array satisfies the following conditions:

a. *It is either type–A or type–B.*
b. *It consists either of only even columns or only odd columns.*
c. *The overall k-parity is even if the array is type–A, and odd otherwise.*
d. *The overall h-parity is even if the array columns are even, and odd otherwise.*
e. *The projection of the array is a codeword of H_6.*

Note that by restricting condition (**a**) of the foregoing definition, that is taking only the type–A arrays, the Leech *half-lattice* H_{24} is obtained. Further restricting condition (**b**), that is taking only the even columns, produces the Leech *quarter-lattice* Q_{24} as defined in [12].

In the next section we describe the decoding algorithm and show that it decodes correctly at least up to the guaranteed error-correction radius of the Leech lattice. Bounded-distance decoding of the binary Golay code is discussed in Section 3. In Section 4 we compute the effective error-coefficient for our Leech lattice decoder. Finally, Section 5 presents the simulation results.

2. Bounded-distance decoding of the Leech lattice

For the sake of simplicity we shall assume an AWGN channel model, even though the algorithm in the sequel extends to other channels as well. Suppose that a point $\lambda \in \Lambda_{24}$ was transmitted and a vector $\rho \in R^{24}$ was observed at the channel output. We shall write $\rho = (\rho_1, \rho_2, \ldots \rho_{12})$, where $\rho_1, \rho_2, \ldots \rho_{12} \in R^2$. In fact in most practical situations, such as QAM signaling, the channel output would be specified explicitly in terms of the sequence $\rho_1, \rho_2, \ldots \rho_{12}$ of two-dimensional symbols.

2.1. The decoding algorithm

As in [12] the decoder for Λ_{24} consists of four separate decoders for the cosets of Q_{24}, operating concurrently. Since the four decoders are essentially identical, we describe only the decoder for Q_{24} which operates on type–A arrays consisting of even columns.

Precomputation: For $n = 1, 2, \ldots 12$ find in each A_{ijk}-subset of D_2 a point $\hat{A}_{ijk}(n)$ which is closest to ρ_n, and set this point as a representative of the entire subset. Let $d_{ijk}(n)$ denote the squared Euclidean distance (SED) between the representative point $\hat{A}_{ijk}(n)$ and the received symbol ρ_n. Further, set

$$d_{ij}(n) := \min\{d_{ij1}(n), d_{ij0}(n)\} \qquad for \quad n = 1, 2, \ldots 12 \tag{3}$$

As in [2, 5, 6, 12] the 48 metrics $d_{ij}(n)$ are assumed to be available directly from the geometry of the signaling constellation. These metrics constitute the input to our decoder.

Step 1: Computing the confidence values of the characters.

As in [12] we compute for each character $x \in GF(4)$ and for each of the six coordinates of the hexacode $l = 1, 2, \ldots 6$, the *confidence value* of x in the l-th coordinate. The decoder for Q_{24} computes only the even-interpretation confidence values, given by

$$\mu_l^e(x) := \min\{\, d_{i_1 j_1}(2l-1) + d_{i_2 j_2}(2l),\; d_{\bar{i}_1 \bar{j}_1}(2l-1) + d_{\bar{i}_2 \bar{j}_2}(2l)\,\} \tag{4}$$

where (i_1, j_1, i_2, j_2), $(\bar{i}_1, \bar{j}_1, \bar{i}_2, \bar{j}_2)$ are the two even interpretations of x, and $\bar{\alpha} = 1 \oplus \alpha$ is the complement of $\alpha \in GF(2)$. The decoder for the coset of Q_{24} in H_{24} computes the odd-interpretation confidence values $\mu_l^o(x)$ (cf. [12]).

Step 2: Computing the confidence values of the blocks.

For each $\underline{x} = (x_1, x_2, x_3, x_4, x_5, x_6) \in GF(4)^6$ we compute

$$\begin{aligned} S_1^e(x_1, x_2) &:= \mu_1^e(x_1) + \mu_2^e(x_2) \\ S_2^e(x_3, x_4) &:= \mu_3^e(x_3) + \mu_4^e(x_4) \end{aligned} \tag{5}$$

and subsequently sort the 16 values of $S_1^e(x_1, x_2)$, respectively $S_2^e(x_3, x_4)$, in ascending order. The decoder for the coset of Q_{24} in H_{24} will compute and sort $S_j^o(x_{2j-1}, x_{2j})$ for $j = 1, 2$, using the odd-interpretation confidence values.

Step 3: Minimization over the hexacode.

For each $\underline{x} = (x_1, x_2, x_3, x_4, x_5, x_6) \in H_6$ we compute the (even type–A) metric of $\underline{x}$, as follows

$$M(\underline{x}) \;:=\; S_1^e(x_1, x_2) + S_2^e(x_3, x_4) + \mu_5^e(x_5) + \mu_6^e(x_6) \tag{6}$$

and subsequently select $\widehat{\underline{x}} \in H_6$, such that $M(\widehat{\underline{x}})$ is minimum among the 64 hexacodewords.

We presently construct an (even type–A) image of $\widehat{\underline{x}}$, denoted by $\mathcal{I}(\widehat{\underline{x}})$, such that the squared Euclidean distance between $\mathcal{I}(\widehat{\underline{x}})$ and the received signal ρ is equal to the metric $M(\widehat{\underline{x}})$. Although this procedure does not require any decoding operations,* it demands some explanation. Assume w.l.o.g. that $\mu_1^e(x) = d_{i_1j_1}(1) + d_{i_2j_2}(2)$. Set $k_1 := 0$ if $d_{i_1j_1}(1) = d_{i_1j_10}(1)$, and $k_1 := 1$ otherwise. Let k_2 be similarly defined. Then the SED between the column $[\hat{A}_{i_1j_1k_1}(1), \hat{A}_{i_2j_2k_2}(2)]^t \in R^4$ and $(\rho_1, \rho_2) \in R^4$ is equal to $\mu_1^e(x)$. Moreover this column is closer to (ρ_1, ρ_2) than any other even column that projects on x. Hence, as in [12], this column is called the *preferable* (even type–A) *representation* of the character x in the first coordinate. The discussion extends in the obvious way to the other coordinates as well.

It is now evident that $\mathcal{I}(\widehat{\underline{x}})$ may be reconstructed from $\widehat{\underline{x}}$ by taking the preferable even type–A representations of the characters of $\widehat{\underline{x}}$ in the six coordinates. Furthermore, it is easy to see that among all the even type–A arrays that project on the hexacode, $\mathcal{I}(\widehat{\underline{x}})$ is the closest to ρ. However, the point of R^{24} described by $\mathcal{I}(\widehat{\underline{x}})$ does not necessarily belong to Λ_{24} as conditions (**c**) and (**d**) of the definition may be violated. In the next two steps we coerce $\mathcal{I}(\widehat{\underline{x}})$ into a point of $Q_{24} \subset \Lambda_{24}$.

Step 4: Resolving the *h*-parity.

Let η denote the overall h-parity of $\mathcal{I}(\widehat{\underline{x}})$. If $\eta \neq 0$ we compute for $l = 1, 2, \ldots 6$ the h-penalties

$$\Delta(\widehat{x}_l) \;:=\; d_{\bar{\imath}_1\bar{\jmath}_1}(2l-1) \;+\; d_{\bar{\imath}_2\bar{\jmath}_2}(2l) \;-\; d_{i_1j_1}(2l-1) \;-\; d_{i_2j_2}(2l) \tag{7}$$

where $(\widehat{x}_1, \widehat{x}_2, \widehat{x}_3, \widehat{x}_4, \widehat{x}_5, \widehat{x}_6) = \widehat{\underline{x}}$ and (i_1, j_1, i_2, j_2), $(\bar{\imath}_1, \bar{\jmath}_1, \bar{\imath}_2, \bar{\jmath}_2)$ are the even interpretations of $\widehat{x}_l$ as in (4), with (i_1, j_1, i_2, j_2) being the preferable of the two. Let $s \in \{1, 2, \ldots 6\}$ be such that $\Delta(\widehat{x}_s)$ is the smallest of the six penalties. We shall modify $\mathcal{I}(\widehat{\underline{x}})$ by replacing $[\hat{A}_{i_1j_1k_1}(2s-1), \hat{A}_{i_2j_2k_2}(2s)]^t$ with $[\hat{A}_{\bar{\imath}_1\bar{\jmath}_1k_1}(2s-1), \hat{A}_{\bar{\imath}_2\bar{\jmath}_2k_2}(2s)]^t$. The overall h-parity of the resulting array $\mathcal{I}'(\widehat{\underline{x}})$ is even and the

---

*In accordance with a well established tradition (cf. [2, 3, 5, 6, 7, 11, 12]) we count only operations on real values. As in [2, 3, 5, 6, 7, 11, 12] various logic operations are neglected, though none of these is allowed to be excessive.

corresponding metric is $M'(\hat{\underline{x}}) := M(\hat{\underline{x}}) + \Delta(\hat{x}_s)$. If $\eta = 0$ then $\mathcal{I}(\hat{\underline{x}})$ agrees with condition (**d**), and hence we set $\mathcal{I}'(\hat{\underline{x}}) := \mathcal{I}(\hat{\underline{x}})$ and $M'(\hat{\underline{x}}) := M(\hat{\underline{x}})$.

Step 5: Resolving the *k*-parity.

Let κ be the overall k-parity of $\mathcal{I}'(\hat{\underline{x}})$. If $\kappa \neq 0$ we find the minimum $\delta(t) := \min_{1\leq n\leq 12}\{\delta(n)\}$ among the twelve k-penalties given by

$$\delta(n) \;=\; d_{ij\bar{k}}(n) - d_{ijk}(n) \qquad \qquad for \quad n = 1, 2, \ldots 12 \qquad (8)$$

where i, j, k are the indices of the corresponding points in $\mathcal{I}'(\hat{\underline{x}})$. Subsequently we construct $\mathcal{I}''(\hat{\underline{x}})$ from $\mathcal{I}'(\hat{\underline{x}})$ by replacing a single point $\hat{A}_{ijk}(t)$ with $\hat{A}_{ij\bar{k}}(t)$, and set $M''(\hat{\underline{x}}) := M'(\hat{\underline{x}}) + \delta(t)$. If $\kappa = 0$ we set $\mathcal{I}''(\hat{\underline{x}}) := \mathcal{I}'(\hat{\underline{x}})$ and $M''(\hat{\underline{x}}) := M'(\hat{\underline{x}})$. Evidently $\mathcal{I}''(\hat{\underline{x}}) \in Q_{24}$. The array $\mathcal{I}''(\hat{\underline{x}})$ along with the corresponding metric $M''(\hat{\underline{x}})$ constitute the output of the Q_{24} decoder. □

The number of real addition-equivalent operations performed by the Q_{24} decoder may be easily counted. The computation of both $\mu_l^e(x)$ and $\mu_l^o(x)$ requires 8 comparisons and 8 additions per coordinate. Thus the the complexity of the first step for the Q_{24} decoder may be taken as $6{\cdot}8 = 48$ operations. The second step requires $2{\cdot}16 = 32$ additions to compute the values of $S_1^e(x_1, x_2)$ and $S_2^e(x_3, x_4)$. Using a partition of H_6 given in [11], the minimization over the hexacode in the third step may be now performed with only $16{\cdot}8{+}4{\cdot}4{+}3 = 147$ operations. As a by-product of the comparisons in (4) the computation of $\Delta(\hat{x}_l)$ in (7) requires no operations at all in the most favorable case, 6 operations in the worst case, and 3 operations on the average. On top of that the fourth step requires 5 comparisons to find the minimum among the six penalties, and one more addition to update the metric $M'(\hat{\underline{x}})$. Note that the computation of $\delta(n)$ in (8) does not require additional operations as the values of $\delta(n)$ are known to the decoder from the comparison in (3). Thus the complexity of the fifth step is 11 comparisons to find $\delta(t)$ and a single addition to compute the final metric. Evidently the operations performed at the last two steps are only required if the corresponding parity checks fail. Hence the cumulative complexity of the Q_{24} decoder is $48{+}32{+}147{+}12{+}12 = 251$ real operations in the worst case, $48{+}32{+}147 = 227$ operations in the most favorable case, and $227{+}0.5{\cdot}9{+}0.5{\cdot}12 = 237.5$ operations on the average.

Given the Q_{24} decoder described above the decoding of Λ_{24} is straightforward. Repeat steps 1 through 5 of the foregoing algorithm for each of the four cosets of Q_{24} in Λ_{24}, while using the appropriate h- and k-parities. Among the outputs of the four Q_{24} decoders select the one with the minimal metric $M''(\hat{\underline{x}})$ as the output of the Leech lattice decoder. The total complexity of such Λ_{24} decoder is $4{\cdot}251 + 3 = 1007$ real addition-equivalent operations in the worst case,

$4{\cdot}237.5+3=953$ operations on the average, and 911 operations in the most favorable case.

2.2. The bounded-distance property

We presently prove that the proposed algorithm decodes correctly at least up to the error-correction radius of the Leech lattice. Let d be the minimum distance between the points of the checkerboard lattice D_2. It is well-known (see for instance [4, 5]) that the minimum SED between the points of Λ_{24} is $16d^2$ and the error-correction radius of the Leech lattice is $R=2d$. In other words, the maximum-likelihood decoder for the Leech lattice guarantees correct decoding provided $N(\rho-\lambda)<R^2=4d^2$, where $N(\cdot)$ is the Euclidean norm (cf. [4, p.41]).

Theorem 1. *If ρ is within an open sphere of radius R about λ, that is to say $N(\rho-\lambda)<4d^2$, then the proposed algorithm decodes ρ to λ.*

Proof. Assume for the time being that $\lambda\in Q_{24}$ and consider the Q_{24} decoder of Section 2.1. We start with a few simple facts.

Fact 1: For any $i,j,k_1,k_2\in GF(2)$, the minimum SED between A_{ijk_1} and $A_{\bar{i}\bar{j}k_2}$ is $4d^2$.

Fact 2: For any $i,j,k\in GF(2)$, the minimum SED between A_{ijk} and $A_{ij\bar{k}}$ is $8d^2$.

Fact 3: The minimum SED between the even representations of a character $x\in GF(4)$ is $8d^2$.

Fact 4: The minimum SED between the even type–A images of the codewords of H_6 is $16d^2$.

The first two facts follow immediately by scrutinizing the 16-point constellation of (1). Since the two even interpretations of a character $x\in GF(4)$ are complements of each other, Fact 3 is an easy corollary of Fact 1. To see that Fact 4 is true note that the minimum Hamming distance between distinct hexacodewords $\underline{x}$ and $\underline{x}'$ is 4 (cf. [1]). Hence w.l.o.g. let $\underline{x}=(x_1,x_2,x_3,x_4,0,0)$ and $\underline{x}'=(x_1',x_2',x_3',x_4',0,0)$ with $x_l\neq x_l'$ for $l=1,2,3,4$. Let (i_1,j_1,i_2,j_2) and (i_1',j_1',i_2',j_2') be the interpretations of x_l and x_l' respectively. Since the minimum SED between the A_{ijk} points is $2d^2$, if $(i_1,j_1)\neq(i_1',j_1')$ and $(i_2,j_2)\neq(i_2',j_2')$ the SED between the representations of x_l and x_l' is at least $4d^2$. Yet if $(i_1,j_1)=(i_1',j_1')$ then $(i_2',j_2')=(\bar{i}_2,\bar{j}_2)$, since both interpretations must be even. In view of Fact 1, the SED between the representations of x_l and x_l' is at least $4d^2$ in this case as well. Thus the SED between the even type–A images of $\underline{x}$ and $\underline{x}'$ is at least $4{\cdot}4d^2=16d^2$.

Now let $\underline{x}_\lambda \in H_6$ be the projection of λ on the hexacode. Assume that $\widehat{\underline{x}} \neq \underline{x}_\lambda$. Then in view of Fact 4 the distance between $\mathcal{I}(\widehat{\underline{x}})$ and λ is at least $4d$, and therefore the distance between $\mathcal{I}(\widehat{\underline{x}})$ and ρ is at least $2d$. But this is impossible, since $\mathcal{I}(\widehat{\underline{x}})$ must be closer to ρ than any other even type–A array that projects on the hexacode (λ being one of these). Hence $\widehat{\underline{x}} = \underline{x}_\lambda$. Write $\underline{x}_\lambda = \widehat{\underline{x}} = (\widehat{x}_1, \widehat{x}_2, \widehat{x}_3, \widehat{x}_4, \widehat{x}_5, \widehat{x}_6)$ as in (7). Since the preferable representation of $\widehat{x}_l$ is closer to (ρ_{2l-1}, ρ_{2l}) than any other even representation of $\widehat{x}_l$, Fact 3 implies that at most one of the preferable representations of the characters of $\widehat{\underline{x}}$ is in error. Note that at Step 4 the Q_{24} decoder essentially applies the Wagner decoding rule of [10] to the interpretations of $(\widehat{x}_1, \widehat{x}_2, \widehat{x}_3, \widehat{x}_4, \widehat{x}_5, \widehat{x}_6)$. Since the Wagner rule is known [10] to be optimal in the single-error case, Step 4 necessarily restores the correct interpretations of the characters of $\widehat{\underline{x}}$. In other words, the ij-subscripts of all the 12 points in $\mathcal{I}'(\widehat{\underline{x}})$ are correct. Moreover, Fact 2 implies that at most one of the k-subscripts is in error. Finally the Wagner rule is applied again at Step 5, this time to the k-subscripts of the twelve points of $\mathcal{I}'(\widehat{\underline{x}})$. This corrects the single bit in error (if at all), and hence $\mathcal{I}''(\widehat{\underline{x}}) = \lambda$.

Now assume that λ belongs to any one of the four cosets of Q_{24} in Λ_{24}. An argument similar to the above shows that the decoder for the corresponding coset necessarily produces $\mathcal{I}''(\widehat{\underline{x}}) = \lambda$ together with the metric $M''(\widehat{\underline{x}}) = N(\rho - \lambda) < 4d^2$. The other decoders produce some three Leech lattice points $\lambda_1, \lambda_2, \lambda_3 \in \Lambda_{24}$ along with the corresponding metrics $N(\rho - \lambda_i)$ for $i = 1, 2, 3$. Since $N(\lambda_i - \lambda) \geq 16d^2$ while $N(\rho - \lambda) < 4d^2$, it follows that $N(\rho - \lambda_i) > N(\rho - \lambda) = M''(\widehat{\underline{x}})$ and hence $\lambda = \mathcal{I}''(\widehat{\underline{x}})$ will be selected as the output of the Leech lattice decoder. ∎

3. Bounded-distance decoding of the binary Golay code

An efficient algorithm for bounded-distance decoding of the binary Golay code C_{24} may be obtained immediately as a "restriction" of the Leech lattice decoder presented in the previous section to $GF(2)^{24}$. Notably, this algorithm achieves the guaranteed error-correction radius of the Golay code on the AWGN channel.

We shall represent the binary Golay code as $C_{24} = C_{24}^e \cup C_{24}^o$, where C_{24}^e is the set of all the Golay codewords obtained by taking the even interpretations of the hexacodewords, and C_{24}^o is the coset of C_{24}^e obtained by taking the odd interpretations. Note that C_{24}^e is generated by the first 11 rows of the Golay generator matrix given in [12, Eq.(2)]. The decoder for C_{24} will consist of two independent (except at Step 1) decoders for C_{24}^e and C_{24}^o, operating concurrently. Again, since the two decoders are essentially identical we describe only the decoder for C_{24}^e. As in [11] the input to the decoder is assumed to consist of the 24 confidence values $\mu_1, \mu_2, \ldots \mu_{24} \in R$.

Step 1: For each character $x \in GF(4)$ and for each of the six coordinates of the hexacode find the preferable interpretation of x and the corresponding confidence value $\mu_i^e(x)$.

Step 2: As in (5) compute and sort the confidence values of the blocks $S_1^e(x_1, x_2)$ and $S_2^e(x_3, x_4)$.

Step 3: As in (6) compute the metrics $M(\underline{x})$ of the 64 hexacodewords and select $\hat{\underline{x}} \in H_6$ with the maximum metric $M(\hat{\underline{x}})$.

Step 4: Reconstruct the binary image of $\hat{\underline{x}}$ denoted $\mathcal{B}(\hat{\underline{x}})$, which is a 4×6 array with entries from $GF(2)$, by taking the preferable even interpretations of all the characters of $\hat{\underline{x}}$. If the parity of the top row of $\mathcal{B}(\hat{\underline{x}})$ is odd use the Wagner rule, — that is to say complement the character with the lowest confidence value, say $\mu_s^e(\hat{x}_s)$. The resulting array $\mathcal{B}'(\hat{\underline{x}})$ along with the corresponding metric $M'(\hat{\underline{x}}) = M(\hat{\underline{x}}) - 2\mu_s^e(\hat{x}_s)$ constitute the output of the C_{24}^e decoder.

The final decision is reached by comparing the metrics at the output of the two decoders and selecting the one with the higher metric as the output of the decoder for C_{24}.

The number of operations required by the proposed algorithm may be calculated as follows. Using the Gray code as in [11] the complexity of the first step is 60 real operations for both decoders. The second step require $2 \cdot 16 = 32$ operations and the third step takes $16 \cdot 8 + 4 \cdot 4 + 3 = 147$ operations. Finally the fourth step takes at most 6 operations. Thus the worst-case complexity of our decoder is $60 + 2 \cdot (32 + 147 + 6) + 1 = 431$ real operations, as compared to 651 operations in [11].

To see that the proposed decoder achieves the maximum-likelihood error-correction radius of C_{24} note that the minimum SED between the Golay codewords is $8d^2$, where d now denotes the Euclidean distance between the signals representing 0 and 1. Thus the maximum-likelihood decoder for C_{24} decodes correctly provided the vector observed at the channel output is at distance less than $R_{ML} = \sqrt{2}d$ from the transmitted codeword. However, in this case at most one of the characters may be assigned a wrong interpretation at the first step, since the SED between the complementary interpretations of a character is $4d^2$. This single error is successfully corrected by the Wagner decoder at the last step of the algorithm.

Note that the hard-decision decoder for C_{24} guarantees correct decoding only up to $R_{HD} = d$.

4. The effective error-coefficient

The performance of the proposed Leech lattice decoder on the AWGN channel is evaluated in two ways in this correspondence. First, the effective error-coefficient is calculated in this section. Since our decoder achieves the error-correction radius of the Leech lattice, the 'error-exponent' is the same as in maximum-likelihood decoding and any loss in the coding gain relative to the maximum-likelihood decoder is due primarily to the increase in the effective error-coefficient (cf. [6]). It is estimated in [6] that increasing the error-coefficient by a factor of about 1.5 should result in a coding-gain loss of only about 0.1 dB at $BER \approx 10^{-6}$. The next section presents the simulation results which confirm this estimate.

The Voronoi cell of $\lambda \in \Lambda_{24}$, denoted $V(\lambda)$, is the set of all $\rho \in R^{24}$ that are at least as close to λ as to any other point of the Leech lattice. As in [6] we define the *decision region* of λ, denoted $D(\lambda)$, as the set of all $\rho \in R^{24}$ that are decoded to λ using the proposed algorithm. Both $V(\lambda)$ and $D(\lambda)$ are convex polyhedra in R^{24}. Further, Theorem 1 shows that a sphere of radius R is the largest sphere that can be inscribed in either $V(\lambda)$ or $D(\lambda)$. Hence the points on the boundary of either $V(\lambda)$ or $D(\lambda)$ that are the closest to λ are at distance exactly R from λ. The number of points at distance exactly R from λ on the boundary of $V(\lambda)$ is just the kissing number $\mathcal{N}_0(\Lambda_{24}) = 196560$ (see [4]). As in [6] the number of such points on the boundary of $D(\lambda)$ is said to be the *effective error-coefficient* $\mathcal{N}_{0,eff}(\Lambda_{24})$ of our algorithm. It is shown in the sequel that our algorithm increases the effective error-coefficient by a factor of less than 1.72 relative to the maximum-likelihood decoder, and a factor of less than 1.15 relative to the bounded-distance decoding algorithm of Forney [6].

Before going into the computation of the effective error-coefficient, it would be helpful to express the relevant lattices in the code-formula notation of Forney [5]. Assume w.l.o.g. that $d = 1/\sqrt{2}$ and hence the minimum nonzero norm of the points of Λ_{24} is 8. With this assumption the Leech lattice may be written as $\Lambda_{24} = H_{24} \cup (\xi + H_{24})$, where $\xi \in R^{24}$ is the 24-tuple $\frac{1}{2}(-3, 1^{23})$. The code-formula for H_{24} may be found in [4, 5]. The code-formula for Q_{24} is given by

$$Q_{24} = C_{24}^{e} + 2(24,23,2) + 4Z^{24} = C_{24}^{e} + 2D_{24} \tag{9}$$

where Z^{24} is the 24-dimensional integer lattice. Let F_{24} be the $(24, 12, 4)$ binary code consisting of all the even interpretations of the 64 hexacodewords. Note that F_{24} is a self-dual code, which is classified in [8] and further studied in [9]. We define the lattice Φ_{24} by the code-formula

$$\Phi_{24} \;=\; F_{24} + 2Z^{24} \tag{10}$$

Obviously $C_{24}^{e} \subset F_{24}$ and $Q_{24} \subset \Phi_{24}$. Furthermore, with this notation the decoder of Section 2.1 is equivalent to the following. Steps 1 through 3 amount to maximum-likelihood decoding of the lattice Φ_{24}. In Step 4 we employ the Wagner rule to obtain a codeword of C_{24}^{e} from a codeword of F_{24}. In Step 5 we employ this rule again to get a point of $2D_{24}$ form a point of $2Z^{24}$. It is evident from (9) and (10) that this produces a point of Q_{24}.

Lemma 2. *The effective error-coefficient of the Q_{24} decoder is $\mathcal{N}_{0,eff}(Q_{24}) =$ 189264.*

Proof. W.l.o.g. we may assume that $\lambda = \underline{0}$ where $\underline{0}$ is the origin of R^{24}. In view of the foregoing discussion the effective error-coefficient of the Q_{24} decoder is equal to the number of points with Euclidean norm 8 in the lattice Φ_{24}. Indeed, all the points of norm 4 in Φ_{24} will be necessarily modified either to $\underline{0}$ or to a point of norm 8 when the Wagner rule is applied at the last two steps of the algorithm. On the other hand, applying the Wagner rule to points of norm 8 cannot map these points onto $\underline{0}$. Thus the "effective nearest neighbors" of the Q_{24} decoder are precisely the points of norm 8 in Φ_{24}. There are exactly 735 codewords of Hamming weight 8 in F_{24}. These arise from the $45{\cdot}2^4 = 720$ even interpretations of the 45 hexacodewords of weight 4 and another $\binom{6}{2} = 15$ interpretations of $(\mathbf{0},\mathbf{0},\mathbf{0},\mathbf{0},\mathbf{0},\mathbf{0})$. Thus there are $735{\cdot}2^8 = 188160$ points of the type $(0^{16} \pm 1^8)$ in Φ_{24}. In addition to these there are also $4{\cdot}\binom{24}{2} = 1104$ points of the type $(0^{22} \pm 2^2)$. Since these are the only points of Euclidean norm 8 in Φ_{24}, we have $\mathcal{N}_{0,eff}(Q_{24}) = 188160 + 1104 = 189264$. ∎

Theorem 3. *The effective error-coefficient of the Λ_{24} decoder is $\mathcal{N}_{0,eff}(\Lambda_{24}) =$ 336720.*

Proof. Given $\mathcal{N}_{0,eff}(Q_{24}) = 189264$, in order to find the effective error-coefficient of the Leech lattice decoder it remains to enumerate the points of norm 8 in the three cosets of Q_{24} in Λ_{24}. The number of these is given by $\mathcal{N}_0(\Lambda_{24}) - \mathcal{N}_0(Q_{24})$. The kissing number of Q_{24} may be easily calculated as follows. There are exactly 375 codewords of weight 8 in C_{24}^{e} arising from the 15 interpretations of the all-zero hexacodeword and $45{\cdot}2^3 = 360$ even interpretations of the 45 weight 4 hexacodewords which satisfy the h-parity constraint.

Therefore Q_{24} contains $375 \cdot 2^7 = 48000$ points of the type $(0^{16} \pm 1^8)$ along with $4 \cdot \binom{24}{2} = 1104$ points of the type $(0^{22} \pm 2^2)$. Hence $\mathcal{N}_0(Q_{24}) = 49104$ and we have $\mathcal{N}_{0,eff}(\Lambda_{24}) = \mathcal{N}_{0,eff}(Q_{24}) + \mathcal{N}_0(\Lambda_{24}) - \mathcal{N}_0(Q_{24}) = 336720$. ∎

Remark. One might argue that all the points of norm 8 in the three cosets of Φ_{24} corresponding to the decoders for the three cosets of Q_{24} in Λ_{24}, should be taken as the effective nearest neighbors of $\underline{0}$. Instead, in the proof of Theorem 3 we have taken into account only those points of these cosets that belong to Λ_{24}. This may require some explanation. Let φ be a point of norm 8 with $\varphi \notin \Phi_{24}$ and $\varphi \notin \Lambda_{24}$. The point φ is an effective nearest neighbor of $\underline{0}$ iff $\varphi' = \varphi/2$ is on the boundary of $D(\underline{0})$. We presently show that φ' is *within* the decision region of $\underline{0}$ rather than on the boundary thereof. First note that the distance between φ' and $\underline{0}$ is $\sqrt{2}$. Thus the decoder for Q_{24} when presented with φ' yields $\underline{0}$ with the corresponding metric $N(\varphi' - \underline{0}) = 2$, unless there exists a point $\phi \in \Phi_{24}$ of norm 8 which is at distance exactly $\sqrt{2}$ from φ'. Evidently, this point ϕ lies on the sphere of radius $2R = 2\sqrt{2}$ about $\underline{0}$. Since the straight line $\underline{0}\,\varphi'$ intersects this sphere at φ and $N(\varphi - \varphi') = N(\varphi' - \underline{0}) = 2$ the only point on the sphere at distance $\sqrt{2}$ from φ' is φ. As $\varphi \notin \Phi_{24}$ it follows that the Q_{24} decoder decodes φ' to $\underline{0}$. Now the other three decoders produce some three points $\lambda_i \in \Lambda_{24} \backslash \{\underline{0}\}$ along with the metrics $N(\varphi' - \lambda_i)$ for $i = 1, 2, 3$. By the same argument if $N(\varphi' - \lambda_i) \leq 2$ for some i, we must have $\lambda_i = \varphi$ which is impossible since $\varphi \notin \Lambda_{24}$. Hence $N(\varphi' - \underline{0}) < N(\varphi' - \lambda_i)$ and the point φ' is necessarily decoded to $\underline{0}$. It follows that the set of the effective nearest neighbors of $\underline{0}$ is equal to the union of the first shell of Λ_{24} and the second shell of Φ_{24}.

5. Simulation results

A comprehensive computer simulation has been performed for both the proposed algorithm and the algorithm of Forney [6]. The simulation assumed a signaling scheme based on a 64-QAM square constellation and employed an additive white Gaussian noise channel model. The results of the simulation for the Leech lattice are presented in Figures 1 and 2. The solid line represents the proposed Leech lattice decoder, the dashed line is the maximum-likelihood decoder, and the dotted line is the decoder of [6]. Finally the dot-dashed line stands for the uncoded QAM transmission.

The simulation results show a loss in coding-gain relative to the maximum-likelihood decoder that is uniformly *less* than 0.1 dB over the whole range of BER from 10^{-1} to 10^{-7}. The loss in coding-gain relative to the decoder of

Forney [6] is less than 0.025 dB over the same range. At $BER \approx 10^{-6}$ the loss in coding-gain versus the maximum-likelihood decoder is approximately 0.075 dB. Since our algorithm increases the effective error-coefficient by a factor of about 1.71, this is even less than might be expected from the estimate of [6]. The reason for the small difference lies in the fact that in practice only a finite subset of the lattice points, as defined by the 64-QAM constellation, is utilized. The lattice points that are close to the outer boundary of this finite subset have fewer effective neighbors than the inner points. Due to this boundary effect the average effective error-coefficient is reduced.

References

1. E.F. Assmus, Jr. and H.F. Mattson, Jr., *Algebraic Theory of Codes,* Report No.1, Contract No. F19628-69C0068, Air Force Cambridge Research Labs., Bedford, MA., 1969.
2. Y. Be'ery, B. Shahar, and J. Snyders, "Fast decoding of the Leech lattice," *IEEE J. Select. Areas Comm.*, vol. SAC-7, pp. 959–967, 1989.
3. J.H. Conway and N.J.A. Sloane, "Soft decoding techniques for codes and lattices, including the Golay code and the Leech lattice," *IEEE Trans. Inform. Theory*, vol. IT-32, pp. 41–50, 1986.
4. J.H. Conway and N.J.A. Sloane, *Sphere Packings, Lattices and Groups*, New York: Springer-Veralg, 1988.
5. G.D. Forney, Jr., "Coset codes II: Binary lattices and related codes," *IEEE Trans. Inform. Theory*, vol. IT-34, pp. 1152–1187, 1988.
6. G.D. Forney, Jr., "A bounded distance decoding algorithm for the Leech lattice, with generalizations," *IEEE Trans. Inform. Theory*, vol. IT-35, pp. 906–909, 1989.
7. G.R. Lang and F.M. Longstaff, "A Leech lattice modem," *IEEE J. Select. Areas Comm.*, vol. SAC-7, pp. 968–973, 1989.
8. V. Pless and N.J.A. Sloane, "On the classification of self dual codes," *J. Comb. Theory*, vol. 18A, pp. 313–335, 1975.
9. M. Ran and J. Snyders, "On maximum likelihood soft decoding of some binary self dual codes," to appear in *IEEE Trans. Comm.*, 1993.
10. R.A. Silverman and M. Balser, "Coding for a constant data rate source," *IRE Trans. Inform. Theory*, vol. 4, pp. 50–63, 1954.
11. A. Vardy and Y. Be'ery, "More efficient soft-decision decoding of the Golay codes," *IEEE Trans. Inform. Theory*, vol. IT-37, pp. 667–672, 1991.
12. A. Vardy and Y. Be'ery, "Maximum-likelihood decoding of the Leech lattice," to appear in *IEEE Trans. Inform. Theory*, 1993.

Some restrictions on distance distribution of optimal binary codes

Sergei I. Kovalov

St.- Petersburg Academy of Airspace Instrumentation,
Bolshaia Morskaia str.,67, St.-Petersburg,190000, Russia,
e-mail : liap@sovam.com

Abstract. Some new restrictions on the last element of a distance spectrum for optimal binary block codes are presented.

Let us consider a binary block code G of length N and size M in which any two words have distance at least d. The code is referred to as an optimal code if it has the maximum number of codewords for given N and d . A pair of codewords x and y located at distance N are called antipodes. Let B_N denote the proportion of the codewords having antipodes in G .

Evidently, $0 \leq B_N \leq 1$. It is easy to see that $B_N \leq 1 - 1/M$ for odd M. In this correspondence we obtain an improvement on this bound. Restrictions on the elements of the weight spectrum may be useful for search problems or problems of classification for optimal codes or in LP bound [1 ,chapter 17.4].

Let us represent the codewords of G as nodes of an undirected graph and mark (x) the node corresponding to the codeword x . The terms "the node of graph" and "the codeword" are equivalent in further consideration. We define the edges of the graph as follows. Let x and y be connected by an edge if the Hamming distance $D(x, y)$ between codewords x and y satisfies the condition : $N - d < D(x, y) < N$. This graph we call the code graph.

Proposition 1 *Every codeword having an antipode in G corresponds to an isolated node of the code graph.*

The validity of Proposition 1 follows from the observation that there are no two codewords $\bar{x}$ and y such that $D(x, \bar{x}) = N$ and $D(x, y) > N - d$ for some $x \in G$; otherwise $D(\bar{x}, y) < d$.

Let us delete all nodes having antipodes from the code graph. The resulting graph will be referred to as a conflict graph. Let S be the number of nodes of the conflict graph . Note that S and the size of the code have the same parity. Evidently, $B_N = 1 - S/M$. We examine various changes of the input code connected with deleting some codewords that are associated with the nodes of the conflict graph. In some cases it gives an opportunity to add antipodes of some codewords to the code. These actions are based on the propositions given below.

Proposition 2 *If there is an isolated node in the conflict graph then one can add an antipode of this node to the code without decreasing the minimal distance.*

Proof. Let (x) be an isolated node of the conflict graph and $\bar{x}$ be the inverse of x. It follows from the definition of the conflict graph that the code does not contain a codeword at distance less than d from $\bar{x}$. Hence we can add this vector to the code without decreasing the minimal distance.

Proposition 3 *Let (x) be an isolated node of the conflict graph. An addition of the vector $\bar{x}$ to the code does not cause an appearance of new edges in this graph.*

Proof. Suppose that the addition of $\bar{x}$ to the code causes an appearance of a new edge. Without loss of the generality we suppose that the new edge connects $(\bar{x})$ and (y) i.e. $N-d < D(\bar{x}, y) < N$. Then $0 < D(x, y) < d$. It contradicts the fact that $x, y \in G$. We can admit that after adding an antipode of some codeword to the conflict graph we construct the conflict graph anew. It follows from the previous proposition that the change of the graph represents the deletion of the node having an antipode.

Proposition 4 *If there is a node of valency 1 in the conflict graph then a code with the same parameters N, M, d and $S' < S - 1$ exists.*

Proof. We delete one node (the codeword) with the edges associated to it and create an isolated node. Then we add the antipode of this node using proposition 2.

Proposition 5 *If the parameter M of an optimal code is odd then $S \geq 3$.*

Proof. Evidently, $S \neq 1$ for any optimal code (see proposition 2).

Proposition 6 *If the parameter M of an optimal code is even then either $S \geq 4$ or an optimal code with $S = 0$ exists.*

Proof. Using the proposition 4 for the case $S = 2$ we obtain the statement of the proposition.

Now, we wish to improve propositions 5 and 6. The following considerations are based on a more detailed analysis of the structure of the conflict graph. We focus on the presence or the absence of cycles in the graph . We will show that the absence of any cycles in the graph leads to $S' = 0$ by deleting some codewords and adding some antipodes. If a cycle exists then we show that its length is not too small. Estimating the number of nodes included in the shortest cycle we obtain a lower bound on the nonzero S'. At first examine the case when M is odd (we know that $S > 0$ in this case) and suppose that an optimal code of size M exists.

Proposition 7 *The conflict graph of an optimal code having an odd size contains at least one cycle.*

Proof. Suppose that the graph is a tree or a collections of trees. Then using proposition 4 several times we generate a sequence of optimal codes and conflict graphs corresponding to them and the sequence $S_1 > S_2 > \ldots$. Since the last element of this sequence is equal to 1 it contradicts proposition 5.

Corollary. There exists a conflict graph of an optimal code of odd size such that every node is included into a cycle.

Proposition 8 *The conflict graph of an optimal code of odd size has at least one cycle of an odd length.*

Proof. Suppose that all cycles have even lengths. Then the graph may be represented as bipartite graph [2]. The numbers of nodes in the two parts are not equal to each other. Let us delete all nodes from the smaller part and all corresponding codewords from the code. Using propositions 2, 3 we can add the antipodes of all remained nodes to obtain a code of greater size than the input code. This contradicts the supposition that the input code is an optimal one.

Thus, the minimal configuration of a conflict graph for odd S is a cycle of odd length. Let us estimate its size.

Proposition 9 *Let S_m be the minimal number of nodes in any odd cycle of the conflict graph. Then $S_m > 1 + (N - d)/(d - 1)$.*

Proof. Consider an arbitrary node (x_0) belonging to a cycle. It is connected with a node (x_1) of this cycle. The last node is connected with a node (x_2) . Then

$$D(x_0, x_1) \geq N - d + 1 \quad ; \quad D(x_1, x_2) \geq N - d + 1$$

and

$$D(x_0, \bar{x}_1) \leq d - 1, \quad D(\bar{x}_1, x_2) \leq d - 1, \Rightarrow D(x_0, x_2) \leq 2d - 2.$$

This means that (x_0) and (x_2) cannot be connected by the edge if $2d - 2 \leq N - d$. Let us increase the chain under consideration by 2 edges. Let the node (x_2) be connected with (x_3) and (x_3) be connected with (x_4). Then similar observations lead to the following inequalities :

$$D(x_0, x_2) \leq 2d - 2 \quad ; \quad D(x_2, x_4) \leq 2d - 2, \Rightarrow D(x_0, x_4) \leq 4d - 4.$$

In general, we have $D(x_0, x_i) \leq id - i$ where i is even. Therefore, the minimal number of nodes included into a cycle is lowerbounded by S_m that is defined as the minimal odd number satisfying the condition $(S_m - 1)(d - 1) > N - d$.

When M is even the statements given below may be proved similarly. Suppose an optimal code contains an even number of codewords. Then

i. an optimal code exists with the conflict graph having either $S = 0$ or a cycle of odd length.

ii. the proposition 9 is valid for any conflict graph having a cycle of odd length.

Using i, ii and proposition 9 we obtain following

Theorem 1 *The minimal number* $S_m{}^o$ *of nodes for the conflict graph associated with an optimal code satisfies the following conditions :*

$$\begin{cases} S_m{}^o > 1 + (N-d)/(d-1) \;\; for\ odd\ M; \\ S_m{}^o = 0 \;\; or \;\; S_m{}^o > 1 + (N-d)/(d-1) \;\; for\ even\ M. \end{cases}$$

Note that the conflict graph determines the sum of the several last elements of the distance spectrum in a unique way. It is easy to see that

$$\sum_{i>N-d} B_i \;=\; 1 - S/M + 2R/M \;,$$

where R is the number of edges of the conflict graph.

References

[1] F.J. MacWilliams and N.J.A.Sloane. The Theory of Error-Correcting Codes. North Holland,Amsterdam,1976.

[2] F.Harary. Graph Theory. Addison-Wesley. 1969.

Two new upper bounds for codes of distance 3*

Simon Litsyn **
Tel-Aviv University, Department of Electrical Engineering – Systems
Ramat-Aviv 69978, Tel-Aviv, Israel

Alexander Vardy ***
IBM Research Division, Almaden Research Center
650 Harry Road, San Jose, CA 95120

Abstract. We prove that the (10,40,4) code found by Best is unique. We then employ this fact to show that $A(10,3) = A(11,4) \leq 78$ and $A(11,3) = A(12,4) \leq 156$.

1. Introduction

An (n,d) binary code is a set of binary vectors of length n such that any two of them differ in at least d positions. The maximum number of codewords in any (n,d) binary code is denoted by $A(n,d)$. If a binary (n,d) code C consists of M codewords it is said to be an (n,M,d) code. If in addition all the codewords of C contain exactly w nonzero positions then C is an (n,M,d,w) constant-weight code. An (n,M,d) binary code such that $M = A(n,d)$ is said to be optimal. Many optimal codes are known to be unique. For instance, the uniqueness of the (24,4096,8) and the (23,2048,7) binary Golay codes is proved in [11, 12, 6]. The (13,32,6) extended Nadler code is shown to be unique in [3]. For the uniqueness of the (16,256,6) Nordstrom-Robinson code see [12]. The optimal (15,128,6) and (14,64,6) are also known to be unique [10, p.75]. Less is known regarding the uniqueness of the optimal $(n,4)$ codes. It is easy to show that the first three (6,4,4), (7,8,4) and (8,16,4) codes are unique. However, the (9,20,4) code is not [10, p.72].

The optimal (10,40,4) code consisting of ten cyclic shifts of the following affine square

$$\begin{matrix} 1\,0\,1\,0\,0\,0\,0\,0\,0\,1 \\ 1\,1\,0\,0\,1\,0\,1\,1\,0\,0 \\ 0\,0\,0\,1\,0\,1\,0\,1\,1\,1 \\ 0\,1\,1\,1\,1\,1\,1\,0\,1\,0 \end{matrix} \tag{1}$$

---

*The paper is a part of [9] submitted to IEEE Transactions on Information Theory.

**Research supported by a grant from the Israeli Ministry of Science and Technology and the Guastallo Fellowship.

***Research supported in part by the Rothschild Fellowship.

has been found by Best [1] in 1980. The Best code may be also described as the union of five affine cubes, while a simple representation of this code over $\mathbb{Z}_4$ has been recently found in [5]. It has long been conjectured that the Best code is unique. In this correspondence we prove that this is indeed true. Since the Best code is cyclic, our proof also implies that the (9,40,3) code is unique.

Along with the interest in the uniqueness of the known optimal codes a considerable effort has been devoted to deriving upper bounds on $A(n,d)$. In fact for $n \leq 16$, all the values of $A(n,d)$ are presently known (see [4, p.248]), except four: $A(10,3), A(11,3), A(11,4)$, and $A(12,4)$. Since $A(n+1,2t) = A(n,2t-1)$, there are essentially only two unknown values. It is shown in Julin [7] that $A(12,4) = A(11,3) \geq 144$, and $A(11,4) = A(10,3) \geq 72$. The upper bounds $A(10,3) \leq 78$ and $A(11,3) \leq 154$ have been claimed by Wax [13] in 1959. However, Best et al. [2] showed in 1978 that the results of Wax are erroneous. In the same paper the authors prove that $A(8,3) = 20$, which implies $A(10,3) \leq 80$ and $A(11,3) \leq 160$. Two years later Best [1] improved this result to $A(10,3) \leq 79$ and $A(11,3) \leq 158$. In this correspondence we employ the uniqueness of the (10,40,4) code to show that $A(10,3) \leq 78$ and $A(11,3) \leq 156$. This rules out the only odd upper bound on $A(n,3)$ in the table of [4, p.248], which is in accord with the well-known conjecture that $A(n,d)$ must be even.

We use the following scheme of proof. First we show that if a (10,40,4) code consists of only even-weight codewords then one of its translates contains a $(10,13,4,3)$ or a $(10,12,4,3)$ constant-weight subcode. Then we classify all such constant-weight codes, demonstrating that there exist exactly two nonisomorphic (10,13,4,3) codes and exactly twenty-eight nonisomorphic (10,12,4,3) codes. We subsequently show that only one of these codes may be extended to a (10,40,4) code. Furthermore the extension may be accomplished in a unique way. Analysis of all the (9,40,3) codes obtained by puncturing the resulting (10,40,4) code completes the proof.

2. The Best code is unique

We start with some notation. The set of all the binary n-tuples is denoted $\mathbb{F}_2^n$. The subset of $\mathbb{F}_2^n$ consisting of all the n-tuples which contain exactly w ones is denoted $\mathbb{F}_2^{n,w}$. For $\underline{x}, \underline{y} \in \mathbb{F}_2^n$ the Hamming distance between $\underline{x}$ and $\underline{y}$, denoted $d(\underline{x}, \underline{y})$, is the number of positions in which $\underline{x}$ and $\underline{y}$ differ. Given $\underline{x} \in \mathbb{F}_2^n$ and a subset $S \subset \mathbb{F}_2^n$ we shall write $d(\underline{x}, S) = \min_{\underline{y} \in S} d(\underline{x}, \underline{y})$. The Hamming weight of $\underline{x} \in \mathbb{F}_2^n$ is given by $wt(\underline{x}) = d(\underline{x}, \underline{0})$, where $\underline{0}$ denotes the all-zero n-tuple. The Hamming sphere of radius r centered at $\underline{x} \in \mathbb{F}_2^n$ is $\mathcal{B}_n(\underline{x}, r) = \{\underline{y} \in \mathbb{F}_2^n : d(\underline{x}, \underline{y}) \leq r\}$. We let $\mathcal{B}_n(S, r) = \cup_{\underline{x} \in S} \mathcal{B}_n(\underline{x}, r)$, where S is any subset of $\mathbb{F}_2^n$. A subset of $\mathbb{F}_2^n$ is a binary code of length n, and a subset of $\mathbb{F}_2^{n,w}$ is a constant-weight code. The maximum number of codewords in any (n, M, d, w) constant-weight code is denoted $A(n, d, w)$. Two codes $C_1, C_2 \subset \mathbb{F}_2^n$ are said to be isomorphic if there exists a permutation π, such that $C_2 = \{\pi(\underline{c}) : \underline{c} \in C_1\}$. They are said to be equivalent if there exist a vector $\underline{a}$ and a permutation π, such that $C_2 = \{\underline{a} + \pi(\underline{c}) : \underline{c} \in C_1\}$.

For a code $C \subset \mathbb{F}_2^n$, the weight distribution of C is a vector $(W_0, W_1, \ldots W_n)$, where W_i is the number of codewords of weight i in C. The distance spectrum of C is given by $(B_0, B_1, \ldots B_n)$ where $B_i = (1/|C|) \sum_{\underline{c} \in C} W_i(\underline{c})$, and $W_i(\underline{c})$ is the number of vectors of weight i in the coset $\underline{c} + C$.

Given a set $S \subset \mathbb{F}_2^n$ we denote by $A(S,d)$ the maximum number of codewords in any (n,d) subcode of S. If S is not too large it is often possible to determine $A(S,d)$ by means of a direct computer search. We have employed a program based on a simple backtracking algorithm (see for instance [8]) for the evaluation of $A(S,d)$. Whenever we use this program we shall indicate the size of the search space S. It is worth mentioning that since the largest set S that we ever need to search has cardinality 72, the total computational effort involved in our proof amounts to less than an hour of computing time on a contemporary PC.

Proposition 1. *Let C be a $(10,40,4)$ consisting of only even-weight codewords. Then C is equivalent to a $(10,40,4)$ code, whose weight distribution satisfies*

$$W_0 = W_1 = W_2 = W_4 = W_6 = W_8 = W_{10} = 0, \quad 12 \le W_3 \le 13,$$

Proof. Denote by H the set of holes of C, that is the set of all $\underline{h} \in \mathbb{F}_2^{10}$ such that $d(\underline{h}, C) = 3$. Evidently $wt(\underline{h})$ is odd for all $\underline{h} \in H$. Since the Hamming spheres of radius 1 about the codewords of C do not intersect, and all the vectors $\underline{x} \in \mathbb{F}_2^{10} \backslash C$ contained in these spheres are of odd weight, it follows that

$$|H| = \frac{|\mathbb{F}_2^{10}|}{2} - 40 \cdot |\mathcal{B}_{10}(\underline{0}, 1)| + |C| = 512 - 440 + 40 = 112$$

Let $\eta_i(\underline{x})$ denote the number of codewords of C at distance i from a vector $\underline{x} \in \mathbb{F}_2^{10}$. Consider the following sum $\sum_{\underline{h} \in H} \eta_3(\underline{h})$. Clearly, the number of times a codeword $\underline{c} \in C$ is counted in the foregoing sum is equal to the number of holes at distance three from $\underline{c}$. The latter is given by $\binom{10}{3} - 4\eta_4(\underline{c})$ since a vector at distance three from $\underline{c}$ is a hole if and only if it is not at distance one from a codeword $\underline{c}' \in C$ with $d(\underline{c}', \underline{c}) = 4$. Thus we have

$$\sum_{\underline{h} \in H} \eta_3(\underline{h}) = \sum_{\underline{c} \in C} \Big(120 - 4\eta_4(\underline{c}) \Big)$$

Let $(B_0, B_1, \ldots B_{10})$ be the distance spectrum of C. After a few simple calculations the foregoing expression transforms into

$$B_4 = 30 - \frac{1}{160} \sum_{\underline{h} \in H} \eta_3(\underline{h}) \tag{2}$$

We now estimate the value of B_4. The distance spectrum of C satisfies the MacWilliams – Delsarte inequalities (cf. [2, 10])

$$K_k(0) + B_4 K_k(4) + B_6 K_k(6) + B_8 K_k(8) + B_{10} K_k(10) \ge 0$$

for $k = 1, 2, \ldots 5$, where the Krawtchouk polynomials $K_k(x)$ are given by

$$K_k(x) = \sum_{j=0}^{k} (-1)^j \binom{x}{j} \binom{n-x}{k-j}.$$

Actually we only need the inequality for $k = 3$ which is of the form

$$8B_4 - 8B_6 + 8B_8 + 120B_{10} \leq 120. \tag{3}$$

Lemma 6 of [1] gives

$$6B_4 + 4B_6 + 2B_8 \leq 190. \tag{4}$$

This inequality counts the number of zeros in the code in two ways while taking into account that the number of zeros in each position cannot exceed $A(9,4) = 20$. In fact, since the number of ones is also upper bounded by $A(9,4) = 20$ there must be exactly 20 zeros in each position, and (4) turns out to be an equality in our case. Obviously

$$B_4 + B_6 + B_8 + B_{10} = 39, \tag{5}$$

and

$$B_6 \geq 0, \quad B_8 \geq 0, \quad B_{10} \geq 0. \tag{6}$$

Solving the linear programming problem for the maximum of B_4 subject to (3) - (6) yields

$$B_4 \leq 22. \tag{7}$$

Substituting (7) into (2) we obtain

$$\frac{1}{|H|} \sum_{\underline{h} \in H} \eta_3(\underline{h}) \geq \frac{(30-22) \cdot 160}{112} = 11.4285\ldots .$$

Thus there exists a vector $\underline{h} \in H$ such that $\eta_3(\underline{h}) > 11$. Clearly $\eta_3(\underline{h}) \leq A(10,4,3) = 13$. Since $\underline{h}$ is of odd weight while all the codewords of C are of even weight, we have $\eta_i(\underline{h}) = 0$ for $i = 0, 2, 4, 6, 8, 10$. Finally $\eta_1(\underline{h}) = 0$ by the definition of the set H. Hence the weight distribution of the translate $C' = \underline{h} + C$ satisfies the claim of the proposition.

Q.E.D.

Thus in order to check the variants of constructing a (10,40,4) code we have to find all the nonequivalent (10,13,4,3) and (10,12,4,3) constant-weight codes. It turns out that the classification of these codes up to a permutation of coordinates is a tractable problem.

Proposition 2. *There exist exactly two nonequivalent* (10, 13, 4, 3) *and exactly* 28 *nonequivalent* (10, 12, 4, 3) *codes.*

Proof. For the complete proof see [9]. Actually there are two nonequivalent codes with 13 words:

$$
T_0 = \begin{array}{c}
1110000000\\
1001100000\\
1000011000\\
1000000110\\
0101010000\\
0100100100\\
0100001010\\
0011001000\\
0010100010\\
0010010100\\
0000110001\\
0001000011\\
0000001101
\end{array}
\qquad
T_1 = \begin{array}{c}
1110000000\\
1001100000\\
1000011000\\
1000000110\\
0101010000\\
0100100100\\
0100001010\\
0001001100\\
0011000010\\
0010110000\\
0010001001\\
0000100011\\
0000010101
\end{array}
\tag{8}
$$

There are 28 nonisomorphic (10, 12, 4, 3) codes. These are given below in a hexadecimal notation.

D_0 :	380	260	218	206	150	124	10A	0C8	0B0	085	043	029
D_1 :	380	260	218	206	150	124	10A	0C8	0B0	085	029	013
D_2 :	380	260	218	206	150	124	10A	0C8	0B0	043	029	015
D_3 :	380	260	218	206	150	124	10A	0C8	0A2	094	045	031
D_4 :	380	260	218	206	150	124	10A	0C8	0A2	094	045	029
D_5 :	380	260	218	206	150	124	10A	0C8	0A2	091	045	029
D_6 :	380	260	218	206	150	124	10A	0C8	0A2	045	029	013
D_7 :	380	260	218	206	150	124	10A	0C8	0A2	043	029	015
D_8 :	380	260	218	206	150	124	10A	0C8	0A1	094	045	032
D_9 :	380	260	218	206	150	124	10A	0C8	0A1	094	045	013
D_{10} :	380	260	218	206	150	124	10A	0C8	0A1	094	043	032
D_{11} :	380	260	218	206	150	124	10A	0C8	0A1	094	043	00D
D_{12} :	380	260	218	206	150	124	10A	0C8	0A1	043	032	015
D_{13} :	380	260	218	206	150	124	10A	0C8	0A1	043	032	00D
D_{14} :	380	260	218	206	150	124	10A	0C8	094	045	029	013
D_{15} :	380	260	218	206	150	124	10A	0C8	085	043	032	029
D_{16} :	380	260	218	206	150	124	10A	0C8	043	032	029	015
D_{17} :	380	260	218	206	150	124	10A	0C2	0A8	094	04C	032
D_{18} :	380	260	218	206	150	124	10A	0C2	0A8	094	04C	031
D_{19} :	380	260	218	206	150	124	10A	0C2	0A8	091	04C	023
D_{20} :	380	260	218	206	150	124	10A	0C2	0A8	091	049	032
D_{21} :	380	260	218	206	150	124	10A	0C2	0A1	049	032	015
D_{22} :	380	260	218	206	150	128	103	0C4	0A2	091	049	025
D_{23} :	380	260	218	206	150	128	103	0C4	0A1	092	049	015
D_{24} :	380	260	218	206	150	122	109	0C4	0A8	091	043	025
D_{25} :	380	260	218	206	150	122	109	0C4	0A1	092	02C	015
D_{26} :	380	260	218	206	150	122	109	0C4	091	08A	043	025
D_{27} :	380	260	218	206	150	122	109	0A1	094	045	02C	013

(9)

It remains to be seen which of the two codes in (8) and the twenty-eight codes in (9) can be completed to a (10,40,4) code.

Proposition 3. *A* $(10,40,4)$ *code with weight distribution satisfying*

$$W_0 = W_1 = W_2 = W_4 = W_6 = W_8 = W_{10} = 0, \quad W_3 = 13$$

does not exist..

Proof. Evidently $W_7 \leq A(10,4,7) = 13$ and $W_9 \leq A(10,4,9) = 1$. Further, if $W_9 = 1$ then $W_7 \leq A(9,4,6) = 12$. Hence $W_7 + W_9 \leq 13$, and therefore

$$W_5 \geq 40 - W_3 - W_7 - W_9 \geq 14 . \tag{10}$$

For $i = 0,1$ let S_i be the set of vectors of weight 5 at distance 4 from the code T_i given in (8), that is $S_i = \mathbb{F}_2^{10,5} \backslash (\mathbb{F}_2^{10,5} \cap \mathcal{B}_{10}(T_i,2))$. We have $|S_i| = 36$ and $A(S_i,4) = 13$ for $i = 0,1$ which contradicts (10). Q.E.D.

Proposition 4. *The* $(10,40,4)$ *code with weight distribution satisfying*

$$W_0 = W_1 = W_2 = W_4 = W_6 = W_8 = W_{10} = 0, \quad W_3 = 12$$

is unique.

Proof. Let C be such a code, and let us write C as an array with 10 columns and 40 rows. Since $A(9,4) = 20$ each column of the array contains exactly 20 zeros and 20 ones. Counting the total number of ones in the array gives

$$5W_5 + 7W_7 + 9W_9 = 200 - 3W_3 = 164 . \tag{11}$$

Evidently,

$$W_5 + W_7 + W_9 = 40 - W_3 = 28 . \tag{12}$$

Taking into account that $W_9 \leq 1$ and solving (11),(12) for W_5 and W_7 we obtain the following two possibilities for the weight distribution of the code

$$W_5 = 16, \quad W_7 = 12, \quad W_9 = 0 \tag{13}$$

$$W_5 = 17, \quad W_7 = 10, \quad W_9 = 1 \tag{14}$$

As before, we let

$$S_i = \mathbb{F}_2^{10,5} \Big\backslash \left(\mathbb{F}_2^{10,5} \cap \mathcal{B}_{10}(D_i,2)\right) \qquad i = 0,1,\ldots 27 \tag{15}$$

denote the set of vectors of weight 5 at distance ≥ 4 from the code D_i given by (9). The following is a list of values of $|S_i|$ and $A(S_i,4)$ for the 28 sets defined in (15):

i	0	1	2	3	4	5	6	7	8	9	10	11	12	13
$\lvert S_i \rvert$	48	48	48	49	49	48	48	48	49	48	49	48	48	48
$A(S_i,4)$	13	14	13	16	16	16	15	15	17	16	15	15	15	15

(16)

i	14	15	16	17	18	19	20	21	22	23	24	25	26	27
$\lvert S_i\rvert$	48	48	49	54	51	49	50	48	48	48	48	48	48	48
$A(S_i,4)$	14	15	16	18	17	16	18	15	15	15	15	15	16	14

Since $W_5 \geq 16$ by (13) and (14), this rules out 17 out of the 28 codes in (9). To rule out the code D_{17} note that this code is essentially a Steiner triple system on the first nine positions, with the last position being entirely zero. Thus it is easy to see that all the $\binom{9}{4} - 12\cdot\binom{6}{1} = 54$ vectors of weight 5 that are at distance ≥ 4 from D_{17} must have one in the last position. Assuming weight distribution (13), we must have exactly $20 - 16 = 4$ vectors of weight 7 with one in the last position. This means that $12 - 4 = 8$ vectors of weight 7 have zero in the last position. But this is impossible since $A(9,4,7) = 4 < 8$. Similarly, if we assume weight distribution (14) then the vector of weight 9 cannot have zero in the last position, since otherwise all the 10 vectors of weight 7 must have one in the last position and the weight of the last column becomes $17 + 10 = 27 > 20$. Hence exactly $20 - 17 - 1 = 2$ out of the 10 vectors of weight 7 contain one in the last position, and again $10 - 2 = 8$ of them have zero in the last position, which is impossible. The code D_{20} may be ruled out in a similar fashion. It is the unique $(10,12,4,3)$ code consisting of 8 columns of weight 4 and two columns of weight 2. W.l.o.g. assume that the two columns of weight 2 are in the last two positions, and partition the 50 vectors in S_{20} according to their values in these positions. Namely, we define

$$S'_{00} = \{\, \underline{s} = (s_0, s_1, \ldots s_9) \in S_{20} \;:\; s_8 = 0, s_9 = 0 \,\}$$
$$S'_{01} = \{\, \underline{s} = (s_0, s_1, \ldots s_9) \in S_{20} \;:\; s_8 = 0, s_9 = 1 \,\}$$
$$S'_{10} = \{\, \underline{s} = (s_0, s_1, \ldots s_9) \in S_{20} \;:\; s_8 = 1, s_9 = 0 \,\}$$
$$S'_{11} = \{\, \underline{s} = (s_0, s_1, \ldots s_9) \in S_{20} \;:\; s_8 = 1, s_9 = 1 \,\}$$

It is easy to see by simple counting arguments that $|S'_{01}| = |S'_{10}| = 13$, $|S'_{11}| = 24$, and $S'_{00} = \emptyset$. Further, we have $A(S'_{01} \cup S'_{10}, 4) = 10$. Now assume that D_{20} may be completed to a $(10,40,4)$ code whose weight distribution is given by (13). Consider the first 28 rows consisting of the 12 vectors of weight 3 and 16 vectors of weight 5. The total number of ones seen in the last two positions of these 28 rows is at least $4 + 10 + 6\cdot 2 = 26$. Hence the total number of zeros is at most $28\cdot 2 - 26 = 30$. Yet the (10,40,4) code contains exactly $2\cdot 20 = 40$ zeros in the last 2 columns. Therefore the last 12 rows, which constitute a $(10,12,4,7)$ constant-weight code, must have at least $40 - 30 = 10$ zeros in the last two positions. But this is impossible since each column of a (10,12,4,7) code contains at most $A(9,4,7) = 4$ zeros. Assuming weight distribution (14) leads to a similar counting argument. The total number of ones seen in the last two positions of the first 30 rows is at least $4 + 10 + 7\cdot 2 + 1 = 29$. Hence the total number of zeros is at most $30\cdot 2 - 29 = 31$, and the $(10,10,4,7)$ subcode in the last 10 rows must contain at least $40 - 31 = 9$ zeros in the last two columns. Since $2\cdot A(9,4,7) = 8$, this is again impossible.

With the remaining nine codes we proceed as follows. For a vector $\underline{v} \in \mathbb{F}_2^{10,7}$ set $S_i(\underline{v}) = \mathbb{F}_2^{10,5} \backslash (\mathbb{F}_2^{10,5} \cap \mathcal{B}_{10}(D_i \cup \{\underline{v}\}, 2))$ and define $V_i = \{\underline{v} \in \mathbb{F}_2^{10,7} :$

$A(S_i(\underline{v}),4) \geq 16\}$. Thus V_i is the set of all $\underline{v} \in \mathbb{F}_2^{10,7}$ that do not preclude the existence of a $(10,16,4,5)$ code at distance ≥ 4 from $D_i \cup \{\underline{v}\}$. The sets V_i for the nine codes at hand are easy to compute, and since none of them is of large cardinality, we may readily find $A(V_i,4)$:

i	3	4	5	8	9	16	18	19	26
$\lvert V_i \rvert$	24	15	14	67	20	26	63	19	24
$A(V_i,4)$	7	7	8	11	8	7	9	7	12

Since $W_7 \geq 10$ by (13) and (14), this test rules out seven out of the remaining nine codes. We are left with the codes D_8 and D_{26}. Note that $A(V_8,4) = 11$, and hence if D_8 may be completed to a $(10,40,4)$ code, the weight distribution of this code must be given by (14). We therefore define $V_8' = \{\underline{v} \in \mathbb{F}_2^{10,7} : A(S_8(\underline{v}),4) \geq 17\}$. Computing V_8' shows $|V_8'| = 16$ and $A(V_8',4) = 6$. Hence the code D_8 is thus ruled out.

We now proceed with the only remaining candidate D_{26}. Let $U_{26} = S_{26} \cup V_{26}$. We have $|U_{26}| = 48 + 24 = 72$ and $A(U_{26},4) = 28$, which demonstrates the existence of a (10,40,4) code. The resulting code C is given by $D_{26} \cup \mathcal{X} \cup \mathcal{Y}$, where $\mathcal{X}$ consists of sixteen vectors of weight 5 and $\mathcal{Y}$ consists of twelve vectors of weight 7, as shown below:

$$D_{26} = \begin{array}{c} 1110000000 \\ 1001100000 \\ 1000011000 \\ 1000000110 \\ 0101010000 \\ 0100100010 \\ 0100001001 \\ 0011000100 \\ 0010010001 \\ 0010001010 \\ 0001000011 \\ 0000100101 \end{array} \qquad \mathcal{X} = \begin{array}{c} 0101101100 \\ 0100011110 \\ 0011111000 \\ 0010110110 \\ 0001011101 \\ 0000111011 \\ 0111100001 \\ 0110000111 \\ 1010100011 \\ 1010101100 \\ 1011001001 \\ 1011010010 \\ 1100010011 \\ 1100110100 \\ 1101000101 \\ 1101001010 \end{array} \qquad \mathcal{Y} = \begin{array}{c} 0011101111 \\ 0101110111 \\ 0110111101 \\ 0111011011 \\ 1001111110 \\ 1010011111 \\ 1011110101 \\ 1100101111 \\ 1101111001 \\ 1110111010 \\ 1111011100 \\ 1111100110 \end{array}$$

Note that the complement of $\mathcal{Y}$ forms a $(10,12,3,4)$ code isomorphic to D_{26} under the permutation $(1,3)(2,8)(4,5)(6,7)$ fixing positions 0 and 9. This indeed has to be so since D_{26} is the unique $(10,12,3,4)$ code that may be completed to a $(10,40,4)$ code C. To prove that C itself is unique we verify that $A(U_{26} \backslash \{\underline{x}\},4) = A(U_{26} \backslash \{\underline{y}\},4) = 27$ for every $\underline{x} \in \mathcal{X}$ and every $\underline{y} \in \mathcal{Y}$. Q.E.D.

Theorem 5. *The* $(10,40,4)$ *Best code is unique.*

Proof. It is an immediate corollary of propositions 1, 3 and 4 that any $(10,40,4)$ code, consisting either of only even-weight codewords or only odd-weight codewords, is equivalent to the code C given by (17). In particular, since the weight of all the four vectors in (1) is odd, the Best code $\mathcal{B}$ and the code (17) must be equivalent. This may be also seen directly by verifying $C = \{\underline{a} + \pi(\underline{c}) : \underline{c} \in \mathcal{B}\}$, where $\underline{a} = (0000010001)$ and $\pi = (0,2,1,6,5)(4,9,7)$. Thus in order to complete the proof of the uniqueness of the Best code, it remains to show that a $(10,40,4)$ code containing both even-weight and odd-weight codewords does not exist.

Assume to the contrary that C' is such a code. We may puncture C' to obtain a $(9,40,3)$ code C^*, and subsequently extend C^* by a parity-check coordinate to obtain a $(10,40,4)$ code C'' consisting of only even-weight codewords. By the preceding argument the code C'' must be equivalent to the Best code $\mathcal{B}$. Consider a graph $G(C^*)$ whose vertices are the 40 codewords of C^*, with two vertices connected by an edge iff the corresponding codewords are at distance 3 from each other. We claim that the graph $G(C^*)$ cannot be connected. Indeed, appending an additional coordinate to C^* such that the resulting code has minimum distance 4 is equivalent to finding a 2-coloring of $G(C^*)$ in such a way that no two vertices incident to the same edge are of the same color. It is easy to see by successive coloring that if $G(C^*)$ is connected such 2-coloring of $G(C^*)$ is unique, and corresponds to appending either even or odd parity-check coordinate. As C^* was obtained by puncturing C', the fact that C' contains both even- and odd-weight codewords implies that $G(C^*)$ is not connected. Now, since C^* is also a punctured version of C'', and C'' is equivalent to the Best code $\mathcal{B}$, it follows that there exists a way to puncture $\mathcal{B}$ such that the graph corresponding to the resulting code $\mathcal{B}^*$ is not connected. As the Best code is cyclic, puncturing in any coordiante produces equivalent codes $\mathcal{B}^*$. It is a matter of straightforward verification that $G(\mathcal{B}^*)$ is connected, and hence C' does not exist.

Q.E.D.

Corollary 6. *The* $(9,40,3)$ *code* $\mathcal{B}^*$ *is unique.*

Proof. This follows immediately from the uniqueness of the $(10,40,4)$ code $\mathcal{B}$ and the fact that $\mathcal{B}$ is cyclic. Q.E.D.

3. New bounds on $A(n,d)$

The uniqueness of the (10,40,4) and (9,40,3) codes makes it possible to improve the known upper bounds on $A(n,d)$ for $n = 10, 11, 12$.

Proposition 7. *A* $(10,79,3)$ *code does not exist.*

Proof. Let C be such a code. Consider an arbitrary position in C, say the last position, and define the codes A and B as follows

$$A = \{\underline{a} : (\underline{a}|0) \in C\}$$
$$B = \{\underline{b} : (\underline{b}|1) \in C\}$$

where $(\cdot|\cdot)$ denotes concatenation. Obviously the minimum distance of each of these codes is at least three, and $d(A,B) = \min_{\underline{a}\in A, \underline{b}\in B} d(\underline{a},\underline{b}) \geq 2$. Since $A(9,3) = 40$ we have $|A| \leq 40$ and $|B| \leq 40$. Furthermore since $|A| + |B| = 79$, we may assume without loss of generality that $|A| = 39$ and $|B| = 40$. Hence B must be equivalent to the unique (9,40,3) code $\mathcal{B}^*$.

Thus to establish the claim of the proposition, it remains to prove the non-existence of a (9,39,3) code A at distance ≥ 2 from $\mathcal{B}^*$. Let H be the set of holes of $\mathcal{B}^*$, that is the set of all $\underline{h} \in \mathbb{F}_2^9$ such that $d(\underline{h}, \mathcal{B}^*) = 2$. Evidently $|H| = 512 - 40 \cdot 10 = 112$ and $A \subset H$. Therefore it would suffice to show that $A(H,3) < 39$. To facilitate the search we partition the set H as follows

$$H_0 = \{\, \underline{h} = (h_0, h_1, \ldots h_8) \in H \; : \; h_8 = 0 \,\}$$
$$H_1 = \{\, \underline{h} = (h_0, h_1, \ldots h_8) \in H \; : \; h_8 = 1 \,\}$$

It is easy to see that $|H_0| = |H_1| = 56$, and we have $A(H_0,3) = A(H_1,3) = 16$. Hence $A(H,3) \leq A(H_0,3) + A(H_1,3) = 32$.

Q.E.D.

Theorem 8.

(a). $A(11,4) = A(10,3) \leq 78$

(b). $A(12,4) = A(11,3) \leq 156$

Proof. The second inequality follows by $A(n+1,d) \leq 2A(n,d)$. Q.E.D.

References

1. M.R. Best, "Binary codes with a minimum distance of four," *IEEE Trans. Inform. Theory*, vol. IT-26, pp. 738–742, 1980.
2. M.R. Best, A.E. Brouwer, F.J. MacWilliams, A.M. Odlyzko, and N.J.A. Sloane, "Bounds for binary codes of length less than 25," *IEEE Trans. Inform. Theory*, vol. IT-24, pp. 81–93, 1978.
3. J.-M. Goethals, "The extended Nadler code is unique," *IEEE Trans. Inform. Theory*, vol. IT-23, pp. 132–135, 1977.
4. J.H. Conway and N.J.A. Sloane, *Sphere Packings, Lattices and Groups*, New York: Springer-Veralg, 1988.
5. J.H. Conway and N.J.A. Sloane, "Quaternary constructions for the binary codes of Julin, Best, and others," preprint.
6. P. Delsarte and J.-M. Goethals, "Unrestricted codes with the Golay parameters are unique," *Discr. Math.*, vol. 12, pp. 211–224, 1975.
7. D. Julin, "Two improved block codes," *IEEE Trans. Inform. Theory*, vol. IT-11, p. 459, 1965.
8. D.E. Knuth, *The Art of Computer Programming: Sorting and Searching*, Reading: Addison-Wesley, 1973.
9. S. Litsyn and A. Vardy, "The uniqueness of the Best code," submitted to *IEEE Trans. Inform. Theory*.
10. F.J. MacWilliams and N.J.A. Sloane, *The Theory of Error-Correcting Codes*, New York: North-Holland, 1977.
11. V. Pless, "On the uniqueness of the Golay codes," *J. Comb. Theory*, vol. 5, pp. 215-228, 1968.
12. S. Snover, "The uniqueness of the Nordstrom-Robinson and the Golay binary codes," *Ph.D. Thesis*, Michigan State University, 1973.
13. N. Wax, "On upper bounds for error-detecting and error-correcting codes of finite length," *IEEE Trans. Inform. Theory*, vol. IT-5, pp. 168–174, 1959.

On Plotkin-Elias Type Bounds for Binary Arithmetic Codes

Gregory Kabatianski
Institute for Problems of Information Transmission
Ermolovoy 19, Moscow GSP-4
Russia

and

Antoine Lobstein
Centre National de la Recherche Scientifique
Télécom Paris, Dpt INF
46 rue Barrault, 75634 Paris Cedex 13
France

Abstract. We establish a new upper bound for binary arithmetic codes, which is asymptotically better than previously known bounds. We also discuss possible "candidates" such as Plotkin and Elias bounds for arithmetic codes over an arbitrary alphabet.

1 Arithmetic Distance, "Volume" of Arithmetic Sphere (Ball) and an Asymptotic Form of Hamming and Varshamov-Gilbert bounds for Arithmetic Codes

The binary arithmetic weight $w_a(x)$ of an integer x is defined (see [1]) as the minimal number of nonzero coefficients in the following representations of x:

(1) $$x = \sum_{i \geq 0} x_i \cdot 2^i, \quad x_i \in \{-1,0,1\} \text{ for all } i \geq 0.$$

The arithmetic distance $d_a(x,y)$ between integers x and y is defined by the usual formula

(2) $$d_a(x,y) = w_a(x-y).$$

It is known that for any integer x, there exists the so-called nonadjacent representation:

(3) $$x = \sum_{i \geq 0} x_i \cdot 2^i, \quad x_i \in \{-1,0,1\}, \ x_i \cdot x_{i+1} = 0 \text{ for all } i \geq 0.$$

The second author is greatly indebted to the first one for inviting him for a 6-week stay at the Institute for Problems of Information Transmission.

The number of nonzero coefficients in the nonadjacent representation of x is minimal, *i.e.*, is equal to $w_a(x)$.

Any set C of integers is called an arithmetic code. The minimal arithmetic distance, $d_a(C)$, of code C is defined as usual by:

$$d_a(C) = \min_{x\in C, y\in C, x\neq y} d_a(x,y). \tag{4}$$

There are different well-known ways of extending the above definitions to the case of the ring Z_N of residues of integers mod N (see [1,2]). For the most important cases, $N = 2^n, 2^n \pm 1$, these definitions coincide and, in fact, are given by the same formulas (1)-(3), but with equalities within the ring Z_N.

We say that the length of code C is equal to n if $C \subseteq A_n$, where A_n is either one of the rings Z_N, $N = 2^n, 2^n \pm 1$, or a set $B_n = \{0,1,\ldots,2^n-1\}$.

Now the number $S_{n,w}$ of elements of A_n with given arithmetic weight w is "approximately" equal to $\binom{n-w}{w}\cdot 2^w$, or more exactly:

$$\lim_{n\to\infty,\ w/n=\omega} 1/n \log_2 S_{n,w} = \omega + (1-\omega)\, H_2\left(\frac{\omega}{1-\omega}\right), \tag{5}$$

where H_2 is the binary entropy function (see [3] for more detail).

Of course, equality (5) holds also for $\omega \leq 1/3$ if one replaces "volume" $S_{n,w}$ of arithmetic sphere $A_{n,w}$ of radius w by "volume" $\tilde{S}_{n,w} = \sum_{0\leq i\leq w} S_{n,i}$ of arithmetic ball $\tilde{A}_{n,w}$ of same radius. The application of usual arguments of packings and coverings gives corresponding Hamming and Varshamov-Gilbert (V-G) bound respectively. We need to make one remark: for the space B_n, part of the points of a ball with its centre inside B_n can be outside B_n. But this cannot change the asymptotic form of Hamming and V-G "arithmetic" bounds.

Denote by $m_a(n,d)$ the maximal cardinality of a code of length n and minimal arithmetic distance at least d, and by $R_a = R_a(n,d) = 1/n \log_2 m_a(n,d)$ the rate of a corresponding maximal code. Then the following asymptotic inequalities hold (see [3]):

$$R_a \lesssim (1-\partial/2)\left(1-H_2\left(\frac{\partial/2}{1-\partial/2}\right)\right), \tag{6}$$ - Hamming arithmetic bound;

$$R_a \gtrsim (1-\partial)\left(1-H_2\left(\frac{\partial}{1-\partial}\right)\right), \tag{7}$$ - V-G arithmetic bound,

where $\partial = d/n$. Note that V-G arithmetic bound guarantees the existence of codes with asymptotically nonzero rate for $\partial < 1/3$; on the other hand, Hamming arithmetic bound shows that rate tends to zero if $\partial \geq 2/3$ when $n\to\infty$. It can be immediately improved (1/2 instead of 2/3), for nonadjacent forms show that the arithmetic weight of any $x \in A_n$ cannot be larger than $(n+1)/2$.

Another simple remark is that arithmetic distance between x and y cannot be larger than Hamming distance between their binary representations. Hence, one can apply any known upper bound for codes in binary Hamming space, in particular the best-known asymptotic upper bound - the McEliece-Rodemich-Rumsey-Welch (MRRW) bound ([4]).

In next Section, we derive a new upper bound for binary arithmetic codes, which is better than the above mentioned application of MRRW bound.

2 A New Upper Bound for Binary Arithmetic Codes

We shall use a well-known idea due to Elias-Bassalygo (see [5]) to estimate the cardinality of the best code in some metric space, by the cardinality of the best code in some suitably chosen subspace.

Let X be an abelian group with some distance $d(x,y)$, such that $d(x+a,y+a) = d(x,y)$ for any $a \in X$. Any subset $Y \subset X$ can be considered as a metric space, too. For any metric space T let $m(T,d)$ be the maximal cardinality of a code with minimal distance d within space T.

Lemma [6]. For any $Y \subset X$ the following inequality holds:

$$\frac{m(X,d)}{|X|} \leq \frac{m(Y,d)}{|Y|}. \tag{8}$$

For completeness, we give here the proof of this lemma. Let C be a maximal code in X with distance d. For any $a \in X$, code $C_a = \{a + c : c \in C\}$ has same distance and cardinality. Consider now code $D_a = C_a \cap Y$, which has distance at least d. As $D_a \subseteq Y$, $m(Y,d) \geq |D_a|$ for any $a \in X$. Now the usual permutation of summations

$$\sum_{a\in X} |D_a| = \sum_{a\in X}\sum_{c\in C} |\{c+a\} \cap Y| = \sum_{c\in C}\sum_{a\in X} |\{c+a\} \cap Y| = \sum_{c\in C} |Y|$$

shows that there exists $a \in X$ such that: $|D_a| \geq \frac{|C||Y|}{|X|}$. ■

Spontaneous extension of Elias-Bassalygo bound to binary arithmetic codes means that we must let $X = A_n$ and Y be the arithmetic sphere $A_{n,w}$ of radius w. But we have not enough skill to use $A_{n,w}$ directly, *i.e.*, we cannot prove an analogue of Johnson bound for arithmetic constant weight codes.

Therefore we use the larger set $J^{(3)}_{n,w}$ of ternary vectors of length n and Hamming weight w; more exactly, we embed sphere $A_{n,w}$ into ternary Hamming sphere $J^{(3)}_{n',w}$ of same radius, as the nonadjacent representation exists for any element in A_n, where $n' = n$ for rings Z_N and $n' = n + 1$ for B_n.

Using the above Lemma together with the following obvious inequality:

$$d_a(x,y) \le d_H(\tilde{x}, \tilde{y}),$$

where $\tilde{x}$ and $\tilde{y}$ are ternary vectors of corresponding minimal (nonadjacent) binary representations of x and y, we get:

$$\frac{m_a(n,d)}{2^n} \le \frac{m(A_{n,w},d)}{S_{n,w}} \le \frac{m_H(n,w,d;q=3)}{S_{n,w}}, \tag{9}$$

where $m_H(n,w,d;q=3)$ is the maximal cardinality of a ternary constant weight code of length n, weight w and distance d. From Johnson bound for ternary codes,

$$m_H(n,w,d;q=3) \le \frac{\frac{2}{3}nd}{w^2 - \frac{4}{3}nw + \frac{2}{3}nd},$$

it follows that

$$m_H(n,w,d;q=3) \le \frac{2}{3}nd \quad \text{for } w \ge w(n,d) = 1 + \left\lceil \frac{2}{3}n - \sqrt{\frac{4}{9}n^2 - \frac{2}{3}nd} \right\rceil.$$

Substituting $w = w(n,d)$ in (9), one gets a new bound, with the following asymptotic form:

$$R_a \lesssim (1-\rho)(1 - H_2(\tfrac{\rho}{1-\rho})), \tag{10}$$

where $\rho = \frac{2}{3} - \sqrt{\frac{4}{9} - \frac{2}{3}\partial}$ and $\partial = d/n$.

This bound is better than the MRRW bound (see Table below).

∂	New upper bound	MRRW bound
0.05	0.8042	0.8251
0.1	0.6571	0.6927
0.2	0.4138	0.4613
0.25	0.3100	0.3537
0.3	0.2170	0.2502
0.33	0.1667	0.1933
0.35	0.1357	0.1581
0.4	0.0685	0.0814
0.45	0.0199	0.0252
0.47	0.0077	0.0104
0.49	0.00093	0.00147

3 Concluding Remarks on Elias-Plotkin Type Bounds for Arithmetic Codes

In our opinion, the main barrier against the generalization of even such simple bounds as Plotkin is that arithmetic space is not an association scheme. We are not the first to note it (see [7]); we just draw attention that even "first polynomial" ([8,9]) does not exist for arithmetic space. On the other hand, we have a good candidate for arithmetic Plotkin bound, as a formula for the sum of arithmetic weights of all codewords of cyclic *AN*-codes is known:

(11) $$\sum_{0\le i\le A^*-1} w_a(Ai) = n\cdot\left(\left\lfloor\frac{qA^*}{q+1}\right\rfloor - \left\lfloor\frac{A^*}{q+1}\right\rfloor\right),$$

where $AA^* = q^n - 1$ (see [2] for $q = 2$ and [10] for the general case). Hence, at least for cyclic *AN*-codes, the following analogue of Plotkin bound (see [10]) holds:

(12) $$\frac{d}{n} \lesssim \frac{q-1}{q+1}.$$

Let us suppose that an analogue of (11) can be proved for arbitrary arithmetic codes, *i.e.*, the following inequality is true:

$$\frac{1}{|C|\cdot(|C|-1)}\sum_{c,c'\in C} d_a(c,c') \le n\,\frac{q-1}{q+1},$$

or that any weaker inequality can be proved - but sufficient for asymptotic bound (12), *i.e.*, the number of codewords in q-ary arithmetic codes with distance $n\,\frac{q-1}{q+1}$ cannot increase exponentially fast with length n.

Then, applying Lemma with n-"dimension" q-ary arithmetic space $X = A_n^{(q)} = \{0,1,\dots,q^n-1\}$ and l-"dimension" q-ary arithmetic space $Y = A_l^{(q)}$ with $l = \frac{q+1}{q-1}\cdot d$ would give the desired arithmetic Plotkin bound

(12) $$R_a \lesssim 1 - \partial\,\frac{q+1}{q-1}.$$

Certainly, if some analogue of Johnson bound for arithmetic constant weight codes is proved, then one can get an analogue of Elias bound by considering q-ary arithmetic sphere with suitable radius $w = w_q(n,d)$. Let us look at an opportunity of generalizing results of Section 2. There are different generalizations of nonadjacent representations to the q-ary case ([11,3]). We consider one of them (see [3]), which seems more natural for dealing with the volume of arithmetic sphere.

Any integer x has exactly one so-called constant-sign representation

(13) $$x = \sum_{i\ge 0} x_i\cdot q^i,\quad x_i\in\{-(q-1),\dots,-1,0,1,\dots,q-1\}\text{ where}$$

(14) $x_i \cdot x_{i+1} \geq 0$, and if $|x_{i+1}| = q-1$ then $x_i = 0$ (for all $i \geq 0$).

The number of nonzero coefficients in this representation is minimal among all possible representations (13), *i.e.*, is equal to the arithmetic weight of x. The constant-sign representation has a simple structure - elements x_i are divided into blocks of consecutive nonzero elements of the same sign, any two blocks are separated by at least one zero element, and an element equal to $\pm(q-1)$ can occur only on the "left" position of a block.

As in Section 2, we can apply Lemma to $X = A_n^{(q)}$ and $Y = J_{n,w}^{(Q)}$, the sphere of radius w in the Q-ary n-dimensional Hamming space, with $Q = 2q-1$.

Unfortunately, the resulting asymptotic upper bound becomes equal to zero at point

$$\partial = \frac{(q-1)(2q+5)}{2(q+1)^2}$$

which is larger for $q > 2$ than point $\frac{q-1}{q}$, point where the best-known upper bounds for codes in q-ary Hamming space become equal to zero. This means that simply the trivial application of known upper bounds for Hamming space gives better results for $q > 2$.

In [12] was presented another upper bound, which is better than known q-ary upper bounds for Hamming space. But there is a gap in the proof given in [12]: namely the assumption that, without loss of generality, all codewords do not contain, in their usual q-ary representation, coefficients equal to $q-1$. In fact, this is a very strong restriction (look at the binary case!), and considering only such codes involves replacing in Lemma the q-ary sphere $A_{n,w}^{(q)}$ by the rather small set $\{x \in A_{n,w}^{(q)} : 0 \leq x_i \leq q-2\}$, which immediately changes the bound. We must note also that Lemma guarantees only the existence of some "shift" $C+a$ of the original code C, with the cardinality of the intersection with set Y not less than average; the author of [12] deals with AN-codes, for which computing the minimal distance is equivalent to computing the minimal arithmetic weight of nonzero codewords. This is no more true for "shifts" of AN-codes, still it is used in [12].

Therefore we have today, for q-ary arithmetic codes, nothing better than the application of the best-known upper bound for q-ary codes in Hamming space. The authors complain of it, but would not bear the blame and just draw attention of people, dealing with upper bounds, to this miserable situation.

References

1. W.W. Peterson, E.J. Weldon, Jr.: Error-correcting codes. Cambridge: MIT Press 1972
2. J.L. Massey, O.N.Garcia: Error-correcting codes in computer arithmetics. In: Advances in Information Systems Science 4, New York-London: Plenum Press 1972, Ch. 5
3. G.A. Kabatianski: Bounds on the number of code words in binary arithmetic codes. Problems of Information Transmission 12-4, 277-283 (1976)
4. R.J. McEliece, E.R. Rodemich, H.C. Rumsey, Jr., L.R. Welch: New upper bounds on the rate of a code via the Delsarte-MacWilliams inequalities. IEEE IT-23, 157-166 (1977)
5. E.R. Berlekamp (ed.): Key papers in the development of coding theory. New-York: IEEE Press 1974
6. L.A. Bassalygo: New upper bounds for codes correcting errors. Problems of Information Transmission 6-4, 41-45 (1965), in Russian
7. P. Solé: A Lloyd theorem in weakly metric association schemes. Europ. J. Combinatorics 10, 189-196 (1989)
8. Ph. Delsarte: An algebraic approach to the association schemes of coding theory. Philips Research Reports Suppl. 10, 1973
9. V.I. Levenshtein: Designs as maximum codes in polynomial metric spaces. Acta Applicandae Mathematicae 25, 1-I, 1992
10. I.M. Boyarinov, G.A. Kabatianski: Arithmetic (n, A)-codes over an arbitrary base. Sov. Phys. Dokl. 20-4, 247-249 (1975)
11. W.E. Clark, J.J. Liang: On modular weight and cyclic nonadjacent forms for arithmetic codes. IEEE-IT 20, 767-770 (1974)
12. S. Ernvall: On bounds for nonbinary arithmetic codes. Ars Combinatoria 29-B, 85-89 (1990)

Bounds on generalized weights

Gérard Cohen[1], Llorenç Huguet[2] and Gilles Zémor[1]

[1] ENST
46 rue Barrault,
75634 Paris 13 France

[2] UIB
Cra. de Valldemossa, km 7.5
07071 Palma, Spain

1 Introduction

The concept of generalized weights, a natural extension of the notion of minimal distance, has received renewed interest lately, [6], [10], [8], [4]. The reasons for that are manyfold: it is a challenging combinatorial problem and has a wealth of more or less unexpected applications; to name but a few:

- In cryptography [11], [7].
- In state complexity of trellis diagram [9].
- In designing codes for the switching Multiple-Acces Channel [12].

The outline of the paper is as follows.

In section 2 we derive upper bounds on the generalized distances of linear codes by adapting the classical bounds of coding theory. Section 3 deals with lower bounds with special emphasis on BCH codes.

Let us denote by $C[n,k,d]$ a linear binary code with length n, dimension k and minimum-Hamming-distance d. A vector v of $\mathbf{F}_2^n$, the n-dimensional vector space over $\mathbf{F}_2$, will be frequently identified with its support $supp(v) = \{j, v(j) \neq 0\}$, i.e. the set of non-zero coordinates of v. We will therefore write $v \cup w$ instead of $supp(v) \cup supp(w)$. The weight of v will be denoted by $|v| = |supp(v)|$.

The i-distance of C, denoted by $d_i(C)$ or simply d_i, is the minimum size of the union of the supports of i linearly independent codewords in C.

2 Upper bounds

Let us denote by $B_i(n,d)$ the maximum size of a linear code of length n and i-distance $d_i \geq d$.

2.1 generalized Plotkin bounds

For any i vectors $c_1, c_2, \ldots, c_i$ of $\mathbf{F}_2^n$ let us define

$$W_i(c_1, c_2, \ldots, c_i) = \# \cup_{h=1}^{i} supp(c_h).$$

Let us now estimate the average value E of W_i over all ordered i-tuples of codewords of rank i in a $[j, k]$ linear code C.

$$E = \frac{1}{(M-1)(M-2)\ldots(M-2^{i-1})} \sum_{\substack{(c_1, c_2, \ldots, c_i) \in C^i \\ rk\{c_1 \ldots c_i\}=i}} W_i(c_1, c_2, \ldots, c_i)$$

where $M = 2^k$ is the number of codewords of C. Set

$$\Sigma = (M-1)(M-2)\ldots(M-2^{i-1})E \tag{1}$$

Consider now the $M \times j$ array made up of all codewords, and evaluate the contribution per column to Σ. An i-tuple of entries in a given column associated to i independent rows does *not* contribute a 1 to Σ iff these i entries are all zeros. Since each column has $2^{k-1} - 1$ zeroes associated to non-zero codewords, the number of such i-tuples is

$$L = (2^{k-1} - 1)(2^{k-1} - 2)\ldots(2^{k-1} - 2^{i-1})$$

and

$$\Sigma = j\left((M-1)(M-2)\ldots(M-2^{i-1}) - L\right).$$

From this and (1) we easily deduce

$$E = \frac{(2^i - 1)}{2^i} \frac{2^k}{(2^k - 1)} j.$$

E is an upperbound to the i-distance d_i of any $[j, k]$ code. Therefore considering an optimal code (i.e. such that $B_i(j, d_i) = 2^k$), we get

Proposition 1. *We have*

$$B_i(j, d_i) \leq \frac{d_i}{d_i - \frac{2^i - 1}{2^i} j}$$

Next we need

Lemma 2. *For all j we have*

$$B_i(n, d_i) \leq 2^{n-j} B_i(j, d_i)$$

Proof. take a maximal $[n,k]$ code with $2^k = B_i(n,d_i)$ and shorten it $n-j$ times, to obtain a $[n, k-(n-j)]$ code with i-distance d_i.

□

Now choosing $j = \lfloor \frac{2^i}{2^i-1} \rfloor - 1$, setting $\delta_i = d_i/n$ $R = n/k$ and taking asymptotics, we get

Proposition 3 (generalized Plotkin bound). *For n large enough, any $[n,k]$ code of i-distance $\delta_i n$ satisfies*

$$R \leq 1 - \frac{2^i}{2^i - 1}\delta_i$$

2.2 A generalized Elias bound

Definition 4. Let C be any subset of $\mathbf{F}_2^n$. We define

$$d_2(C) = \min_{\substack{x,y,t \in C \\ x \neq y, y \neq t, x \neq t}} |(x+t) \cup (y+t)|$$

Note that if C is a linear code, $d_2(C) = \min_{x \neq y, y \neq 0, x \neq 0} |x \cup y|$ coincides with the "usual" generalized distance. The purpose of this generalization is the straightforward lemma.

Lemma 5. *d_2 is invariant under translation. For any set $C \subset \mathbf{F}_2^n$, and any $\tau \in \mathbf{F}_2^n$,*

$$d_2(C + \tau) = d_2(C)$$

Definition 6.

$$A_2(n,d) = \max\{|C|, C \subset \mathbf{F}_2^n, d_2(C) \geq d\}$$
$$A_2(n,d,w) = \max\{|C|, C \subset \mathbf{F}_2^n, d_2(C) \geq d, \forall x \in C, w(x) = w\}$$

Lemma 7. *We have*

$$A_2(n,d) \leq \frac{2^n A_2(n,d,w)}{\binom{n}{w}}$$

Proof. This is essentially the Elias-Bassalygo lemma, [1]. Take a code C realizing $A_2(n,d)$ and consider its 2^n translates $C+\tau, \tau \in \mathbf{F}_2^n$. They constitute a $A_2(n,d)$-covering of $\mathbf{F}_2^n$, and in particular of the set of vectors of weight w. Thus one of these translates - in itself a code with $d_2 \geq d$ by lemma 5 - must contain at least $\binom{n}{w}A_2(n,d)2^{-n}$ such vectors. Hence

$$\binom{n}{w}A_2(n,d)2^{-n} \leq A_2(n,d,w)$$

□

Lemma 8. *If* $d - 3w + \frac{3w^2}{n} > 0$ *then*

$$A_2(n,d,w) \leq \frac{d}{d-3w+\frac{3w^2}{n}}$$

Proof. Let C be a code of length n, constant weight w, size $|C| = M$, with $d_2(C) = d$. We evaluate in two ways the quantity

$$T = \sum_{x\neq y, y\neq t, x\neq t} |x+t \cup y+t|. \tag{2}$$

Since all of the terms inside the sum are bigger than d we have

$$T \geq M(M-1)(M-2)d. \tag{3}$$

Next consider the separate contribution of every coordinate i to T, i.e. write T as

$$T = \sum_{i=1}^{n} \sum_{x\neq y, y\neq t, x\neq t} |x(i)+t(i) \cup y(i)+t(i)|$$

where, abusing notation, $\cup$ stands for a logical "or". Let w_i be the number of vectors of C that have a one in coordinate i. The sum

$$\sum_{x\neq y, y\neq t, x\neq t} |x(i)+t(i) \cup y(i)+t(i)|$$

is equal to

$$3w_i(M-w_i)(M-w_i-1) + 3(M-w_i)w_i(w_i-1) = 3w_i(M-w_i)(M-2).$$

Together with (3) this yields

$$M(M-1)d \leq 3\sum_{i=1}^{n} w_i(M-w_i) \tag{4}$$

Since $\sum_{i=1}^{n} w_i = Mw$, the sum in (4) is maximum if $w_i = Mw/n$. Hence

$$(M-1)d \leq 3wM(1-\frac{w}{n})$$

which reduces to

$$M \leq \frac{d}{d-3w+\frac{3w^2}{n}}.$$

□

Let us now derive an Elias-type asymptotic upper bound on the quantity $R_2 = n^{-1}\log A_2(n, n\delta)$ by minimizing $A_2(n,\delta)$ in lemma 7 over every possible choice of the normalized weight $\omega = w/n$. By lemma 7

$$R_2 \leq \min_{\omega}\{1-H(\omega)+n^{-1}\log A_2(n,n\delta,w)\}.$$

The minimum is attained when ω is the smallest root of $3\omega^2 - 3\omega + \delta = 0$, giving

Proposition 9. *Any code of length n and second distance $n\delta$ satisfies, when n goes to infinity,*

$$R_2 \leq 1 - H\left(1/2 - 1/2\sqrt{1-4\delta/3}\right)$$

The same line of reasoning generalizes to arbitrary d_i and yields the following proposition, which will be elaborated on in a forthcoming paper.

Proposition 10. *Let R_i be the largest possible asymptotic rate of a code with relative i-distance $\delta_i = d_i/n$. We have*

$$R \leq 1 - H(\lambda)$$

where λ is the smallest positive root of the equation

$$\delta_i = 1 - x^{i+1} - (1-x)^{i+1}.$$

3 Lower bounds

The best asymptotic lower bound is the following VG-type bound, which is derived by random coding arguments, and is due to Wei [13], see also [3].

Proposition 11. *Given δ_i, there exist asymptotically $[n, nR]$ linear codes of i-distance $d_i \geq n\delta_i$ with a rate R arbitrarily close to*

$$\mathcal{R} = 1 - \frac{1}{i}(H(\delta_i) + \delta_i \log(2^i - 1)).$$

In what follows we derive some constructive lower bounds on the dual i-distance, with special emphasis on BCH codes.

Let $\overline{C}[n, n-k, \overline{d}]$ be the dual of $C[n, k, d]$. The i-distance of $\overline{C}$ is written $\overline{d}_i$. Denote for short $d_i(m, t)$ (resp. $\overline{d}_i(m, t)$) the i-distance of the t-error-correcting BCH code of length $2^m - 1$ (resp. of its dual).

We use the following straightforward generalization of a result of [5], already implicit in [14].

Proposition 12. *Shortening an $[n, k, d]$ code C on $\overline{d}_i$ positions corresponding to the support of a subcode of dimension i in $\overline{C}$ gives a*

$$C_{\overline{d}_i}[n - \overline{d}_i, k - \overline{d}_i + i, d] \text{ code.}$$

Let us apply the previous proposition to get lower bounds on $\overline{d}_m(m, 2)$.

- $m = 5$. Suppose $\overline{d}_5(5, 2) \leq 24$. The C_{24} would be a $[7, 2, 5]$ code contradicting Griesmer bound. Hence $\overline{d}_5(5, 2) \geq 25$.
- $m = 6$. Suppose $\overline{d}_6(6, 2) \leq 54$. Then C_{54} is an impossible $[9, 3, 5]$ code and $\overline{d}_6(6, 2) \geq 55$.
- $m = 7$. Although $\overline{d}_7(7, 2) = 113$ is not ruled out by the Griesmer bound (a $[14, 7, 5]$ satisfies it), a $[14 - j, 7 - j, 5]$ code does not exists for $j \leq 3$. Hence $\overline{d}_7(7, 2) \geq 116$.

More generally, $\overline{d}_m(m, 2)$ is at least equal to the first integer j, for which there exists a code of parameters $[n = 2^m - 1 - j, 2^m - 1 - 2m - j + i, 5]$. By use of the sphere covering bound, this gives:

$$n^2 < 2^{2m-i+1}$$

and

Proposition 13. *Let C be a 2 error correcting BCH code with length $2^m - 1$ then*

$$\overline{d}_i(m, 2) \geq 2^m - 2^{\frac{2m+1-i}{2}}$$

Using the same line of reasoning, we get the following lower bound on $\overline{d}_i(m, t)$ for t fixed larger than 2 and m large enough:

$$\overline{d}_i(m, t) \geq 2^m - \left(\frac{t}{e}\right) 2^{m - \frac{i}{t}}.$$

Using proposition 12, we get a few improvements on the tables of [4].

$$\overline{d}_6(7, 2) \geq 112$$
$$\overline{d}_7(8, 2) \geq 232$$
$$\overline{d}_8(8, 2) \geq 240$$

Also in [4], $d_8(6,2)$ should read 17 instead of 18. The complete weight hierarchies of two-error correcting BCH codes are known for $m \le 6$ ([4]) and read

$$m=4 \begin{cases} C[15,7].\{d_i(4,2)\} = \{5,8,10,11,13,14,15\} \\ \overline{C}[15,8].\{\overline{d}_i(4,2)\} = \{4,7,9,10,12,13,14,15\} \end{cases}$$

$$m=5 \begin{cases} \{d_i(5,2)\} = \{5,8,10,12,13,[15-19],[21-31]\} \\ \{\overline{d}_i(5,2)\} = \{12,18,21,23,25,26,28,29,30,31\} \end{cases}$$

$$m=6 \begin{cases} \{d_i(6,2)\} = \{5,8,10,11,[13-15],[17-21],[23-27],[29-39],[41-63]\} \\ \{\overline{d}_i(6,2)\} = \{24,36,42,48,52,55,57,58,[60-63]\} \end{cases}$$

where $[i,j]$ means that all elements in the interval are realized as generalized distances.

References

1. L. A. Bassalygo, **Новые верхние границы для кодов, исправляющих ошибки**, Problemy Peredachi Informatsii, 1 (1965), pp. 41–45.
2. H. Chung, *The second generalized Hamming weight of double-error correcting binary BCH codes and their dual codes*, in AAECC 9, Springer-Verlag, Lec. N. Comp. Sci. 539, 1991.
3. G. Cohen and G. Zémor, *Intersecting codes and independent families.* Submitted to IEEE Trans on Inf. Theory.
4. G. L. Feng, K. K. Tzeng, and V. K. Wei, *On the generalized Hamming weights of several classes of cyclic codes*, IEEE Trans. on Inf. Theory, IT-38 (1992), pp. 1125–1130.
5. H. J. Helgert and R. D. Stinaff, *Shortened BCH codes*, IEEE Trans. on Inf. Theory, IT-19 (1973), pp. 818–820.
6. T. Helleseth, T. Kløve, and Ø. Ytrehus, *Generalized Hamming weights of linear codes*, IEEE Trans. on Inf. Theory, IT-38 (1992), pp. 1133–1140.
7. L. Huguet, *Coding scheme for a wire-tap channel using regular codes*, Discrete Math., 56 (1985), pp. 191–201.
8. G. Kabatiansky, *On second generalized Hamming weight*, in Intern. Workshop on Algebraic and Combinatorial Coding Theory, Bulgaria, 1992, pp. 98–100.
9. T. Kasami, T. Takata, T. Fujiwara, and S. Lin, *On the optimum bit order with respect to the state complexity of treillis diagrams for binary linear codes*, IEEE Trans. on Inf. Theory, IT-39 (1993), pp. 242–245.
10. T. Kløve, *Upperbounds on codes correcting asymmetric errors*, IEEE Trans. on Inf. Theory, IT-35 (1989), pp. 797–810.

11. L. H. OZAROW AND A. D. WYNER, *Wire-tap channel II*, AT and T. B.S.T.J., Vol. 63 (1984), pp. 2135–2157.
12. P. VANROOSE, *Code construction for the noiseless binary switching multiple access channel*, IEEE Trans. on Inf. Theory, IT-34 (1988), pp. 1100–1106.
13. V. K. WEI, *Generalized Hamming weights for linear codes*, IEEE Trans. on Inf. Theory, IT-37 (1991), pp. 1412–1418.
14. V. A. ZINOVIEV AND S. N. LITSYN, *Shortening of codes*, Problemy Peredachi Informatsii, 20 (1982), pp. 3–11.

Threshold effects in codes

Gilles Zémor

Ecole Nationale Supérieure des Télécommunications
Dépt. Réseaux
46 rue Barrault, 75634 Paris Cedex 13, France

Abstract. A theorem of Margulis states the existence of a threshold phenomenon in the probability of disconnecting a graph, given that each of its edges is independently severed with some probability p. We show how this theorem can be reinterpreted in the coding context: in particular we study the probability $f_C(p)$ of residual error after maximum likelihood decoding, when we submit a linear code C to a binary symmetric channel with error probability p. We show that the function $f_C(p)$ displays a threshold behaviour i.e. jumps suddenly from almost zero to almost one, and how the acuteness of the threshold effect grows with the minimal distance of C. Similar results for the erasure channel are also discussed.

1 Introduction

Let C be an $[n, k, d]$ linear code, and let us submit its codewords to a binary symmetric channel with error probability p. We are interested in the residual error probability after maximum-likelihood decoding. Let us recall that maximum-likelihood decoding of a received vector v consists of choosing the closest (for the Hamming distance) codeword c, and if there are several codewords equally close to v, choosing one of them according to some arbitrary scheme.

We wish to prove a threshold behaviour of the residual error probability after maximum-likelihood decoding. Let us recall that, for growing n, the average behaviour of this function over all linear codes with rate R is well known. It is an exponentially small function of the block length whenever $p < H^{-1}(1 - R)$, and jumps suddenly from almost 0 to almost 1 around the value $p = H^{-1}(1-R)$. Our purpose is to show that *all* families of linear codes whose minimal distance grows with block length (however slowly) display such a threshold behaviour, i.e. are such that $f_C(p)$ jumps suddenly from almost zero to almost one around some threshold probability θ (that may of course be smaller than $H^{-1}(1 - R)$).

Since we are dealing with linear codes, we shall focus on syndrome decoding. We shall use the following notation and terminology. Let $\mathbf{H}$ be an $r \times n$ parity-check matrix of the code C with $r = n - k$. Let σ be the syndrome function.

$$\sigma : \mathbf{F}_2^n \to \mathbf{F}_2^r$$
$$v \mapsto \mathbf{H}\,{}^t v$$

For any binary vector v, we denote by $|v|$ its weight. For any vector s of $\mathbf{F}_2^r$, define $w(s) = \min\{|v|, \sigma(v) = s\}$, and

$$S_i = \#\{s \in \mathbf{F}_2^r, w(s) = i\}.$$

Any syndrome decoding scheme consists of associating an error pattern to each $s \in \mathbf{F}_2^r$, i.e. it is a function:

$$\epsilon : \mathbf{F}_2^r \to \mathbf{F}_2^n$$

such that $\forall x \in \mathbf{F}_2^r$, $\sigma\epsilon(x) = x$. The set $\epsilon(\mathbf{F}_2^r)$ is the set of decodable error vectors. The residual errror probability associated to the decoding scheme ϵ is the probability that a random error vector is not decodable. Its expression is, as a function of the error probability p,

$$f_C^\epsilon(p) = 1 - \sum_{s \in \mathbf{F}_2^r} p^{|\epsilon(s)|}(1-p)^{n-|\epsilon(s)|}.$$

The function ϵ is a maximum-likelihood decoding scheme if

$$\forall s \in \mathbf{F}_2^r, \quad |\epsilon(s)| = w(s).$$

The residual error probability when ϵ is a maximum likelihood decoding scheme is therefore

$$f_C(p) = 1 - \sum_{s \in \mathbf{F}_2^r} p^{w(s)}(1-p)^{n-w(s)} = 1 - \sum_{i=0}^{n} S_i p^i (1-p)^{n-i}.$$

We shall show that the residual error probability $f_C(p)$ always behaves in a "threshold manner", i.e. jumps suddenly from almost zero to almost one in a way dictated by the minimum distance $d(C)$ of C. More precisely

Theorem 1. *For any $\varepsilon, \alpha > 0$, there exists $\delta > 0$ such that whenever the minimum distance $d(C)$ of a linear code C satisfies $d(C) > \delta$, then*

$$m(\{x, \varepsilon < f_C(x) < 1 - \varepsilon\}) < \alpha.$$

where, $m(X)$ denotes the Lebesgue measure of X.

Theorem 1 makes it therefore natural to define the *threshold* probability $\theta(C)$ associated to C as the real number such that $f_C(\theta) = 1/2$. It can also be proved that whenever $p < \theta$, the residual error probability $f_C(p)$ is upper bounded by an exponentially decreasing function of the minimum distance $d(C)$. More precisely,

Theorem 2. *There exists a positive constant T such that for any linear code C, and for any $p < \theta(C)$, we have*

$$f_C(p) \leq e^{-Td(C)(\theta-p)^2}$$

The paper is organised as follows. In section 2 we state Margulis's threshold theorem, and relevant notions. In section 3, We apply Margulis's theorem to codes in order to study the threshold behaviour of residual error probability after decoding, first on the erasure channel, and then on the binary symmetric channel.

For vectors v and v' of $\mathbf{F}_2^n$, we shall write $v \subset v'$ (respectively $v \supset v'$) if the support of v is contained in (respectively contains) the support of v'. We shall also write $v \cup v'$ to denote the vector whose support is the union of the supports of v and v'.

2 The theorem of Margulis

Let W be a non-empty subset of $\mathbf{F}_2^n$. Denote by $f_W(p)$ the function

$$\begin{aligned} [0,1] &\rightarrow [0,1] \\ p &\mapsto f_W(p) = \sum_{v \in W} p^{|v|}(1-p)^{n-|v|} \end{aligned}$$

Denote by $\overline{W}$ the set of vectors not in W. Denote by $B(v,1)$ the Hamming ball of radius one centered on v. Denote by ∂W the "frontier of W", i.e.

$$\partial W = \{v \in W, B(v,1) \cap \overline{W} \neq \emptyset\}.$$

Margulis's theorem gives a sufficient condition for $f_W(p)$ to behave in a threshold manner. It associates to W the following

$$\Delta(W) = \min_{v \in \partial W} \#(B(v,1) \cap \overline{W}).$$

Call W an increasing (respectively decreasing) set if whenever $v \in W$ and $v \subset v'$ (respectively $v \supset v'$), then $v' \in W$. Call W monotonous if it is either increasing or decreasing. The theorem can be worded as follows.

Theorem 3. *For any $\varepsilon, \alpha > 0$, there exists $\delta > 0$ such that whenever an increasing set W satisfies $\Delta(W) > \delta$, then*

$$m(\{x, \epsilon < f_W(x) < 1-\varepsilon\}) < \alpha.$$

The following easy lemma shows that "increasing" may be replaced by "decreasing" in theorem 3.

Lemma 4. *Let* $\mathbf{1}$ *denotes the all-one vector. The mapping* $x \mapsto x^c = \mathbf{1} - x$ *transforms any increasing (decreasing) set* W *into a decreasing (increasing) set* W^c *such that* $\Delta(W^c) = \Delta(W)$ *and* $f_{W^c}(p) = f_W(1-p)$.

For a proof of theorem 3 see [1] or [2]. The key idea is to study the derivative of $f_W(p)$ and to notice that for any increasing set W,

$$\begin{aligned}\frac{d}{dp} f_W(p) &= \frac{1}{p} \sum_{v \in \partial} \#(B(v,1) \cap \overline{W}) p^{|v|}(1-p)^{n-|v|} \\ &\geq \frac{\Delta(W)}{p} \sum_{v \in \partial W} p^{|v|}(1-p)^{n-|v|}. \end{aligned} \tag{1}$$

Margulis's theorem was qualitatively improved by Talagrand [2]. He obtains

Theorem 5. *For any increasing set* W *we have*

$$\frac{d}{dp} f_W(p) \geq \left(\frac{2\Delta(W)}{p(1-p)} \right)^{1/2} f_W(p)(1 - f_W(p)). \tag{2}$$

Furthermore, there exists a universal constant K *such that*

$$\frac{d}{dp} f_W(p) \geq \frac{1}{K} \left(\frac{\Delta(W)}{p(1-p)} \right)^{1/2} f_W(p)(1 - f_W(p)) \sqrt{\ln \frac{1}{f_W(p)(1 - f_W(p))}}. \tag{3}$$

Note that lemma 4 transforms theorem 5 into an analogous statement about decreasing sets. We shall use the following

Definition 6. For any monotonous non-empty set $W \subset \mathbf{F}_2^n$ such that $W \neq \mathbf{F}_2^n$, let us call the *threshold probability* $\theta(W)$ of W, the real number $\theta \in [0,1]$ such that $f_W(\theta) = 1/2$.

The following application of theorem 5 will be useful to us in obtaining an asymptotic estimation of residual error probabilities.

Proposition 7. *There is a constant* T *such that for any increasing subset* W *of* $\mathbf{F}_2^n$, $W \neq \mathbf{F}_2^n$, *and for any* $p < \theta(W)$, *we have*

$$f_W(p) \leq e^{-T(\theta - p)^2 \Delta(W)}.$$

Proof. Set $F = -\ln f_W$. Remembering that $f_W(p) < 1$, we obtain from (3) that whenever $p < \theta$

$$\frac{d}{dp} F(p) \leq -\frac{1}{2K} \Delta^{1/2} F(p)^{1/2}$$

in other words

$$2 \frac{d}{dp} \sqrt{F(p)} \leq -\frac{1}{2K} \Delta^{1/2}.$$

Noting that F is positive on $]0,\theta]$ and writing

$$\sqrt{F(\theta)}-\sqrt{F(p)}=\int_p^\theta \frac{d}{dt}\sqrt{F(t)}\,dt \leq -\frac{1}{4K}\Delta^{1/2}(\theta-p)$$

we obtain

$$\sqrt{F(p)} \geq \sqrt{F(\theta)}+\frac{1}{4K}\Delta^{1/2}(\theta-p)$$

and

$$F(p) \geq \frac{1}{16K^2}\Delta(\theta-p)^2.$$

Going back to the definition of F we have the result with $T=1/16K^2$.

□

3 Applying theorem 3 to linear codes

Our purpose is to apply theorem 3 to the coding context. Let C be a linear $[n,k,d]$ code. We will associate two monotonous sets W_a and W_b to C such that $\Delta(W_a)$ and $\Delta(W_b)$ are closely related to the minimum distance $d(C)$ of C. Furthermore $f_{W_a}(p)$ and $1-f_{W_b}(p)$ will respectively be interpreted as residual error probabilities after decoding on the erasure channel and on the binary symmetric channel.

3.1 The set $W_a(C)$ and the erasure channel

Let W_a be the increasing set of vectors of $\mathbf{F}_2^n$ that cover some non-zero codeword of C, i.e.

$$W_a=\{v\in\mathbf{F}_2^n,\ \exists c\in C\setminus\{0\},\ c\subset v\}$$

Proposition 8. *We have* $\Delta(W_a)=d(C)$

Proof. We first note that if a vector v is such that its support, $supp(v)$, contains the supports of two different non-zero codewords c and c', then $B(v,1)\subset W_a$, so that $v\notin\partial(W_a)$. This is because if v' is obtained from v by changing a "one" coordinate to a "zero", then either $v'\supset c$ or $v'\supset c'$ or $v'\supset(c+c')$. Therefore, if $v\in\partial W_a$, then $supp(v)$ contains the support of a unique non-zero codeword, and $B(v,1)$ contains exactly $|c|$ elements not in W_a. Hence the proposition.

□

Recall that, when submitted to an erasure channel, any symbol may be blurred - erased - with probability p, and non-blurred symbols are always received correctly, i.e. 0 and 1 symbols are never mistaken for each other. The possibility of unambiguous decoding of the received vector is equivalent to the sent codeword being the only codeword that coincides with the non-erased received symbols. But, because of linearity, this is equivalent to the non-existence of a codeword whose support is totally included in the set of erased positions. In other words, the probability of a decoding ambiguity occuring is equal to the probability that the erasure vector covers a non-zero codeword: i.e. it is equal to $f_{W_a(C)}(p)$. This issue is discussed in more detail in [3].

Margulis's original goal in devising theorem 3 was actually to prove the threshold behaviour of the probability $f_{\mathcal{G}}(p)$ of disconnecting a graph $\mathcal{G}$ given that each edge is severed with probability p. $f_{\mathcal{G}}(p)$ is exactly $f_{W_a(C)}(p)$ where $W_a(C)$ is the cocycle code of the graph $\mathcal{G}$. Given that the minimum distance of the cocycle code of $\mathcal{G}$ is equal to the connectivity of $\mathcal{G}$, proposition 8 recovers Margulis's original theorem.

In the next section we give a lower bound on the threshold probability $\theta(W_a(C))$ of $W_a(C)$ that we denote by $\theta_a(C)$ for short.

3.2 An asymptotic lower bound on $\theta_a(C)$

It is fairly obvious that for any $[n,k,d]$ code C of relative minimum distance $\delta = d/n$, we must have $\theta_a(C) \geq \delta$. We prove in this section that "when n is large", $\theta_a(C)$ must actually exceed 2δ.

For any nonempty set $W \subset \mathbf{F}_2^n$, denote the *interior* of W by $\overset{\circ}{W} = W \setminus \partial W$. Note that if W is increasing, then so is $\overset{\circ}{W}$.

Lemma 9. *Let $(W_n)_{n\geq 0}$ be a sequence of increasing sets with $\Delta(W_n) \xrightarrow[n\to\infty]{} \infty$. Then*

$$\lim_{n\to\infty} (\theta(\overset{\circ}{W}_n) - \theta(W_n)) = 0.$$

Proof. Suppose the contrary. Then there exists some sequence $(W_n)_{n\geq n_0}$ with $\Delta(W_n) \xrightarrow[n\to\infty]{} \infty$ and $\beta > 0$ such that $\forall n,\ \ \theta(\overset{\circ}{W}_n) - \theta(W_n) \geq \beta$. Choose any ε, $\beta > \varepsilon > 0$. When n is big enough, for any $p \in [\theta(W_n) + \varepsilon,\ \theta(\overset{\circ}{W}_n)]$, we have $f_{W_n}(p) \geq 1 - \varepsilon$ and $f_{\overset{\circ}{W}_n}(p) \leq 1/2$. Since $f_W(p) = f_{\partial W}(p) + f_{\overset{\circ}{W}}(p)$, we deduce from the above that we must have $f_{\partial W_n}(p) \geq 1/2 - \varepsilon$ on the whole interval

$[\theta(W_n)+\varepsilon,\ \theta(\overset{\circ}{W}_n)]$. But (1) implies that $\frac{d}{dp}f_{W_n}(p)$ tends to infinity uniformly with n on an interval of length $\beta-\varepsilon$, which contradicts the fact that $f_{W_n}(p)\leq 1$.

□

For any linear code C, define $W_a^t(C)$ as the set obtained from $W_a(C)$ by taking the interior t times. In other words,

$$W_a^1(C)=\overset{\circ}{\widehat{W_a(C)}},\ldots,W_a^{t+1}(C)=\overset{\circ}{\widehat{W_a^t(C)}}\ldots$$

Recall that the (generalized) t-distance d_t of C is the smallest integer m such that C admits a subcode of dimension t on a support of size m. We have the following.

Proposition 10. *$W_a^t(C)$ is the set of vectors v of $\mathbf{F}_2^n$ for which there exists a subcode C' of C of dimension $t+1$, such that $v\supset c$ for any $c\in C'$. In particular,*

$$\Delta(W_a^t(C))=d_{t+1}.$$

Proof. Proceed by induction. Let v be a vector of $\overset{\circ}{\widehat{W_a^{t-1}(C)}}$. By the induction hypothesis, v must cover the codewords of a subcode C' of dimension t. Furthermore v must cover some additional codeword x, otherwise change a coordinate of v on the support of a non-zero codeword of C' and we have a vector not in $W_a^{t-1}(C)$ which would contradict $v\in\overset{\circ}{\widehat{W_a^{t-1}(C)}}$. Therefore, v covers the subcode of dimension $t+1$ generated by x and C'.

Conversely, let v be a vector that covers a subcode C' of dimension $t+1$, and let $v'\subset v$, with $|v'|=|v|-1$. Let i be the coordinate on which v' and v differ. Then v' must cover the subcode C'' of C' which consists of those codewords of C' that have a zero on coordinate i. Since C'' is of codimension at most 1 in C', we have $v'\in W_a^{t-1}(C)$, so that $v\in\overset{\circ}{\widehat{W_a^{t-1}(C)}}$.

□

Proposition 11. *Let $(C_n)_{n\geq n_0}$ be a sequence of linear codes with lengths n and minimal distances $d(C_n)\xrightarrow[n\to\infty]{}\infty$. Let $\delta_t(C_n)=d_t(C_n)/n$, and let $\delta_t=\liminf_{n\to\infty}\delta_t(C_n)$. Let $\theta_a=\liminf_{n\to\infty}\theta_a(C_n)$. Then, for any t,*

$$\theta_a\geq\delta_t$$

and in particular,

$$\theta_a\geq 2\delta_1.$$

Proof. This is a straightforward application of proposition 10 and lemma 9, and the fact that for $t \geq 2$ we have (Griessmer bound),

$$\delta_t \geq \delta_1(1 + \frac{1}{2} + \ldots + \frac{1}{2^{t-1}}).$$

□

3.3 The set $W_b(C)$ and the binary symmetric channel

To define the set W_b, we introduce an order $\preceq$ on the vectors of $\mathbf{F}_2^n$. Let us define $x \preceq y$ if

$$\begin{cases} \text{either } |x| < |y| \\ \text{or } |x| = |y| \text{ and } supp(x) \text{ precedes } supp(y) \text{ in the lexicographic order.} \end{cases}$$

Now for any $s \in \mathbf{F}_2^r$, define $\epsilon(s)$ as the lowest vector in $\sigma^{-1}(s)$ for the order $\preceq$. This defines a maximum-likelihood decoding scheme ϵ. Let $W_b(C) \subset \mathbf{F}_2^n$ be the set $W_b(C) = \epsilon(\mathbf{F}_2^r)$.

We have

Proposition 12. *for any linear code C, $W_b(C)$ is a decreasing set of vectors of $\mathbf{F}_2^n$, such that*

$$\Delta(W_b) \geq \lfloor \frac{d(C) - 1}{2} \rfloor.$$

Proof. First let us see that W_b is decreasing. It will be enough to prove that for any $x \in W_b$ and any $y \subset x$ with $|y| = |x| - 1$ we have $y \in W_b$. Let us pick such vectors x and y, so that $x = y \cup u = y + u$ where u is a vector of weight one. If $y \notin W_b$, then there is a vector $y' \preceq y$ such that $\sigma(y') = \sigma(y)$. This implies that $\sigma(y' + u) = \sigma(x)$. But since $x \in W_b$ we must have $x \preceq y' + u$. From this we first deduce y' and u have disjoint supports, and then obtain a contradiction by applying the straightforward

Lemma 13. *If y and y' both have their support disjoint from that of u, and if $y' \preceq y$, then $y' \cup u \preceq y \cup u$.*

Next, we must evaluate $\Delta(W_b)$. Let us pick a vector x in ∂W_b. This means that $x \in W_b$ and that there is a vector u of weight one, whose support is disjoint from that of x, such that $x+u$ is not in W_b. This means that there is a vector $y \neq x+u$, such that $y \preceq x + u$ and $\sigma(y) = \sigma(x) + \sigma(u)$. First notice that $u \not\subset y$, otherwise lemma 13 would imply $y+u \preceq x$ and contradict $x \in W_b$ (since $\sigma(y+u) = \sigma(x)$). Also, since $x \preceq y + u$, it is easily checked that either $|y| = |x| - 1$ or $|y| = |x|$ or $|y| = |x| + 1$.

Let V be the set of vectors v of weight one, such that $v \subset y$ and $v \not\subset x$. We will distinguish two cases.

1. $|y| = |x| - 1$ or $|y| = |x|$. Then $\forall v \in V, |y + v + u| < |x + v|$ so that $y + v + u \preceq x + v$. In other words $\forall v \in V, x + v \notin W_b$.
2. $|y| = |x| + 1$. Then $y \preceq x + u$ implies the existence of a $v^\star \in V$ whose support precedes the support of $x + u$ in the lexicographic order. Then $\forall v \in V \setminus \{v^\star\}, y + v + u \preceq x + v$. In other words $\forall v \in V \setminus \{v^\star\}, x + v \notin W_b$.

To conclude we now need to estimate the size $\#V$ of V. Note that $\sigma(x+y+u) = 0$, so that $x + y + u \in C$ and $|x + y + u| \geq d(C)$. If $|y| = |x|$ or $|y| = |x| - 1$, then $|V| = \lfloor |x + y|/2 \rfloor \geq \lfloor (d-1)/2 \rfloor$. If $|y| = |x| + 1$ then $\#V = \lfloor |x + y|/2 \rfloor + 1$ and $\#(V \setminus \{v^\star\}) \geq \lfloor (d-1)/2 \rfloor$. This proves the proposition.

□

Theorems 1 and 2 follow therefore, by applying proposition 12 to theorem 3 and proposition 7.

References

1. G. Margulis, *Probabilistic characteristics of graphs with large connectivity*, Problemy Peredachi Informatsii, 10 (1974), pp. 101–108.
2. M. Talagrand, *Isoperimetry, logarithmic Sobolev inequalities on the discrete cube, and Margulis' graph connectivity theorem*, Geometric and Functional Analysis, 3 (1993), pp. 295–314.
3. G. Zémor and G. Cohen, *The threshold probability of a code*. Submitted to IEEE Trans on Inf Theory.

Decoding a bit more than the BCH bound *

Josep Rifà Coma†
Universitat Autònoma de Barcelona

Abstract

The concept and characterization of ϵ-best rational approximations (ϵ-BRA) are given in this paper. And, by using this concept, a decoding algorithm for some cyclic codes is presented.

The conventional algorithms (Berlekamp-Massey, Continued Fraction, Extended Euclidean, ...) allows us to correct up to $e_{BCH} \leq \frac{d-1}{2}$ errors where d is the designed minimum distance of the cyclic code. However our algorithm will be able to correct more than $\frac{d-1}{2}$ errors in case that the true distance δ be greater than d.

The Expurged Golay Code is a very good example of the algorithm presented which allows us to correct up to three errors. This code $G(23,11)$ is 3-error correcting but, by using the conventional algorithms we can only correct up to two errors.

Keywords - Cyclic Codes, Expurged Golay Code, Continued Fraction Expansion, Convergents, ϵ-BRA, Rational Approximations.

---

*This work was partially supported by Spanish Grant TIC91-0472.

†Author's Address: Dept. d'Informàtica, Universitat Autònoma de Barcelona, 08193-Bellaterra (Spain). Fax number: 3-5812478. E-mail: j.rifa@ieee.org

1 Introduction

Our decoding algorithm is based on rational approximations and consists of the computation of the convergents in the continued fraction expansion of received code-word polynomial divided by a predefined polynomial.

It is known that there exists an equivalence between the Extended Euclidean algorithm, Berlekamp-Massey algorithm and the computation of convergents of the continued fraction expansion of a rational fraction [WeSc79], so the complexity of our decoding algorithm will be the same as in Extended Euclidean or Berlekamp-Massey algorithms.

The referred algorithms were conceived to correct $e_{BCH} = [(d-1)/2]$ or fewer errors, where d is the designed minimum distance. If δ is the true minimum distance of the code, the method presented, in case $d < \delta$, can correct $[d/2]$ errors. When d is odd no improvement is done, but when d is even we can correct one error more than the conventional scheme.

Section 2 deals with the continued fraction expansion of a rational fraction, we quote some propositions about convergents and we define the concept of ϵ-best rational approximations (ϵ-BRA). The main theorem in this section is the characterization of ϵ-BRA, by using the convergents to a given rational fraction.

In section 3 we give a coding and decoding method for cyclic codes based on the computation of ϵ-BRA. The main theorem (see theorem 12) shows, in case $e \leq d/2$, that we can correct the errors by using an ϵ-BRA. This theorem allows us to correct one error more than the designed minimum distance determines and we give an algorithm to do this which has the same complexity as the Extended Euclidian or Berlekamp-Massey algorithms.

Several authors (see [Elia87], [BoJa90], [FeTz91]) have described algorithms that decode beyond the BCH bound. The difference between our algorithm and the others is that that have succeeded for particular codes. In certain cases, as in [FeTz91], is possible to decode up to e_{Roos}, a lower bound for the error correcting capability coming from the Roos bound (see [Roos83]). In other cases, as in [Elia87] for the binary Golay code, the method is only applied to a particular code.

Our method is more general and can be applied to all the cyclic codes with even designed minimum distance $d < \delta$, where δ is the true minimum distance, but can only correct one error more than e_{BCH}. In [LiWi86] there is a table with all the binary cyclic codes of length up to 63 which have the true minimum distance δ greater than the designed minimum distance d; there are 59 codes with d even, and with our algorithm we can decode 41 of them up to their actual minimum distance, the other 18 codes can be decoded by correcting one error more than e_{BCH} but not until the true error correcting capability.

Section 4 deals with an application of the proposed algorithm. We use the algorithm to correct up to 3 errors by using the Expurged Golay code and we give a numerical example. We emphasize that with the conventional algorithms it is only possible to correct up to two errors by using this code.

2 Continued fraction expansion of a rational fraction

Let K be a finite field, $K[x]$ the polynomial ring and $K(x)$ its fractions field. We define a norm $\|\ \|$ in $K(x)$ as

$$\left\|\frac{a(x)}{b(x)}\right\| = 2^{deg(a(x))-deg(b(x))}$$

Assuming $deg(0) = -\infty$ it is not difficult to prove that $\|\ \|$ is a ***non-archimedian*** norm, that is for all $f(x), g(x) \in K(x)$, the following holds:

1. $\|f(x)+g(x)\| \leq max\{\|f(x)\|, \|g(x)\|\}$ *(ultrametric inequality)*
2. $\|f(x)g(x)\| = \|f(x)\|\|g(x)\|$
3. $\|f(x)\| \geq 0$ and $\|f(x)\| = 0$ if and only if $f(x) = 0$

$K[x]$ is an Euclidean Domain, so we can define the *integer part* of $\frac{a(x)}{b(x)} \in K(x)$, as the quotient obtained by dividing $a(x)$ into $b(x)$. Clearly, by division properties the choice is unique.

We construct the continued fraction expansion of $\alpha(x) = \frac{a(x)}{b(x)} \in K(x)$ as follows:

$$a_0(x) = Quotient(\frac{a(x)}{b(x)})$$

where $\quad a(x) = b(x)a_0(x) + r_0(x) \quad$ and $\quad \|r_0(x)\| < \|b(x)\|$;

$$a_1(x) = Quotient(\frac{b(x)}{r_0(x)}),$$

where $\quad b(x) = r_0(x)a_1(x) + r_1(x) \quad$ and $\quad \|r_1(x)\| < \|r_0(x)\|$;
In general, for $i > 1$

$$a_i(x) = Quotient(\frac{r_{i-2}(x)}{r_{i-1}(x)}),$$

where $\quad r_{i-2}(x) = r_{i-1}(x)a_i(x) + r_i(x) \quad$ and $\quad \|r_i(x)\| < \|r_{i-1}(x)\|$;

Remark: Since $\|r_i(x)\| < \|r_{i-1}(x)\|$ the process will finish.

It is not difficult to verify that

$$\alpha(x) = a_0(x) + \cfrac{1}{a_1(x) + \frac{1}{a_2(x)+\dots}}$$

and we will write it as

$$\alpha(x) = [a_0(x); a_1(x), a_2(x), \dots]$$

Definition 1 *The* **i-th convergent**

$$\frac{P_i(x)}{Q_i(x)}$$

of the continued fraction expansion is defined as

$$\frac{P_i(x)}{Q_i(x)} = [a_0(x); a_1(x), \ldots, a_i(x)]$$

The following matricial relation (see [EgKo91]) is useful to compute the convergents

$$\begin{pmatrix} P_k(x) & P_{k-1}(x) \\ Q_k(x) & Q_{k-1}(x) \end{pmatrix} = \begin{pmatrix} a_k(x) & 1 \\ 1 & 0 \end{pmatrix} \begin{pmatrix} a_{k-1}(x) & 1 \\ 1 & 0 \end{pmatrix} \cdots \begin{pmatrix} a_0(x) & 1 \\ 1 & 0 \end{pmatrix} \tag{1}$$

So

$$\left.\begin{array}{l} P_k(x) = a_k(x)P_{k-1}(x) + P_{k-2}(x) \\ Q_k(x) = a_k(x)Q_{k-1}(x) + Q_{k-2}(x) \end{array}\right\} \tag{2}$$

With the initial conditions $P_{-1}(x) = 1$, $Q_{-1}(x) = 0$, $P_0 = a_0$ and $Q_0 = 1$.

2.1 Some facts about convergents

Here we give some propositions about convergents. The proofs are straightforward and we have not included them in the paper. The reader interested in a more detailed explanation can see the reference [WeSc79].

Let $\frac{P_i(x)}{Q_i(x)}$ be the continued fraction expansion of a given $\alpha(x) = \frac{a(x)}{b(x)} \in K(x)$

Proposition 2 *Every two* $P_i(x), Q_i(x)$ *are relatively prime for* $i \geq 1$.

Proposition 3 *The sequences* $\{P_i(x)\}$ *and* $\{Q_i(x)\}$ *are strictly increasing, and the sequence* $\{r_i(x)\}$ *is strictly decreasing.*

Proposition 4 $P_i(x)b(x) - Q_i(x)a(x) = (-1)^{i+1}r_i(x)$.

Proposition 5 *The sequence* $\left\{\frac{P_i(x)}{Q_i(x)}\right\}$ *satisfies*

$$\left\|\frac{P_{i+1}(x)}{Q_{i+1}(x)} - \alpha(x)\right\| < \left\|\frac{P_i(x)}{Q_i(x)} - \alpha(x)\right\| = \frac{1}{\|Q_i(x)Q_{i+1}(x)\|}$$

Proposition 6 $\|r_{i-1}(x)\| = \frac{\|b(x)\|}{\|Q_i(x)\|}$

2.2 Rational approximations

As far as we know the concepts and proofs of this section are absolutely new.

Let K be a field (possibly finite), $K[x]$ the polynomial ring and $K(x)$ its fractions field.

Given $a, b \in K[x]$, relatively prime, and $\alpha \in K(x)$, we say that $\frac{a}{b}$ is a ϵ-rational approximation (ϵ-RA) to α if

$$\left\| \alpha - \frac{a}{b} \right\| \leq \epsilon$$

and we will define the concept of ϵ-BRA (ϵ-*Best Rational Approximation*) as:

Definition 7 *$\frac{a}{b}$ is a ϵ-BRA to α if $\frac{a}{b}$ is a ϵ-RA to α and for all ϵ-RA $\frac{c}{d} \neq \frac{a}{b}$ to α we have $\|b\| \leq \|d\|$.*

We give the next technical lemma, that will be used in the characterization of ϵ-BRA to a rational fraction.

Lemma 8 *Let $\frac{r(x)}{s(x)} \neq \alpha(x)$ in $K(x)$ be an ϵ-RA to $\alpha(x)$. If $\frac{P_i(x)}{Q_i(x)}$ is the convergent to $\alpha(x)$ such that*

$$\left\| \frac{P_i(x)}{Q_i(x)} - \alpha(x) \right\| \leq \epsilon < \left\| \frac{P_{i-1}(x)}{Q_{i-1}(x)} - \alpha(x) \right\|$$

then the following identity holds

$$\|r(x)Q_{i-1}(x) - s(x)P_{i-1}(x)\| = \frac{\|s(x)\|}{\|Q_i(x)\|}$$

Proof: Using proposition 5 we can write

$$\left\| \frac{r(x)}{s(x)} - \frac{P_{i-1}(x)}{Q_{i-1}(x)} \right\| = \left\| \frac{r(x)}{s(x)} - \alpha(x) + \alpha(x) - \frac{P_{i-1}(x)}{Q_{i-1}(x)} \right\| =$$

$$= \left\| \frac{P_{i-1}(x)}{Q_{i-1}(x)} - \alpha(x) \right\| = \frac{1}{\|Q_i(x)Q_{i-1}(x)\|}$$

Putting left side over a common denominator and multiplying through by $\|Q_{i-1}(x)\|\|s(x)\|$ we have

$$\|r(x)Q_{i-1}(x) - s(x)P_{i-1}(x)\| = \frac{\|s(x)\|}{\|Q_i(x)\|}$$

∎

Next theorem characterizes the ϵ-BRA to a rational fraction

Theorem 9 *(Characterization of ϵ-BRA)*
Assume $\alpha(x) = A(x)/B(X) \in K(x)$. Then $\frac{r(x)}{s(x)} \in K(x)$ is a ϵ-BRA to $\alpha(x)$ if and only if, up to a constant factor,

$$r(x) = P_i(x) + \tau(x)P_{i-1}(x)$$

$$s(x) = Q_i(x) + \tau(x)Q_{i-1}(x)$$

where $\frac{P_i(x)}{Q_i(x)}$ is the first convergent to $\alpha(x)$ such that

$$\left\|\frac{P_i(x)}{Q_i(x)} - \alpha(x)\right\| \leq \epsilon < \left\|\frac{P_{i-1}(x)}{Q_{i-1}(x)} - \alpha(x)\right\|$$

and

$$0 \leq \|\tau(x)\| \leq \epsilon\|Q_i(x)\|^2 < \frac{\|Q_i(x)\|}{\|Q_{i-1}(x)\|}$$

Proof:

- First: Let $\frac{r(x)}{s(x)}$, be an ϵ-BRA, $r(x)$ and $s(x)$ are relatively prime, and $\frac{P_i(x)}{Q_i(x)}$ is the convergent that satisfies lemma 8, then by definition of ϵ-BRA

$$0 < \|s(x)\| \leq \|Q_i(x)\| \tag{3}$$

and by lemma 8 we have

$$\|r(x)Q_{i-1}(x) - s(x)P_{i-1}(x)\| = \frac{\|s(x)\|}{\|Q_i(x)\|}$$

So from (3), $\frac{\|s(x)\|}{\|Q_i(x)\|} = 1$.
Hence $deg(s(x)) = deg(Q_i(x))$ and

$$\|r(x)Q_{i-1}(x) - s(x)P_{i-1}(x)\| = 1 \tag{4}$$

From the definition of norm,

$$r(x)Q_{i-1}(x) - s(x)P_{i-1}(x) = c \tag{5}$$

where $c \in K$ is a constant. But we know by proposition 2 that

$$P_i(x)Q_{i-1}(x) - Q_i(x)P_{i-1}(x) = (-1)^{i+1} \tag{6}$$

Then substracting (6) from (5) we obtain

$$\left(\frac{r(x)}{\pm c} - P_i(x)\right) Q_{i-1}(x) = \left(\frac{s(x)}{\pm c} - Q_i(x)\right) P_{i-1}(x) \tag{7}$$

Also, by proposition 2 we know that $P_{i-1}(x)$ and $Q_{i-1}(x)$ are relatively prime, so

$$\left.\begin{array}{ll}\left(\frac{r(x)}{\pm c} - P_i(x)\right) & \text{is a multiple of } P_{i-1}(x) \\ \left(\frac{s(x)}{\pm c} - Q_i(x)\right) & \text{is a multiple of } Q_{i-1}(x)\end{array}\right\} \text{the same multiple !}$$

If $\tau(x)$ is the common multiple then

$$\left.\begin{array}{l}\frac{r(x)}{\pm c} = P_i(x) + \tau(x)P_{i-1}(x) \\ \frac{s(x)}{\pm c} = Q_i(x) + \tau(x)Q_{i-1}(x)\end{array}\right\} \tag{8}$$

Dividing the equations of (8) we can conclude that

$$\frac{r(x)}{s(x)} = \frac{P_i(x) + \tau(x)P_{i-1}(x)}{Q_i(x) + \tau(x)Q_{i-1}(x)}$$

If $\tau(x) = 0$ we have $\frac{r(x)}{s(x)} = \frac{P_i(x)}{Q_i(x)}$

If $\tau(x) \neq 0$, using proposition 4 it follows that

$$\left\|\alpha(x) - \frac{P_i(x) + \tau(x)P_{i-1}(x)}{Q_i(x) + \tau(x)Q_{i-1}(x)}\right\| = \tag{9}$$

$$= \left\|\frac{A(x)Q_i(x) - B(x)P_i(x) + \tau(x)\{A(x)Q_{i-1}(x) - B(x)P_{i-1}(x)\}}{B(x)\{Q_i(x) + \tau(x)Q_{i-1}(x)\}}\right\| =$$

$$= \left\|\frac{(-1)^{i+1}r_i(x) + (-1)^i\tau(x)r_{i-1}(x)}{B(x)\{Q_i(x) + \tau(x)Q_{i-1}(x)\}}\right\| =$$

$$= \left\|\frac{(-1)^{i+1}r_i(x) + (-1)^i\tau(x)r_{i-1}(x)}{B(x)Q_i(x)}\right\| \tag{10}$$

We are in the case $\tau(x) \neq 0$, so by proposition 3 we can write

$$\left\|\alpha(x) - \frac{P_i(x) + \tau(x)P_{i-1}(x)}{Q_i(x) + \tau(x)Q_{i-1}(x)}\right\| = \left\|\frac{\tau(x)r_{i-1}(x)}{B(x)Q_i(x)}\right\| \tag{11}$$

Now, by using proposition 6

$$\left\|\alpha(x) - \frac{P_i(x) + \tau(x)P_{i-1}(x)}{Q_i(x) + \tau(x)Q_{i-1}(x)}\right\| = \frac{\|\tau(x)\|}{\|Q_i(x)\|^2}$$

But $\frac{P_i(x) + \tau(x)P_{i-1}(x)}{Q_i(x) + \tau(x)Q_{i-1}(x)}$ is an ϵ-BRA to $\alpha(x)$, so

$$\frac{\|\tau(x)\|}{\|Q_i(x)\|^2} \leq \epsilon < \frac{1}{\|Q_i(x)Q_{i-1}(x)\|}$$

and, multiplying all by $\|Q_i(x)\|^2$

$$\|\tau(x)\| \leq \epsilon\|Q_i(x)\|^2 < \frac{\|Q_i(x)\|}{\|Q_{i-1}(x)\|}$$

- Second: Suppose that

$$\frac{P_i(x) + \tau(x)P_{i-1}(x)}{Q_i(x) + \tau(x)Q_{i-1}(x)}$$

which satisfies the theorem hypothesis is not a ϵ-BRA to $\alpha(x)$, then we can find a ϵ-RA $\frac{r(x)}{s(x)}$, $r(x)$ and $s(x)$ relatively prime, such that

$$0 < \|s(x)\| < \|Q_i(x) + \tau(x)Q_{i-1}(x)\| < \|Q_i(x)\|$$

$\frac{P_i(x)}{Q_i(x)}$ satisfies lemma 8 for $\frac{r(x)}{s(x)}$, so

$$\|r(x)Q_{i-1}(x) - s(x)P_{i-1}(x)\| = \frac{\|s(x)\|}{\|Q_i(x)\|} \tag{12}$$

where

$$0 < \frac{\|s(x)\|}{\|Q_i(x)\|} < 1$$

But this goes to a contradiction because it is not possible to have a polynomial the norm of which fulfils $0 < \|r(x)Q_{i-1}(x) - s(x)P_{i-1}(x)\| < 1$.

∎

3 A method to decoding cyclic codes

3.1 Cyclic Codes

Let $C(n,k)$ be a cyclic code over $GF(q)$, with generator polynomial $g(x)$. To encode we use the non systematic way, consisting of multiplying the polynomial associated to the information vector by the generator polynomial, that is, if the information vector is $i = (i_0, i_1, \ldots, i_{k-1})$, and its associate polynomial is $i(x) = i_0 + i_1x + \cdots + i_{k-1}x^{k-1}$ then the associate code-word is

$$c(x) = i(x)g(x) = c_0 + c_1x + \cdots + c_{n-1}x^{n-1}$$

We have $deg(g(x)) = n - k$ and in its decomposition field, $g(x)$ has $n - k$ roots. Since by construction $g(x)$ divides $x^n - 1$, the roots of $g(x)$ must be n-th roots of the unity. We can choose a primitive n-th root of unity, say α, and every root of $g(x)$ can be expressed as a power of α. Suppose that $g(x)$ has $d-1$ roots which have consecutive exponents expressed as powers of α, that is

$$\alpha^b, \alpha^{b+1}, \ldots, \alpha^{b+d-2}$$

are roots of $g(x)$. We say that d is the designed minimum distance of cyclic code.

In this situation, given a vector $c = (c_0, c_1, \ldots, c_{n-1}) \in GF(q)^n$ and the polynomial $c(x) = c_0 + c_1 x + \cdots + c_{n-1}x^{n-1}$, we can compute:

$$\sum_{i=0}^{n-1} \frac{c_i \alpha^{ib}}{x - \alpha^i} = \frac{C(x)}{g_n(x)}$$

where $g_n(x) = \prod_{i=0}^{n-1}(x - \alpha^i) = x^n - 1$.

Polynomial $C(x)$ is a kind of *Discrete Fourier Transform* of $c(x)$.

Next proposition shows how to compute the polynomial $C(x)$.

Proposition 10 *Let* $C(x) = C_0 + C_1 x + C_2 x^2 + \cdots + C_{n-1}x^{n-1}$, $r_i = n-1-i+b$ *and* $i = 0 \ldots n-1$. *Then the following identities hold:*

- $C_i = c(\alpha^{r_i})$
- $c_i = \dfrac{C(\alpha^i)}{n\alpha^{i(n-1-b)}}$

Proof: From $c = (c_0, c_1, \cdots, c_{n-1}) \in (GF(q))^n$ we can compute

$$\sum_{i=0}^{n-1} \frac{c_i \alpha^{ib}}{x - \alpha^i} = \frac{C(x)}{g_n(x)} \tag{13}$$

Let $D(x) = D_0 + D_1 x + D_2 x^2 + \ldots + D_{n-1}x^{n-1}$, where $D_i = c(\alpha^{r_i})$ and $r_i = n - 1 - i + b$.

To see that $D(x) = C(x)$ we will only need to prove that polynomials $D(x)$ and $C(x)$ have the same values on the n points $x = \alpha^i$ $(0 \le i \le n-1)$, that is to say we must prove that $C(\alpha^i) = D(\alpha^i)$.

From equation (13) we can write $C(x) = \sum_{i=0}^{n-1} c_i \alpha^{ib} \prod_{j=0}^{n-1}(x - \alpha^j)$.

Then $C(\alpha^i) = \sum_{k=0}^{n-1} c_k \alpha^{kb} \prod_{j=0}^{n-1}(\alpha^i - \alpha^j) = c_i \alpha^{ib} \prod_{j=0 j\neq i}^{n-1}(\alpha^i - \alpha^j) = c_i \alpha^{ib} g_n'(\alpha^i)$,

where $g_n(x) = \prod_{i=0}^{n-1}(x-\alpha^i) = x^n - 1$, so $g_n'(x) = nx^{n-1}$
Then $C(\alpha^i) = c_i\alpha^{ib}g_n'(\alpha^i) = c_i\alpha^{ib}n\alpha^{i(n-1)}$
On the other hand,

$$D(\alpha^i) = \sum_{j=0}^{n-1} c(\alpha^{r_j})\alpha^{ij} = \sum_{j=0}^{n-1}\sum_{k=0}^{n-1} c_k\alpha^{r_jk}\alpha^{ij} =$$

$$\sum_{j=0}^{n-1}\sum_{k=0}^{n-1} c_k\alpha^{(n-1-j+b)k}\alpha^{ij} = \sum_{k=0}^{n-1} c_k\alpha^{(n-1+b)k}\sum_{j=0}^{n-1}\alpha^{j(i-k)}$$

It is well know that if γ is a n-th root of unity, then

$$\sum_{j=0}^{n-1}\gamma^j = \begin{cases} 0 & \text{if } \gamma \neq 1 \\ n & \text{if } \gamma = 1 \end{cases}$$

So $D(\alpha^i) = nc_i\alpha^{(n-1+b)i}$
∎

Proposition 11 *If $c = (c_0, c_1, \ldots, c_{n-1}) \in (GF(q))^n$ is a codeword then $deg(C(x)) \leq n-d$.*

Proof: If $c = (c_0, c_1, \ldots, c_{n-1}) \in (GF(q))^n$ is a codeword then $\alpha^b, \alpha^{b+1}, \ldots, \alpha^{b+d-2}$ are roots of polynomial $c(x) = c_0 + c_1x + c_2x^2 + \ldots + c_{n-1}x^{n-1}$, and $c(\alpha^{r_i}) = 0$ where $r_i = n-1-i+b$ and $i = n-1, n-2, \ldots, n-d+1$, so the last $(d-1)$ coefficients in $C(x)$ are zero and $deg(C(x)) \leq n-d$. ∎

As we stated at the beginning of this section, starting from the information vector $i = (i_0, i_1, \ldots, i_{k-1})$ and its associated polynomial $i(x) = i_0 + i_1x + i_2x^2 + \ldots + i_{k-1}x^{k-1}$, we can construct the codeword

$$c(x) = i(x)\cdot g(x) = c_0 + c_1x + c_2x^2 + \ldots + c_{n-1}x^{n-1}$$

We consider the transmission process of a codeword $(c_0, c_1, c_2, \ldots, c_{n-1})$ in an additive memoryless channel, so the received vector will be:

$$(v_0, v_1, \ldots, v_{n-1}) = (c_0 + e_0, c_1 + e_1, \ldots, c_{n-1} + e_{n-1})$$

where $e = (e_0, e_1, \ldots, e_{n-1})$ is the error vector.
Consider:

$$\sum_{i=0}^{n-1}\frac{c_i\alpha^{ib}}{x-\alpha^i} = \frac{C(x)}{g_n(x)}$$

$$\sum_{i=0}^{n-1} \frac{e_i \alpha^{ib}}{x - \alpha^i} = \frac{F(x)}{g_n(x)}$$

$$\sum_{i=0}^{n-1} \frac{v_i \alpha^{ib}}{x - \alpha^i} = \frac{V(x)}{g_n(x)}$$

Notice that if e is the number of nonzero coordinates of the error vector, the fraction $\frac{F(x)}{g_n(x)}$ will simplify to $\frac{E(x)}{g_e(x)}$, where $deg(g_e(x)) \leq e$.

Theorem 12 *From the received vector $v = (v_0, v_1, \ldots, v_{n-1})$ we can compute* $\sum_{i=0}^{n-1} \frac{v_i \alpha^{ib}}{x - \alpha^i} = \frac{V(x)}{g_n(x)}$.

If $e \leq \frac{d}{2}$ then the fraction $\sum_{i=0}^{n-1} \frac{e_i \alpha^{ib}}{x - \alpha^i} = \frac{E(x)}{g_e(x)}$ is an ϵ-BRA to $\frac{V(x)}{g_n(x)}$, where $\epsilon = 2^{-d}$.

Proof: We have

$$\frac{V(x)}{g_n(x)} = \frac{C(x)}{g_n(x)} + \frac{E(x)}{g_e(x)}$$

and $deg(C(x)) \leq n - d$ so $\|\frac{C(x)}{g_n(x)}\| \leq 2^{-d}$.

Hence $\|\frac{V(x)}{g_n(x)} - \frac{E(x)}{g_e(x)}\| = \|\frac{C(x)}{g_n(x)}\| \leq 2^{-d}$

If $\frac{E(x)}{g_e(x)}$ is not a 2^{-d}-BRA to $\frac{V(x)}{g_n(x)}$, then we can find a 2^{-d}-RA, $\frac{\phi(x)}{\sigma(x)} \neq \frac{E(x)}{g_e(x)}$ such that $\|\sigma(x)\| < \|g_e(x)\|$.

Then

$$\begin{aligned}\left\|\frac{\phi(x)}{\sigma(x)} - \frac{E(x)}{g_e(x)}\right\| &= \left\|\frac{\phi(x)}{\sigma(x)} - \frac{V(x)}{g_n(x)} + \frac{V(x)}{g_n(x)} - \frac{E(x)}{g_e(x)}\right\| \leq \\ &\leq max\left\{\left\|\frac{\phi(x)}{\sigma(x)} - \frac{V(x)}{g_n(x)}\right\|, \left\|\frac{V(x)}{g_n(x)} - \frac{E(x)}{g_e(x)}\right\|\right\} \leq 2^{-d}\end{aligned} \tag{14}$$

So

$$2^{-d} \geq \left\|\frac{\phi(x)}{\sigma(x)} - \frac{E(x)}{g_e(x)}\right\| = \left\|\frac{\phi(x)g_e(x) - E(x)\sigma(x)}{\sigma(x)g_e(x)}\right\| > \frac{1}{\|g_e(x)\|^2} = \frac{1}{2^{2e}} \geq \frac{1}{2^d} = 2^{-d}$$

and this goes to a contradiction. ∎

Corollary 13 *Let δ be the true minimum distance and suppose $d < \delta$.*

If $e \leq \frac{d}{2}$ then the fraction $\frac{E(x)}{g_e(x)}$ is an ϵ-BRA to $\frac{V(x)}{g_n(x)}$ and is the unique ϵ-BRA to $\frac{V(x)}{g_n(x)}$ which satisfies $g_e(x)|g_n(x)$, where $\epsilon = 2^{-d}$.

Proof: Direct from the properties of minimum distance ∎

3.2 Decoding Algorithm

From the above section we know that $\frac{E(x)}{g_e(x)}$ is a 2^{-d}-BRA to the fraction $\frac{V(x)}{g_n(x)}$ if the number of errors produced in the transmission is $e \leq d/2$.

As we stated in theorem 9, $E(x) = P_i(x) + \tau(x)P_{i-1}(x)$ and $g_e(x) = Q_i(x) + \tau(x)Q_{i-1}(x)$, where $deg(Q_i(x)) = e \leq \frac{d}{2}$ and $\|\tau(x)\| \leq 2^{-d}2^d = 1$, so $\tau(x)$ is a constant.

$Q_i(x)$ is the first convergent to $\frac{V(x)}{g_n(x)}$ such that:

$$\left\| \frac{P_i(x)}{Q_i(x)} - \frac{V(x)}{g_n(x)} \right\| \leq 2^{-d}$$

By using proposition 4 we can write:

$$\left\| \frac{P_i(x)}{Q_i(x)} - \frac{V(x)}{g_n(x)} \right\| = \left\| \frac{r_i(x)}{Q_i(x)g_n(x)} \right\| \leq 2^{-d}$$

So the condition to stop the algorithm, in case that $deg(Q_i(x)) = e \leq \frac{d}{2}$, will be

$$\|r_i(x)\| \leq 2^{-d}\|Q_i(x)g_n(x)\|$$

that is to say

$$deg(r_i(x)) \leq n - d + deg(Q_i(x))$$

To compute τ we can notice from theorem 9 that:

$$0 \leq \|\tau\| \leq \varepsilon\|Q_i(x)\|^2 = 2^{-d}2^{2e} = 2^{2e-d}$$

and τ will be $\tau = 0$ in case that $e < \frac{d}{2}$ and τ will be a constant in case that $e = \frac{d}{2}$.

Remark: In case that the number e of errors be $e \leq e_{BCH}$, where e_{BCH} is the designed error capability, we have $\tau = 0$ and the decoding algorithm is reduced the the computation of convergents. The reference [WeSc79] deals with another version of this conventional case.

We know the roots of $g_e(x) = Q_i(x) + \tau Q_{i-1}(x)$ are of type α^j, when j are the error places, so $g_e(\alpha^j) = 0$ for all the places j where there is an error.

We can observe the list $\{\tau_0, \tau_1, \ldots, \tau_{n-1}\}$ of all possible values $\tau_j = -\frac{Q_i(x)}{Q_{i-1}(x)}$ by doing $x = \alpha^j$ for $j = 0, 1, \ldots, n-1$.

τ will be the value in the above list repeated e times. Say it $\tau = \tau_{j_1} = \tau_{j_2}, \ldots, \tau_{j_e}$. If there is not such a value we deduce that it is not possible to decode the received codeword.

At this moment we now how many errors there are and also where they are placed. There are e errors and the error places are $j_1, j_2, \ldots, j_e$.

Now we can write:

$$\frac{E(x)}{g_e(x)} - \frac{V(x)}{g_n(x)} = \frac{P_i(x) + \tau P_{i-1}(x)}{Q_i(x) + \tau Q_{i-1}(x)} - \frac{V(x)}{g_n(x)} =$$

$$\frac{g_n(x)P_i(x) + \tau P_{i-1}(x)g_n(x) - V(x)Q_i(x) - \tau Q_{i-1}(x)V(x)}{g_n(x)(Q_i(x) + \tau Q_{i-1}(x))} =$$

$$\frac{(-1)^{i+1}r_i(x) + (-1)^i \tau r_{i-1}(x)}{g_n(x)(Q_i(x) + \tau Q_{i-1}(x))} = \frac{C(x)}{g_n(x)}$$

So

$$C(x) = \frac{(-1)^{i+1}r_i(x) + (-1)^i \tau r_{i-1}(x)}{Q_i(x) + \tau Q_{i-1}(x)}$$

Summarizing these results we can write an algorithm to decode a cyclic code with the improvement that this algorithm allows us to correct up to $\frac{d}{2}$ errors in case that the designed minimum distance d is less than the true minimum distance δ. The conventional algorithms, like Berlekamp-Massey, Extended euclidean, etc., allow us to correct up to $\frac{d-1}{2}$ errors.

In our algorithm we need to compute the continued fraction expansion to a given rational fraction and this computation is equivalent to the Berlekamp-Massey algorithm as it was stablished in [WeSc79]. So the complexity of our algorithm is $O(n^2)$, the same that in the conventional algorithms quoted above.

The presented algorithm is interesting in particular when d is even, because in this case we can correct one error more than the conventional algorithms allowed.

Here we present the above algorithm in the particular case of Expurged Golay Code $G(23, 11)$.

Our algorithm is more general and this example is only a particular case of it. Others algorithms were conceived to be only used in a particular case, you can see for instance the algorithm in [Elia87] for the binary Golay code.

4 Decoding the Expurged Golay code $G(23, 11)$

We can use the above theory to correct up to 3 errors by using the Expurged Golay Code $G(23, 11)$. The designed minimum distance is $d = 6$ and the true minimum distance $\delta = 8$, so with the conventional BCH-theory and by using the well-known algorithms like Berlekamp-Massey, Extended Euclidean, etc., we can correct up to two errors, but by using our decoding algorithm we will correct up to 3 errors.

The Rational Approximations Algorithm will be:

Begin

Initialization

$P_{-1}(x) = 1;\ P_0(x) = 0;\ Q_{-1}(x) = 0;\ Q_0(x) = 1$

$r_{-2}(x) = V(x); r_{-1}(x) = g_n(x)$

$i = 0$

EndInitialization

While $deg(r_i(x) > 17 + deg(Q_i(x))$ and $deg(Q_i(x)) \leq 3$ *do*

$i = i + 1$

$$a_i(x) = Quotient\left(\frac{r_{i-2}(x)}{r_{i-1}(x)}\right)$$

$r_i(x) = r_{i-2}(x) - r_{i-1}(x)a_i(x)$

$Q_i(x) = a_i(x)Q_{i-1}(x) + Q_{i-2}(x)$

$P_i(x) = a_i(x)P_{i-1}(x) + P_{i-2}(x)$

endwhile

if $deg(Q_i(x)) > 3$ *then* *Write* (Decoding Error: Too many errors)

else

if $deg(Q_i(x) < 3$ *then* $\tau = 0$

compute $Q_i(\alpha^j)$ *for* $j = 0, 1, \ldots, 22$, *and find the roots* α^{j_k} *of* $Q_i(x)$

if such a roots there not exists then *Write* (Decoding Error)

else

$$compute\ \tau_j = -\frac{Q_i(\alpha^j)}{Q_{i-1}(\alpha^j)}\ for\ j = 0, 1, \ldots, 22$$

$\tau =$ *the value repeated* 3 *times* $= \tau_{j_1} = \tau_{j_2} = \tau_{j_3}$

if such a value τ *do not exists then* *Write* (Decoding Error)

endelse

$e = (e_1, e_2, \ldots, e_{22})$ *with all the* $e_i = 0$ *except* $e_{j_k} = 1$

endelse

End

Numerical Example:

We can take $g(x)$ as generator polynomial of the Expurged Golay Code $G(23, 11)$

$$g(x) = 1 + x^2 + x^5 + x^8 + x^9 + x^{10} + x^{11} + x^{12}$$

The roots of $g(x)$ are $\alpha^0, \alpha^1, \alpha^2, \alpha^3, \alpha^4, \alpha^6, \alpha^8, \alpha^9, \alpha^{12}, \alpha^{13}, \alpha^{16}$ and α^{18}, where $\alpha \in GF(2^{11})$ is a 23-th root of unity, $\alpha = \beta^{89}$ where β is a primitive element of $GF(2^{11})$

In this example $b = 0$, and we will see the performance of the *Rational Approximation algorithm* by using a particular example.

Take the information vector:

$$i(x) = 1 + x + x^7$$

Its associate codeword will be

$$c(x) = i(x)g(x) = 1+x+x^2+x^3+x^5+x^6+x^7+x^8+x^9+x^{12}+x^{13}+x^{15}+x^{16}+x^{17}+x^{18}+x^{19}$$

Now we will suppose that the channel has inserted three errors and the error vector is

$$e(x) = x + x^{10} + x^{17}$$

so the received vector will be

$$v(x) = 1 + x^2 + +x^3 + x^5 + x^6 + x^8 + x^9 + x^{10} + x^{12} + x^{13} + x^{15} + x^{16} + x^{18} + x^{19}$$

- The receiver will compute $V(x) = \sum_{i=0}^{n-1} V_i x^i$, where $V_i = v(\alpha^{r_i})$ and $r_i = n - 1 - i + b = 22 - i$.

$$\begin{aligned} V(x) = {} & \beta^{946} + \beta^{1892}x + \beta^{630}x^2 + \beta^{1737}x^3 + \beta^{1390}x^4 + \beta^{1260}x^5 + \beta^{1371}x^6 + \\ & \beta^{1427}x^7 + \beta^{1614}x^8 + \beta^{733}x^9 + \beta^{1770}x^{10} + \beta^{473}x^{11} + \beta^{315}x^{12} + \\ & \beta^{695}x^{13} + \beta^{1709}x^{14} + \beta^{807}x^{15} + \beta^{885}x^{16} + \beta^{1181}x^{17} + \beta^{1878}x^{18} + \\ & \beta^{1466}x^{19} + \beta^{939}x^{20} + \beta^{1493}x^{21} + x^{22} \end{aligned}$$

- Take $g_n(x) = \prod_{i=0}^{n-1}(x - \alpha^i) = x^{23} - 1$.

  Now by using the Rational Approximation algorithm we have the following partial computations:

i=0 $r_{i-2} = V(x)$
$r_{i-1} = g_n(x)$
$r_i(x) = V(x)$
$Q_i(x) = 1$
$P_i(x) = 0$

i=1 $r_{i-2} = g_n(x)$
$r_{i-1} = V(x)$

$$\begin{aligned} r_i(x) = {} & \beta^{1988} + \beta^{887}x + \beta^{511}x^2 + \beta^{163}x^3 + \beta^{524}x^4 + \beta^{892}x^5 + \beta^{947}x^6 + \\ & \beta^{687}x^7 + \beta^{1233}x^8 + \beta^{443}x^9 + \beta^{122}x^{10} + \beta^{717}x^{11} + \beta^{1883}x^{12} + \\ & \beta^{392}x^{13} + \beta^{1614}x^{14} + \beta^{1385}x^{15} + \beta^{376}x^{16} + \beta^{1820}x^{17} + \beta^{1504}x^{18} + \\ & \beta^{1737}x^{19} + \beta^{244}x^{20} \end{aligned}$$

$Q_i(x) = \beta^{1493} + x$
$P_i(x) = 1$

i=2 $r_{i-2} = V(x)$

$r_{i-1} = r_1(x)$

$$r_i(x) = \beta^{1849} + \beta^{748}x + \beta^{528}x^2 + \beta^{172}x^3 + \beta^{1047}x^4 + \beta^{1817}x^5 + \beta^{1521}x^6 + \beta^{852}x^7 + \beta^{2041}x^8 + \beta^{1902}x^9 + \beta^{980}x^{10} + \beta^{638}x^{11} + \beta^{694}x^{12} + \beta^{1864}x^{13} + \beta^{5714}x^{14} + \beta^{46}x^{15} + \beta^{1804}x^{16} + \beta^{1650}x^{17} + \beta^{157}x^{18}$$

$$Q_i(x) = \beta^{1474} + \beta^{398}x + \beta^{1249}x^2 + \beta^{1803}x^3$$

$$P_i(x) = \beta^{398} + \beta^{1803}x^2$$

The algorithm stops when $deg(r_i(x)) \leq n - d + deg(Q_i(x)) = 17 + deg(Q_i(x))$.

At this point we know $deg(Q_i(x)) = 3$, so we wait to correct three errors.

- Compute the list of values $\tau_j = -\dfrac{Q_2(\alpha^j)}{Q_1(\alpha^j)}$ for $j = 0, 1, 2, \ldots, n-1$.

$$\tau_j = \{\beta^{396}, \beta^{1149}, \beta^{1043}, \beta^{186}, \beta^{664}, \beta^{855}, \beta^{1837} \beta^{123}, \beta^{2019}, \beta^{936}, \beta^{1149}, \beta^{1926}, \beta^{977}, \beta^{1555}, \beta^{1761}, \beta^{1955}, \beta^{376}, \beta^{1149}, \beta^{1885}, \beta^{1975}, \beta^{325}, \beta^{595}, \beta^{756}\}$$

and take $\tau = \beta^{1149} = \tau_1 = \tau_{10} = \tau_{17}$

τ is the value repeated three times in the above list.

- In this example we are in the binary case, so at this point we know the errors are located in the coordinates where $\tau = \beta^{1149}$, that is to say the 1st, 10th and 17th.

  The error vector is:

$$e = (01000000001000000100000)$$

and the codeword will be computed as:

$$c = v - e = 1+x+x^2+x^3+x^5+x^6+x^7+x^8+x^9+x^{12}+x^{13}+x^{15}+x^{16}+x^{17}+x^{18}+x^{19}$$

which coincides with the true codeword being transmitted.

Acknowledgement

I would like to express my thanks to the referees for their comments and suggestions. I also thank R.Struik and R.W.Versseput for their kind suggestions and J. Cotrina for made ready a module by using the *Mathematica*© software which allowed to compute the numeric example on the Expurged Golay code.

References

[Berl68] E.R.Berlekamp, *Algebraic Coding Theory* , McGraw-Hill, 1968.

[BoJa90] P.Bours, J.C.M.Janssen, M.Van Asperdt, H.C.A.Van Tilborg, "Algebraic Decoding Beyond e_{BCH} of Some Binary Cyclic Codes, when $e > e_{BCH}$", IEEE Trans., IT-36, pag. 214-222, 1990.

[EgKo91] O.Egecioglu, Ç.K.Koç, J.Rifà, " Fast computation of continued fractions", Computer Math. Applic. vol 21, n. 2-3, pag.167-169, 1991.

[Elia87] M.Elia, "Algebraic Decoding of the (23,12,7) Golay Code", IEEE Transactions on Information Theory, IT-33, n. 1, January 1987.

[FeTz91] G.Feng, K.K.Tzeng, "Decoding Cyclic Codes up to Actual Minimum Distance Using Nonrecurrent Syndrome Dependence Relations", IEEE Trans., IT-37, pag. 1716-1723, 1991.

[LiWi86] J.H. van Lint, R.M.Wilson, "On the Minimum Distance of Cyclic Codes", IEEE TRans. IT-32, pag. 23-41, 1986.

[MaSl77] F.J.MacWilliams, N.J.A.Sloane. *The Theory of Error-Correcting Code*, North-Holland Publishing Company, 1977.

[Mass69] J.L.Massey, "Shift-register synthesis and BCH decoding", IEEE Trans., IT-15,pag.122-127, 1969.

[Pete68] W.W.Peterson, "Encoding and error-correction procedures for Bose-Chaudhuri codes", IEEE Trans. Info. Theory, IT-6,pag.459-470, 1960.

[ReSc78] I.S.Reed, R.A.Scholtz, "The fast decoding of Reed Solomon Codes using Fermat theoretic Transforms and Continued Fractions", IEEE Trans. Inform. Theory, vol IT.24 pag.100-106, 1978.

[Roos83] C.Roos, "A New Lower Bound for the Minimum Distance of a Cyclic Code", IEEE Trans. IT-37, pag.330-332, 1983.

[Sugi75] Y.Sugiyama, M.Kasahara, S.Hirasawa, T.Namekawa, " A method for solving key equation for decoding Goppa codes", Inform. Contr., 21, pag.87-99, 1975.

[WeSc79] LL.R.Welch and R.A.Scholtz, "Continued Fractions and Berlekamp's Algorithm", IEEE Trans. Inform. Theory, vol.IT-25 No.1, pag.19-27, 1979

Product codes and the Singleton bound

Nicolas Sendrier
INRIA, Domaine de Voluceau, Rocquencourt,
BP 105, 78153 Le Chesnay CEDEX, FRANCE

Abstract

Minimum distance is not always the most determinant factor to acheive high performance for error correction. Of course the knowledge of the whole weight distribution of the code is more accurate than the knowledge of the mere minimum distance, and the phenomenon amplifies for a high noise level. Besides this fact, the use of error-correcting codes in practical situations requires a trade-off between the algorithmic complexity and the performance of the decoding procedure. We show here that for low rates a very good trade-off is possible using product codes, although they are known for their poor minimum distance.

1 Introduction

We consider transmission of information through a memoryless q-ary symmetric channel which can model an additive channel for instance. In such a model, the probability of transition from a word to another is a decreasing function of their Hamming distance. This means that finding codes with a large minimum distance is desirable, though not necessary, to acheive good performance. The "best" decoding algorithm for any given code is the maximum likelihood decoder. Unfortunately, this algorithm usually has a prohibitive algorithmic cost. It is thus necessary to find a trade-off between algorithmic complexity and performance.

One of the most efficient pair code/decoder is the Berlekamp-Massey decoding algorithm for alternant codes. This algorithm is bounded by the designed distance and has a relatively low algorithmic complexity. But it cannot correct error patterns of weight larger than half the designed distance.

Some other classes of codes, like product codes, have a relatively bad minimum distance but possess a natural decoding algorithm that can correct many error patterns of weight larger than half the minimum distance. We have obtained very good decoding performance for such codes. Furthermore we will prove that the algorithmic complexity of the product code decoder is less than the complexity of the Berlekamp-Massey algorithm for a code of same length and dimension, that acheives similar performance.

We will give in sections 2 and 3 the definitions and tools that help to highlight the main result, and in section 4 we will give an example of a good product code; the product code $RS(256; 15, 7, 9) \otimes RS(256; 15, 7, 9)$, where $RS(256; 15, 7, 9)$ is the shortened Reed-Solomom code over $GF(256)$, has parameters $(225, 49, 81)$ and acheives a residual error rate of 10^{-6} for a channel error probability of 0.254. For the same channel error probability, the shortened Reed-Solomom code $RS(256; 225, 49, 177)$ over $GF(256)$ using Berlekamp-Massey decoding algorithm has a residual error rate of $2.2\,10^{-6}$ and a higher algorithmic complexity.

2 Error correcting algorithm

Let $C(n, k, d)$ denote a linear code over $GF(q)$, of length n dimension k and minimum distance d.

Definition 1 *An* error-correcting algorithm *for C is a mapping γ from $GF(q)^n$ into $C \cup \{\infty\}$ such that for all x in C, $\gamma(x) = x$. (The symbol ∞ denotes a decoding failure)*

*The error-correcting algorithm γ is said to be C-*additive *if for all y in $GF(q)^n$ and all x in C, $\gamma(y + x) = \gamma(y) + x$ (with the convention $\infty + x = \infty$).*

Definition 2 *An error pattern y in $GF(q)^n$ is said to be* correctable *if for all x in C, $\gamma(x + y) = x$.*

When an error-correcting algorithm γ is C-additive, an error pattern y in $GF(q)^n$ is correctable if and only if $\gamma(y) = 0$. From now on, we will only consider C-additive algorithms. This is the case for all syndrome based decoders.

Let γ be a C-additive decoding algorithm. We will denote by $\mathcal{P}_\gamma(p)$ the probability of correct transmission of a codeword transmitted through a q-ary symmetric channel of error probability p; its complement to one $1 - \mathcal{P}_\gamma(p)$ is called the residual error rate. These probabilities can be computed if we are able to describe the set $\{y \in GF(q)^n, \gamma(y) = 0\}$ of correctable error patterns.

Proposition 1 *Let γ be a C-additive error-correcting algorithm, we call* decoding region *of γ the set*

$$E_\gamma = \{y \in GF(q)^n, \gamma(y) = 0\}$$

of correctable error patterns. The probability of correct transmission of a codeword from C transmitted through a memoryless q-ary symmetric channel of error probability p and decoded by γ is equal to

$$\mathcal{P}_\gamma(p) = \sum_{y \in E_\gamma} \left(\frac{p}{q-1}\right)^{w_H(y)} (1-p)^{n-w_H(y)} = \sum_{i=0}^{n} a_i \left(\frac{p}{q-1}\right)^i p^{n-i},$$

where w_H denotes the Hamming weight over $GF(q)$, and a_i is the number of correctable error patterns of weight i.

Proof: In a q-ary symmetric channel of error probability p, the conditional probability to receive b in $GF(q)$ given that a in $GF(q)$ was sent is equal to

$$P(b \mid a) = \begin{cases} 1-p & \text{if } b = a \\ \dfrac{p}{q-1} & \text{if } b \neq a \end{cases}$$

If we denote by $P_n(y \mid x)$ the conditional probability to receive $y = (y_0, \ldots, y_{n-1})$ in $GF(q)^n$ given that $x = (x_0, \ldots, x_{n-1})$ in $GF(q)^n$ was sent, then we have

$$P_n(y \mid x) = \prod_{i=0}^{n-1} P(y_i \mid x_i),$$

because the channel is memoryless, and thus

$$P_n(y \mid x) = \left(\frac{p}{q-1}\right)^{d_H(x,y)} (1-p)^{n-d_H(x,y)}, \tag{1}$$

where d_H denotes the Hamming distance over $GF(q)$.

Since γ is C-additive, the probability of correct transmission of any codeword x is equal to

$$\sum_{y \in E_\gamma} P_n(y + x \mid x) = \sum_{y \in E_\gamma} P_n(y \mid 0), \tag{2}$$

which is independant of x. From (1) and (2) we obtain the result. □

Thus if the weight distribution of the correctable error patterns is known, it is possible to compute the probability $\mathcal{P}_\gamma(p)$ for all p.

3 Closed algorithm – Equivalent diameter

In all of this section we will denote by C a linear code over $GF(q)$ of length n, dimension k and minimum distance d, and by γ a C-additive error-correcting algorithm for C.

Definition 3 *The error-correcting algorithm γ for C is said to be* bounded *by an integer e if for all y in $GF(q)^n$ and all x in C,*

$$d_H(y, x) < \frac{e}{2} \Rightarrow \gamma(y) = x.$$

If also $\gamma(y) = x \neq \infty \Rightarrow d_H(y, x) < e/2$, then γ is said to be closed *by e (or e-closed).*

The best possible bound for an error-correcting algorithm is the minimum distance of the code. A maximum likelihood decoder is always bounded by the minimum distance.

The Berlekamp-Massey decoding algorithm for BCH codes, for instance, is closed by the designed distance. An e-closed algorithm can correct exactly the error patterns of weight $(e-1)/2$ or less, if we denote by ϕ_e such an algorithm, we have

$$\mathcal{P}_{\phi_e}(p) = \mathcal{V}_e(p) = \sum_{i=0}^{(e-1)/2} \binom{n}{i} p^i (1-p)^{n-i} \tag{3}$$

We wish to compare the performance of γ to the performance of an e-closed algorithm ϕ_e for a code of the same length. If $\mathcal{V}_e(p) \leq \mathcal{P}_\gamma(p)$, then γ has better performance than ϕ_e for the channel error probability p. For a given probability p we are interested in the largest value of e such that γ is better than ϕ_e.

Definition 4 *The* equivalent diameter of the decoding region *of an error-correcting algorithm γ for the channel error probability p is the unique odd integer $d^*(p)$ such that :*

$$\mathcal{V}_{d^*(p)}(p) \leq \mathcal{P}_\gamma(p) < \mathcal{V}_{d^*(p)+2}(p).$$

For an e-closed algorithm, we have $d^*(p) = e$ for all p. For an e-bounded algorithm we have $d^*(p) \geq e$, and for low noise level $d^*(p) = e$.

Proposition 2 *Let γ be an error-correcting algorithm, and let e be its largest odd bound. We have*

1. $d^*(p) \geq e$,

2. $\lim_{p \to 0} d^*(p) = e$.

Proof:

1. We have $\mathcal{P}_\gamma(p) \geq \mathcal{V}_e(p)$ because E_γ contains all the words of weight smaller or equal to $(e-1)/2$, thus $d^*(p) \geq e$.

2. For any integer i let a_i denote the number of error patterns of weight i correctable by γ. We have

$$\mathcal{P}_\gamma(p) - \mathcal{V}_{e+2}(p) = \left(a_{t+1} - \binom{n}{t+1}(q-1)^{t+1}\right) p^{t+1} + o(p^{t+1}),$$

where $t = (e-1)/2$. Since $e+2$ is not a bound, there exists uncorrectable error patterns of weight $t+1$, and we have

$$a_{t+1} - \binom{n}{t+1}(q-1)^{t+1} < 0.$$

This proves that when p is small enough $d^*(p) \leq e$.

□

However, for larger values of p we may have $d^*(p) > e$. Of course, a large value of $d^*(p)$ has very little meaning if the performance is poor for the channel error probability p. In practice we are interested in the values of p such that simultaneously $d^*(p) > e$ and the residual error rate $\epsilon = 1 - \mathcal{P}_\gamma(p)$ is small (say smaller than 10^{-4}).

Remark. Let's consider the measure defined for all y in $GF(q)$ by

$$\mu_p(y) = \left(\frac{p}{q-1}\right)^{w_H(y)} (1-p)^{n-w_H(y)}.$$

Then the volume of any subset E of $GF(q)^n$ is

$$\mu_p(E) = \sum_{y \in E} \mu_p(y)$$

which is equal to the probability of receiving a word in E given that the all-zero word of $GF(q)^n$ was sent through a q-ary symmetric channel of error probability p. In particular, we have $\mu_p(E_\gamma) = \mathcal{P}_\gamma(p)$ and $\mu_p(GF(q)^n) = 1$. The equivalent diameter of E_γ is the diameter of the largest open ball centered on the all-zero word whose volume is smaller or equal to the volume of the decoding region E_γ.

The value of the equivalent diameter of E_γ depends only on the real number $\mu_p(E_\gamma) = \mathcal{P}_\gamma(p)$, and this dependence is monotonous. Thus for a given channel the most relevant measure of the goodness of an error-correcting algorithm is its equivalent diameter rather than its bound. Furthermore, $d^*(p)$ is comparable to a distance since an e-closed algorithm (metric condition) has an equivalent diameter exactly equal to e, and even comparable to a minimum distance since the largest possible value of e is the minimum distance of the code.

These remarks will make sense for product codes, which possess error-correcting algorithms with a good equivalent diameter in spite of a poor minimum distance.

4 Product codes

Definition 5 [1] *Let $C_1(n_1, k_1, d_1)$ and $C_2(n_2, k_2, d_2)$ be two linear codes over $GF(q)$. The* product code *of C_1 and C_2, denoted $C_1 \otimes C_2$, is the set of all $n_2 \times n_1$ matrices over $GF(q)$ whose rows are in C_1 and whose columns are in C_2.*

The code $C_1 \otimes C_2$ has length $n_1 n_2$, dimension $k_1 k_2$ and minimum distance $d_1 d_2$. This minimum distance is bad, but there exist decoding algorithms that correct many error patterns of weight larger than half the minimum distance. Two of them, given in [3] and [2], are recalled here.

4.1 Decoding product codes

We give first a "natural" decoding algorithm for the product code $C_1 \otimes C_2$. We then present the more sophisticated Reddy-Robinson algorithm and its iterative version.

We will consider here the case $C_1 = C_2 = C(n, k, d)$, which does not differ from the the general case, but is clearer.

We suppose that we have an error-correcting algorithm γ for C, and we assume that in case of decoding failure we have $\gamma(y) = y$ instead of $\gamma(y) = \infty$. Thus γ will be a mapping from $GF(q)^n$ into itself rather than into $C \cup \{\infty\}$. We denote by $\mathcal{M}_q(n)$ the set of $n \times n$ matrices over $GF(q)$. Let ψ denote the following procedure

$$\begin{array}{cccc} \psi & : & \mathcal{M}_q(n) & \longrightarrow & \mathcal{M}_q(n) \\ & & \begin{pmatrix} y_1 \\ \vdots \\ y_n \end{pmatrix} & \longmapsto & \begin{pmatrix} \gamma(y_1) \\ \vdots \\ \gamma(y_n) \end{pmatrix} \end{array} \quad (4)$$

For a matrix M in $\mathcal{M}_q(n)$ the procedure ψ consists of applying γ to each row of M.

```
Algorithm A_1
begin procedure decode_product_code(M)
        M_0 := M
        SUCCES := false
        for i from 0 to 2n - d + 1 do
                M_{i+1} := ψ(^t M_i)            # ^t M denotes the transpose of M
                if M_{i+1} ∈ C ⊗ C then
                        SUCCES := true
                        break
                else if M_{i+1} = M_i then
                        break
        if i is odd then M :=^t M_i else M := M_i
        return(SUCCES)
end procedure
```

Figure 1: A product code decoder

The "natural" algorithm (Figure 1). This procedure returns **true** if it is succesfull, and, as a side effect the value of its argument is changed to the value of the decoded codeword. The maximum number of iterations allowed is $2n-d+1$. Very few correctable error patterns are concerned by this limit. This is discussed in [3].

This algorithm is very efficient in some cases, as we will see in the following sections. However, it is not minimum distance bounded, there exist error patterns of weight $(t+1)^2$, where $t = (d-1)/2$, that are not correctable. A d^2-bounded decoder would correct all error patterns up to weight $(d^2-1)/2 =$

$2t(t+1)$. On the other hand, it may correct error patterns up to weight $t(2n-t)$ (see [3] for more details).

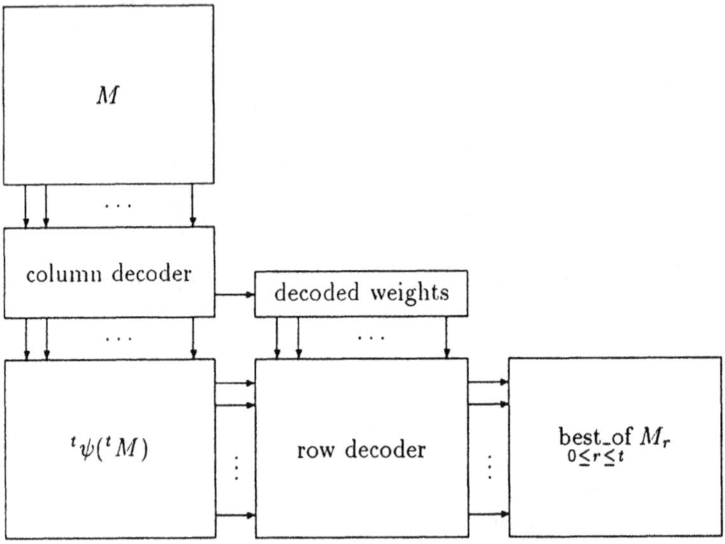

Figure 2: Reddy-Robinson decoder

The Reddy-Robinson algorithm. [2] This decoding algorithm is bounded by the minimum distance d^2. It has only two steps as described in Figure 2, but it requires a decoder for C which is able to handle errors and erasures simultaneously (for instance, the Berlekamp-Massey decoding algorithm for BCH codes is able to correct ν errors and ρ erasures as long as $2\nu + \rho$ is smaller than the designed distance).

The Reddy-Robinson algorithm can be described as follows:

- Given a matrix M in $\mathcal{M}_q(n)$,
  1. using γ, decode each column of M. For all i, $0 \le i < n$, let w_i denote the number of errors corrected in the i-th column. In case of decoding failure of the i-th column, let $w_i = t + 1$, where $t = (d-1)/2$,
  2. make $t + 1$ decoding tentatives of the resulting matrix ${}^t\psi({}^tM)$; for all r, $0 \le r \le t$, do
     (a) erase all columns such that $w_i > r$,
     (b) decode all rows with γ (errors and erasures), store the result as M_r,
  3. among all M_r in the product code, return the one closest to the received matrix.

The iterative Reddy-Robinson algorithm. Using the above algorithm we can describe the following procedure, which we refer to as algorithm $\mathcal{A}_2$:

- Given a matrix M in $\mathcal{M}_q(n)$,

1. try to decode M using $\mathcal{A}_1$, store the result in M',
2. if M' is in the product code then stop, else apply the Reddy-Robinson algorithm to M' and then, if necessary, to ${}^tM'$.

4.2 The simulation

Here and after we will denote by $RS(q;n,k,d)$ the Reed-Solomon code over $GF(q)$ of length n, dimension k and minimum distance $d = n - k + 1$. We have simulated algorithms $\mathcal{A}_1$ and $\mathcal{A}_2$ for a product of two identical Reed-Solomom codes $RS(q;n,k,d)$. The component decoder was the Berlekamp-Massey decoding algorithm, which is minimum-distance-closed.

For all weights i between $(t+1)^2$ and $(2n-t)t$, we have computed a large number of error patterns. We have thus obtained for every i, $(t+1)^2 \leq i \leq (2n-t)t$ a ratio r_i of correctable error patterns of weight i. For $i < (t+1)^2$, we have $r_i = 1$, and for $i > (2n-t)t$ we put $r_i = 0$.

Let a_i be the number of error patterns of weight i correctable by a decoding algorithm $\mathcal{A}$ (here $\mathcal{A}$ can be $\mathcal{A}_1$ or $\mathcal{A}_2$). The value of $\mathcal{P}_{\mathcal{A}}(p)$ is given by

$$\mathcal{P}_{\mathcal{A}}(p) = \sum_{i=0}^{n^2} a_i \frac{p^i}{(q-1)^i}(1-p)^{n^2-i}.$$

The ratio of correctable patterns of weight i is

$$\bar{r_i} = \frac{a_i}{\binom{n^2}{i}(q-1)^i}$$

and thus

$$\mathcal{P}_{\mathcal{A}}(p) = \sum_{i=0}^{n^2} \bar{r_i}\binom{n^2}{i}p^i(1-p)^{n^2-i}.$$

The simulation will give us an estimate r_i of $\bar{r_i}$ for all i, from which we can deduce an estimate of $\mathcal{P}_{\mathcal{A}}(p)$:

$$\tilde{\mathcal{P}}_{\mathcal{A}}(p) = \sum_{i=0}^{n^2} r_i\binom{n^2}{i}p^i(1-p)^{n^2-i}.$$

4.3 Simulating the product of two Reed-Solomom codes of length 15 and dimension 7 over $GF(16)$

We consider here the code $RS(16;15,7,9) \otimes RS(16;15,7,9)$. The decoder of the code $RS(16;15,7,9)$ is the Berlekamp-Massey algorithm, which is able to correct all patterns of ν errors and ρ erasures such that $2\nu + \rho < 9$, and no other.

For all error-weights between 25 and 104, one million random error patterns were generated and then corrected by algorithms $\mathcal{A}_1$ and $\mathcal{A}_2$. The results are given in Table 1 and Table 2.

w	r_w	w	r_w	w	r_w	w	r_w
≤ 70	1.000000	79	0.999694	88	0.940953	97	0.193981
71	0.999998	80	0.999413	89	0.908290	98	0.117210
72	0.999996	81	0.998923	90	0.861178	99	0.063536
73	0.999994	82	0.997871	91	0.797590	100	0.029475
74	0.999991	83	0.996030	92	0.716976	101	0.011363
75	0.999973	84	0.992899	93	0.620352	102	0.003459
76	0.999959	85	0.987346	94	0.512229	103	0.000772
77	0.999911	86	0.978030	95	0.399526	104	0.000104
78	0.999832	87	0.963620	96	0.289376	≥ 105	0

Table 1: Ratios of correctable patterns among one million for the product code $RS(16;15,7,9) \otimes RS(16;15,7,9)$ decoded by $\mathcal{A}_1$

w	r_w	w	r_w	w	r_w	w	r_w
≤ 70	1.000000	79	0.999811	88	0.948679	97	0.244678
71	1.000000	80	0.999612	89	0.919257	98	0.164251
72	1.000000	81	0.999223	90	0.876693	99	0.103916
73	0.999998	82	0.998367	91	0.818453	100	0.060768
74	0.999998	83	0.996916	92	0.744382	101	0.033018
75	0.999987	84	0.994329	93	0.654543	102	0.016812
76	0.999980	85	0.989627	94	0.552942	103	0.008005
77	0.999945	86	0.981540	95	0.446094	104	0.003494
78	0.999896	87	0.968961	96	0.339633	≥ 105	0

Table 2: Ratios of correctable patterns among one million for the product code $RS(16;15,7,9) \otimes RS(16;15,7,9)$ decoded by $\mathcal{A}_2$

By iteration we mean a call to ψ. Each such call needs at most n row decoding. The total number of iterations for each weight was stored. We obtained for a channel error probability of $p = 25\%$ an average number of iterations equal to 2.23 to be compared to a maximum number of $n + k = 22$. Thus the average algorithmic complexity in this case is about one tenth of the maximum algorithmic complexity. This number increases with the channel error probability. It grows to 2.85 for $p = 30\%$, and the maximum of 5.17 is reached for $p = 40\%$, but for this value of p the residual error rate is of 31%.

The increase of complexity for $\mathcal{A}_2$ is negligible, since only a very small portion of the error patterns require additionnal operation. The equivalent diameter of the two algorithms are given in Table 3, for a residual error rate from 10^{-4} to 10^{-7}.

ϵ	10^{-4}		10^{-5}		10^{-6}		10^{-7}	
	$\mathcal{A}_1$	$\mathcal{A}_2$	$\mathcal{A}_1$	$\mathcal{A}_2$	$\mathcal{A}_1$	$\mathcal{A}_2$	$\mathcal{A}_1$	$\mathcal{A}_2$
p	.2846	.2860	.2657	.2673	.2490	.2508	.2338	.2360
$d^*(p)$	179	179	177	179	175	177	175	175

Table 3: Equivalent diameter of the decoding region for the product code $RS(16; 15, 7, 9) \otimes RS(16; 15, 7, 9)$. ($\epsilon$ stands for the residual error rate)

4.4 Simulating the product of two Reed-Solomom codes of length 15 and dimension 7 over $GF(256)$

In order to make a comparison with an MDS code with identical parameters, we have made the same simulation for the product code $RS(256; 15, 7, 9) \otimes RS(256; 15, 7, 9)$. As above one million error patterns for each weight between 25 and 104 were decoded. The results are compiled in Table 4.

ϵ	10^{-4}		10^{-5}		10^{-6}		10^{-7}	
	$\mathcal{A}_1$	$\mathcal{A}_2$	$\mathcal{A}_1$	$\mathcal{A}_2$	$\mathcal{A}_1$	$\mathcal{A}_2$	$\mathcal{A}_1$	$\mathcal{A}_2$
p	.2881	.2891	.2695	.2706	.2530	.2543	.2382	.2396
$d^*(p)$	181	181	179	179	177	179	177	177

Table 4: Equivalent diameter of the decoding region for the product code $RS(256; 15, 7, 9) \otimes RS(256; 15, 7, 9)$. ($\epsilon$ stands for the residual error rate)

The slight improvement compared with Table 3 is due to the very low probability of miscorrecting a row or a column; the density of the spheres of radius 4 centered on the codewords is $3\,10^{-7}$ for $GF(256)$ instead of 0.016 for $GF(16)$. The average number of iterations for $p = 25\%$ is 2.22, very close to the case $q = 16$ (2.23).

Singleton bound. The Singleton bound for the code considered is equal to $n^2 - k^2 + 1 = 177$. Any code of length n^2 and minimum distance k^2 has a minimum distance lower or equal to 177. We see in Table 4 that algorithm $\mathcal{A}_2$ has an equivalent diameter larger or equal to this bound for the values of the residual error rates considered.

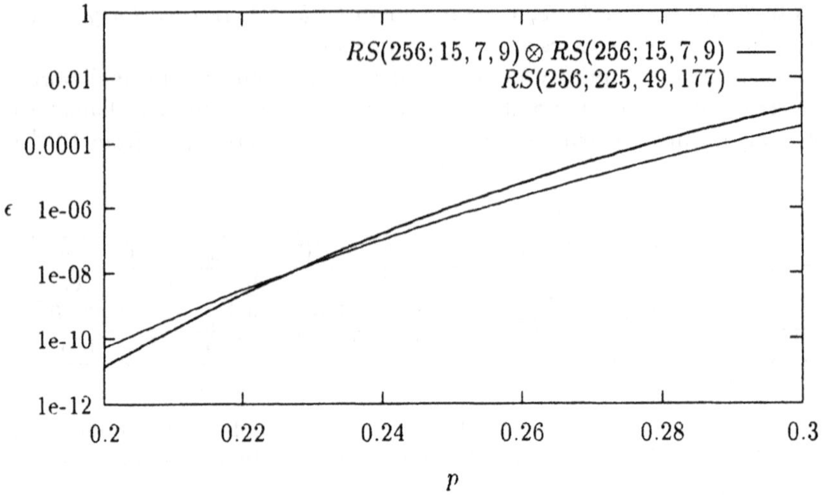

Figure 3: Decoding performance of the code $RS(256;15,7,9) \otimes RS(256;15,7,9)$ versus $RS(256;225,49,177)$

The Reed-Solomom code $RS(256;225,49,177)$ is decoded by the 177-closed Berlekamp-Massey algorithm, its probability of correct transmission for a channel error probability p is given by equation (3)

$$V_{177}(p) = \sum_{i=0}^{88} \binom{225}{i} p^i (1-p)^{225-i}.$$

The residual error rate of both this code and the product code above decoded by $\mathcal{A}_2$ are given in Figure 3 versus the channel symbol error probability. The curves meet at $p = 0.2272$, corresponding to a residual error rate of 10^{-8}. For a larger channel error probability, the product code has better performance, and for a lower channel error probability it is the Reed-Solomom code. But in the last case both codes have a very low residual error rate.

4.5 Algorithmic complexity

We will compare here the algorithmic complexity of algorithm $\mathcal{A}_1$ of a product code $RS(q;n,k,d) \otimes RS(q;n,k,d)$, with the algorithmic complexity of the Berlekamp-Massey algorithm for a Reed-Solomon code $RS(q;n^2,k^2,d')$. All the codes are shortened, if necessary, to be defined on the same alphabet $GF(q)$.

The algorithmic complexity of the Berlekamp-Massey algorithm is $Kn^2(n^2 -$

k^2) for the code $RS(q; n^2, k^2, d')$, and $Kn(n-k)$ for the code $RS(q; n, k, d)$. The constant K depends only on the finite field $GF(q)$.

For the product code we have at most n row-decodings at each iteration. If $\lambda(n, k)$ denotes the average number of iterations, we have a total average complexity of at most

$$\lambda(n, k) K n^2 (n - k).$$

Thus the ratio with the $RS(q; n^2, k^2, d')$ code is at most equal to

$$\frac{\lambda(n, k)}{n + k}.$$

The algorithm we used for the simulation had the number of iterations limited to $n + k$, thus the *extremal* complexity for the product code is the same as the complexity for the Reed-Solomon code.

In practice, for $n = 15$ and $k = 7$, the average number of iterations for the whole simulation is between 2 and 3 (the average value of λ varies with the channel error probability p) instead of a maximal value of 22. In this case the product code has a substantial advantage in average computation time. We have seen above that it has comparable, and even in some case better, decoding performance than the code $RS(256; 225, 49, 177)$.

5 Conclusion

We have shown a product code which has better performance, lower algorithmic complexity and the same transmision rate as the shortened Reed-Solomon code $RS(256; 225, 49, 177)$, for a residual error rate larger than 10^{-8}. Other simulations were made for product codes of higher rates [4]. The results are good, but the equivalent diameter does not beat the Singleton bound.

We did not obtained general result on the "goodness" of product codes, although it seems that the best behavior is obtained for low rates.

References

[1] P. Elias. Error-free coding. *IEEE Transaction on Information Theory*, 4:29–37, 1954.

[2] S.M. Reddy and J.P. Robinson. Random error and burst correction by iterated codes. *IEEE Transaction on Information Theory*, 18(1):182–185, January 1972.

[3] N. Sendrier. Product of linear codes. Rapport de Recherche 1286, INRIA, October 1990.

[4] N. Sendrier. *Codes Correcteurs d'Erreurs à Haut Pouvoir de Correction.* Thèse de doctorat, Université Paris 6, December 1991.

Erasure Correction Performance of Linear Block Codes

Ilya I. Dumer* and Patrick G. Farrell**

* Institute for Problems of Information Transmission
Moscow, Russia

** University of Manchester
United Kingdom

Abstract

We estimate the probability of incorrect decoding of a linear block code, used over an erasure channel, via its weight spectrum, and define the weight spectra that allow us to achieve the capacity of the channel and the random coding exponent. We derive the erasure correcting capacity of long binary BCH codes with slowly growing distance and their duals. Concatenated codes of growing length $n \to \infty$ and polynomial decoding complexity $O(n^2)$, achieving the capacity of the erasure channel (or any other discrete memoryless channel), are considered.

1. Introduction

In this paper we consider the performance of binary linear block codes used over an erasure channel. The channel is defined by the input set $E = \{0,1\}$, the output set $S = \{0,1,*\}$, including the erasure symbol *, and the set of transition probabilities $p(*|0) = p(*|1) = p$, and $p(0/0) = p(1|1) = 1-p$. The capacity of the erasure channel, $C = 1-p$, and the random coding exponent, $E(R,p)$ for codes of rate R are well known (see [1]).

Consider a binary n-dimensional Hamming space $E^n = \{0,1\}^n$ and let A, $A \subseteq E^n$ be a linear (n,k)-code of length n and code rate $R = k/n$, defined by a $(k{\times}n)$ generator matrix $G = (g_{ij})$, $i = 1,\ldots, k$; $j = 1, \ldots, n$. The input information sequence $u=(u_1, u_2, \ldots, u_k)$ is encoded into the codeword $a = (a_1, a_2, \ldots, a_n) = uG$. Let $J = \{j_1, \ldots, j_s\}$ denote an ordered set of s unerased positions in the received vector $z = (z_1, \ldots, z_n) \in S^n$, $1 \leq j_1 \leq \ldots j_s \leq n$. Hereafter $z_J = (z_j | j \in J) \in E^s$ denotes the unerased s-subvector of vector z and $G_J = (g_{ij})$, $i = 1, \ldots, k$; $j \in J$, denotes a $(k{\times}s)$-sub matrix of G, defined on the set J. Obviously $z_J = a_J$.

The maximum likelihood (ML) decoder derives the information sequence $u \in E^k$ by solving the system.

$$uG_J = z_J \tag{1}$$

of s linear equations in k binary variables $u_1, \ldots, u_k$. Note that system (1) has a unique solution, and decoding is unambiguous, if and only if the rank $r(J) = Rk(G_J)$ of matrix G_J is equal to k. Otherwise any solution among the set of $L = 2^{k-r(J)}$ solutions is possible and the probability of correct decoding is equal to 1/L. ML-decoder (1) is known to provide the minimal probability of incorrect decoding (per block) of any linear code A with polynomial complexity $O(n^3)$, $n \to \infty$.

Let α_t denote the fraction of erasure patterns of weight t resulting in ambiguous decoding, for a given code A with code distance d. The probability P of ambiguous decoding is defined by the set $\{\alpha_t,\ t=d, \ldots, n-k\}$ since $\alpha_t = 0$ for $t < d$ and $\alpha_t = 1$ for $t > n-k$. Below we consider so called (n,R)-families of (n_i,k_i)-codes, with rates $R_i = k_i/n_i$, converging to R, $0 \le R \le 1$, when $i \to \infty$. We prove further, that the set of inequalities $\alpha_1 \le \alpha_2 \le \ldots \le \alpha_n$ holds for any linear code and consider (n,R) - families with the following threshold property.

Definition 1 The infinite (n,R) - family is called an (n,R,θ)-family, $0 \le \theta \le 1$, if for any $\epsilon > 0$ and any $t_i \le (1-\epsilon)\theta(n_i-k_i)$ the fraction $\alpha_{t_i} \to 0$, when $i \to \infty$, while for any $t_i \ge (1+\epsilon)\theta(n_i-k_i)$ the fraction $\alpha_{t_i} \ge \gamma$ for some $\gamma > 0$, when $i \to \infty$.

According to proposition 1 below, the parameter θ defines the relative erasure capacity of the (n,R)-family. Namely, an (n,R,θ)-family gives a vanishing probability of incorrect decoding, if used over an erasure channel with transition probability $p < \theta(1-R)$. We are mainly interested in constructing (n,R,1)-families of codes, since only these families achieve the capacity 1-p of the erasure channel for any $R < 1-p$, and all $0 < p < 1$.

In section 2 we estimate the numbers α_t and probability P of ambiguous decoding via the weight spectrum $W = (1, W_d, \ldots, W_n)$ of code A, where $W_i, i=d,\ldots,n$, is the number of codewords of Hamming weight i. We prove that an (n,R,1)-family is obtained, if the restrictions

$$W_i \le \binom{n}{i} \Big/ \binom{n-k}{i} \tag{2}$$

hold. Moreover, the probability P decreases exponentially with distance d for any $p < 1-R$. Restrictions (2) allow any growth of distance d, no matter how slow it is.

On the other hand, much stronger restrictions $W_i \le L_i^{\,1+o(1)}$, where $i = 1, \ldots, n(1-R)$, $o(1) \to 0$, $n \to \infty$ and

$$L_i = \begin{cases} \binom{n}{i} \Big/ 2^{n(1-R)}, & 1 \le i \le n(1-R)/2 \\ \binom{n}{i} \Big/ \binom{n(1-R)}{i}, & n(1-R)/2 \le i \le n(1-R) \end{cases} \tag{3}$$

allow us to achieve the random coding exponent E(R,p), $0 < R < 1-p$. Moreover, the exponent E(R,p) can be achieved, for any $p < 1-R$, only if the distance d satisfies the asymptotical inequality $d/n \gtrsim \delta$, $n \to \infty$, δ being the relative Varshamov-Gilbert distance of the code; $\delta = H^{-1}(1-R)$, $\delta \leq ½$ and $H(x) = - x \log_2 x - (1-x) \log_2 (1-x)$ is the binary entropy function, $0 \leq x \leq 1$

In section 3 we consider the erasure capacity of primitive BCH-codes of growing length $n=2^m-1 \to \infty$ and fixed or slowly growing distance. We prove that these codes correct virtually all erasure patterns of weight $t \sim em = n-k$, where $e = \lfloor (d-1)/2 \rfloor$. Therefore these codes make up an (n,1,1)-family. We also prove that their dual codes correct virtually all erasure patterns of weight $t \sim n-em$ and therefore form an (n,0,1)-family.

Following Forney [2], we consider in section 4 the (n,R)-families of concatenated codes and define the (n,R,1)-families with maximal possible erasure capacity and complexity of order $O(n^2)$ for any $0<R<1$ and $n \to \infty$. These families provide a positive answer to an interesting question brought up by Zemor and Cohen [3]. Namely, codes satisfying the weight restrictions in inequality(3), are proved in [3] to achieve the maximal capacity $\theta=1$, and the question raised there is whether any constructive families achieve the same capacity.

In section 5 we state some open problems resulting from our investigation.

2. Erasure Performance of Linear Block Codes

Consider a binary linear (n,k)-code A, used over a binary erasure channel with erasure probability p. Let the codeword positions be $N=\{1,2,...,n\}$. $I = \{i_1, i_2,...,i_t\}$ is a set of t erased positions, and $J = N \setminus I$ is the complementary set of n - t unerased positions. For any codeword $b = (b_1 \ldots b_n)$ define the set $Supp(b) = \{j | b_j = 1\}$. The set I is defined as a covering set, iff (if and only if) $Supp(b) \subseteq I$ for at least one non-zero codeword $a \in A$. An ML-decoder tries to restore the transmitted codeword a from its unerased subvector $a_J = z_J$ by solving the system (1).

Lemma 1. The following conditions (i), (ii), (iii) are equivalent:

(i) The decoding of erasure pattern I is ambiguous
(ii) I is a covering set
(iii) The rank r(J) of the matrix G_J is less than k

The proof is obvious.

Let $M_t = \{I: |I| = t, r(J) < k\}$ denote the complete set of covering t-subsets. Let $\alpha_t = |M_t| / \binom{n}{t}$ denote the fraction of covering t-subsets among all $\binom{n}{t}$ subsets of cardinality t.

Lemma 2. The numbers α_t form a nondecreasing function of the weight t.

Proof: any covering t-subset can be obtained from t+1 covering subsets of weight t at most. Therefore $|M_{t+1}| \geq |M_t| \cdot (n-t)/(t+1)$ and $\alpha_{t+1} \geq \alpha_t$. QED

According to Lemma 1, the probability P of ambiguous decoding is

$$P = \sum_{t=d}^{n} M_t \, p^t \, (1-p)^{n-t} = \sum_{t=d}^{n} \alpha_t \, \beta(t) \qquad (4)$$

where $\beta(t) = \binom{n}{t} p^t (1-p)^{n-t}$ is the binomial distribution of the probability of t erasures on n positions. The probability P_e of incorrect decoding can be estimated as: $P/2 \leq P_e < P$, since the transmitted codeword is selected from among $2^{k-r(J)} \geq 2$ possible codewords whenever the decoding is ambiguous. Thus the problem of estimating the probabilities P and P_e leads to the problem of estimating the numbers α_t.

Proposition 1. Any (n,R,θ)-family with $0 < R < 1$ has vanishing probability $P \to 0$ of ambiguous decoding for $p < \theta(1-R)$ and non-vanishing probability $P \geq \gamma$ for any $p > \theta(1-R)$, and some $\gamma > 0$, when $n \to \infty$.

Proof: according to definition 1, $\alpha_t \to 0$ for any $\epsilon > 0$ and $t \leq \tau = \lfloor (1-\epsilon)\theta(1-R)n \rfloor$, when $n \to \infty$. From Lemma 2 the inequality follows:

$$P = \sum_{t=d}^{\tau} \alpha_t \, \beta(t) + \sum_{\tau+1}^{n} \alpha_t \, \beta(t) < \alpha_\tau + \sum_{t=\tau+1}^{n} \beta(t)$$

For $p < \theta(1-R)$ and ϵ small enough, the inequality $\tau/n > p$ holds and the probability

$$\sum_{t=\tau+1}^{n} \beta(t)$$

of $\tau+1$ or more erasures tends to zero, when $n \to \infty$. Since $\alpha_\tau \to 0$, the first part of the proposition follows. On the contrary, for any $\epsilon > 0$ and $t \geq T = \lfloor (1+\epsilon)\theta(1-R)n \rfloor$ the inequalities $\alpha_t \geq \gamma$ and

$$P \geq \gamma \sum_{t=T}^{n} \beta(t)$$

hold.

For $p > \theta(1\text{-}R)$ and ϵ small enough $\sum_{T}^{n} \beta(t) \to 1$, and the second part of the proposition follows. QED

According to Proposition 1 (n,R,θ)-families achieve the capacity 1-p of the erasure channel iff $\theta=1$. Hereafter we consider (n,R,1)-families and estimate their weight spectra.

Proposition 2. If the weight spectra of the (n,R)-family satisfy restrictions (2), then the family achieves the maximal erasure capacity $\theta=1$ and has probability P of ambiguous decoding, decreasing exponentially with distance $d \to \infty$ for any $p < 1\text{-}R$.

Proof: following [3], we estimate the numbers α_t via the weight spectrum W_i, $i=d, \ldots, n\text{-}k$. Let $\alpha_t(i)$ be the fraction of t-subsets covering codewords of weight i, $t \geq i$. Obviously,

$$\alpha_t(i) \le W_i \binom{n-i}{t-i} / \binom{n}{t} = W_i \binom{t}{i} / \binom{n}{i}, \alpha_t \ge \sum_{i=d}^{t} W_i \binom{t}{i} / \binom{n}{i} \tag{5}$$

since each codeword of weight i is covered by $\binom{n-i}{t-i}$ subsets of weight t.

If restrictions (2) are satisfied, then

$$\alpha_t(i) \le \binom{t}{i} / \binom{n-k}{i}$$

Therefore $\alpha_t(i) \leq f^i$ and $\alpha_t < f^d/(1\text{-}f)$, where $f = t/(n\text{-}k)$.
For any $\epsilon > 0$ and $t \leq (1\text{-}\epsilon)(1\text{-}R)n$ the estimate

$$\alpha_t < (1-\epsilon)^d/\epsilon$$

holds and therefore α_t decreases exponentially with d. Now the proof follows from proposition 1. QED.

More explicit estimates of the numbers α_t (see (5)) can be obtained in the following way. Let $\alpha_{t,j}$ denote the fraction of t-subsets covering 2^j codewords, $j=1,\ldots,t$. Obviously

$$\alpha_t = \sum_{j=1}^{t} \alpha_{tj}\, , \; and \sum_{j=1}^{t} \alpha_{tj}\, (2^j-1) = \sum_{i=d}^{t} W_i \binom{t}{i} / \binom{n}{i}$$

since any t-subset, covering 2^j-1 non-zero codewords, is counted 2^j-1 times in the right hand side of the last equality. Let $D = \{d_1=d, d_2, \ldots, d_k=n\}$ denote the set of generalised minimum distances [10] of a linear (n,k)-code. Any t-subset, $d_1 \le t < d_2$, covers at most one non-zero codeword and the equality

$$\alpha_t = \sum_{i=d}^{t} W_i \binom{t}{i} / \binom{n}{i} \tag{6}$$

holds for these t. For larger values of t lower estimates of α_t can be obtained, if the sets D and W are known.

Hereafter all unspecified logarithms and exponents are defined over binary base; $o(1) \to 0$, when $n \to \infty$.

Consider now the weight restrictions of (n,R)-families that give the random coding exponent E(R,p) under ML decoding for all $0 < p < 1-R$. Let $T(\gamma) = -\gamma \log p - (1-\gamma)\log(1-p) - H(\gamma)$ denote the limiting exponent of the probability $\beta(\gamma n)$ of erasure patterns of weight γn, $n \to \infty$.

According to [1], the best families of codes satisfy the inequality $(-\log P)/n \ge E(R,p) + o(1)$ for the probability P of ML-decoding, where $n \to \infty$ and

$$E(R,p) = \begin{cases} T(1-R), \ if \ (1-R)/(1+R) \le p \le (1-R) \\ -\log(1+p) + 1-R, \ if \ p \le (1-R)/(1+R) \end{cases} \tag{7}$$

<u>Proposition 3</u>

1. The random coding exponent E(R,p) is achieved for all p, $0 \le p \le 1-R$, by an (n,R)-family, if the weight spectra satisfy the restrictions (3).

2. The random coding exponent E(R,p) is achieved for all p, $0 \le p < 1-R$, by an (n,R)-family, only if the code distances d(n) in the (n,R)-family satisfy the restrictions

$$d(n)/n \ge H^{-1}(1-R) + o(1), \ n \to \infty.$$

Proof: similarly to proposition 2.

Consider now the set of codes, generated by (kxn)-matrices $G = (g_{ij})$, $i = 1, \ldots, k$, $j = 1, \ldots, n$, with rate $R = k/n$, $0 < R < 1$. It is well known that virtually all matrices G have rank k, when $n \to \infty$, and generate (n,R)-codes, satisfying inequalities (3) for all $i=1,\ldots,n$ (see[4]). Therefore we have:

Corollary 1. Virtually all linear (n,R)-codes achieve the random coding exponent E(R,p) of the erasure channel for any $p \leq 1-R$.

3. Performance of BCH-codes over an erasure channel

Below we estimate the performance of long BCH-codes with slowly growing or fixed distance used over an erasure channel. Similarly to the estimates of section 2, the performance can be estimated via their weight spectra (whereas the performance in a binary symmetric channel is defined by the weight distribution of the coset leaders). Still not much is known about the explicit weight spectra of algebraic constructions. The known results include [4] the weight spectra of primitive BCH-codes correcting up to 3 errors, Reed-Muller codes of the second order and some of their subcodes, and the weight spectra of the dual codes. Therefore the performance of all these codes can be estimated from (5).

Let B(n,s) denote the primitive binary BCH-code of length $n = 2^m-1$ and designed distance $d^* = 2s + 1$ with $k \geq n-ms$ information symbols. The asymptotical performance of B(n,s)-codes with $n \to \infty$ and slowly growing (or fixed) distance d^* can be estimated by the following result.

Lemma 4 [5]. The number W_i of codewords of weight i in BCH-code B(n,s) with

$$m \to \infty,\ n = 2^m - 1,\ s \leq 0.2\ \{\ln(n)/\ln(\ln(n))\} \tag{8}$$

can be estimated as

$$W_i = \left(\binom{n}{i} / 2^{n-k} \right) (1 + \epsilon_n) \tag{9}$$

for all $d^* \leq i \leq n-d^*$, where $\varepsilon_n = O(n^{-0.1})$.

According to [4, section 9.3] the equality $k=n-ms$ holds for long BCH-codes with parameters (8). Therefore, unambiguous ML-decoding of these codes in an erasure channel can be done only if the number t of erased symbols satisfies the inequality $t \leq ms \sim s\log n,\ n \to \infty$.

Note also that the actual distance d of long BCH-codes with parameters (8) coincides with the designed distance d^*. The following proposition gives estimates of numbers α_t in the asymptotical interval $d^* \leq t \leq (n-k)(1-o(1))$, when $n \to \infty$.

Proposition 4. BCH-codes with parameters (8) correct virtually all erasure patterns of weight

$$t \leq ms - o(m) \tag{10}$$

where o(m) is any positive function increasing more slowly than m.

Corollary 2. BCH-codes with parameters (8) form an (n,1,1)-family.

Proof: by substituting the weight spectra coefficients (9) into (5).

Consider now the asymptotical performance of the codes $B^{\perp}(n,s)$, dual to the codes B(n,s). We estimate their performance under the following restrictions, with parameter c, $0< c< 1$:

$$m \to \infty\ ,\ n = 2^m - 1\ ,\ s \leq c2^{\lfloor m/2 \rfloor - 1} \tag{11}$$

These restrictions are weaker than restrictions (8). Similarly to B(n,s)-codes, the relation $k = ms < \sqrt{(n)} \log n$ holds for long $B^{\perp}$ (n,s) codes. Moreover, according to the Karlits-Uchiyama bound [4, section 9.9], the inequality:

$$d \geq 2^{m-1} - (s-1)2^{m/2} > 2^{m-1}(1-c) \tag{12}$$

holds.

Proposition 5. $B^{\perp}$ (n,s)-codes with parameters (11) correct virtually all erasure patterns of weight

$$t \leq n-k/(1-\log(1+c)) - o(m) \tag{13}$$

where o(m) is any positive function, increasing more slowly than m.

Proof: according to inequality (5),

$$\alpha_t \leq \sum_{i=d}^{n} W_i \binom{t}{i} / \binom{n}{i} < \sum_{i=d}^{n} W_i \binom{t}{d} / \binom{n}{d} = 2^k \binom{t}{d} / \binom{n}{d}$$

Consider the function $f(t) = \log\{ \binom{t}{d} / \binom{n}{d} \}$, which grows with t. Direct calculations show that the asymptotic equality $f(t) \sim (n - t) \log (1 - d/n)$ holds when $n \to \infty$ and $t \sim n$. According to (12), $d/n > (1-c)/2$. Therefore, the proposition holds since $\alpha_t \to 0$ for $n \to \infty$, and any t satisfying (13). QED

Corollary 3. The family of $B^{\perp}$ (n,s)-codes with parameters (11) form an (n,0,1)-family.

Propositions 4 and 5 show that long BCH-codes B(n,s) with restrictions (8) and dual codes $B^{\perp}(n,s)$ with restrictions (11) achieve the capacity $\theta=1$, correcting virtually all erasure patterns of weight $t \sim n-k$, when $n \to \infty$. The problem of estimating the erasure correcting capacity of BCH-codes of arbitrary rate R, $0 < R < 1$, is still open. In the following section we describe concatenated constructions achieving maximal possible capacity $\theta=1$ for any R, $0 < R < 1$.

4. Concatenated Codes

We consider below the classical Forney construction of concatenated codes [2]. Let A(q,n,m) denote a q-ary code of length n with m codewords A(i), $i=1,\dots,m$, and rate $R_A = (\log_q m)/n$. Let $B(m,\ell,d,M)$ denote an m-ary code of length ℓ, Hamming distance d with M codeword and rate $R_B = (\log_m M)/\ell$. The concatenated code C(q,N,M) of length $N=n\ell$ and rate $R=R_AR_B$ is defined by replacing symbol i_j, $j=1,\dots,\ell$ in any codeword $(i_1,\dots,i_\ell)\in B$ by the corresponding q-ary n-vector $A(i_j)$: $(i_1,\dots,i_\ell),\dots,(A(i_1),\dots,A(i_\ell))$.

Consider the asymptotical performance of q-ary concatenated codes with fixed rate R, $0 < R < 1$, when $n \to \infty$, $\ell \to \infty$, in an arbitrary memoryless channel. We choose inner codes A to be of very short length $n=o(\log N)$, which achieve the random coding exponent of the channel and then we use ML-decoding matched to this channel. Families of $B(m,\ell,d,M)$-codes with linearly growing distance for any rate $R_B<1$ and $\ell \to \infty$, are chosen as outer codes regardless of the channel. Only bounded distance is required for the outer channel, outer decoding providing correction of up to (d-1)/2 errors.

Such families of $B(m,\ell,d,M)$-codes themselves can be constructed as concatenated codes, according to [6]. A complexity of construction, including encoding and decoding with correction of up to (d-1)/2 errors, of $O(\ell^2)$ can be achieved [6].

Proposition 6. For any discrete memoryless channel with capacity C and any rate R, $R < C$, there exist infinite families of concatenated (N,R)-codes, $N \to \infty$, with complexity $O(N^2)$ of construction, encoding and decoding, that provide exponentially decreasing probability of incorrect decoding with length N.

Proof: Similar to Forney's proof [2] for the binary symmetric channel.

Corollary 4. For any rate R, $0 < R < 1$, there exist (N,R,1)-families of concatenated codes with complexity $O(N^2)$.

Notes.

1. An algebraic construction of concatenated codes, based on Justesen concatenated codes, is considered in [7]. This construction provides the conditions of proposition 6 for so called regular channels, forming a subclass of discrete memoryless symmetric channels. Searching algorithms are only used in [7] for constructing large Galois fields, rather than for constructing good short codes, matched to the channel, as above.

2. The exponent E(R) of an erasure channel can be achieved via a class of concatenated codes in the following (non-constructive) way. Let A(2,n,m) be a binary code of growing length $n \to \infty$ and rate R_A, satisfying the weight restrictions (3). Let $B = \{B(m,\ell,d,M)\}$ denote the ensemble of generalised Reed-Solomon codes of length

ℓ=m-1 and rate R_B. Consider the corresponding ensemble {C} of concatenated codes of rate $R=R_AR_B$. According to [8], virtually all codes from ensemble {C} satisfy restrictions (3), if

$$R_A > 1 + \log_2 (1-H^{-1}(1-R)). \tag{14}$$

Therefore according to proposition 2, we have:

Corollary 5. Virtually all linear concatenated codes $C \in \{C\}$ of length $N \rightarrow \infty$ achieve random coding exponent E(R,p) of the erasure channel for any $0 \leq p \leq 1-R$, under restriction (14).

3. Cascaded ML-decoding for the erasure channel is considered in [9] for concatenated codes, where it is carried out by constructing lists of inner codewords for every column. Though the maximal possible inner decoding list is estimated in [9] as the complete list of $M = 2^{nR_A}$ codewords, the decoding complexity is claimed to be $O(N^2)$ for $N \rightarrow \infty$. We should note that the overall number of trials grows for this algorithm as 2^{NR}, whenever complete lists of M codewords are considered. Therefore the maximal complexity of the algorithm in [9] is actually determined by an exhaustive search over all codewords. Therefore the problem of constructing ML-decoding schemes for erasures of complexity $O(N^2)$ is still open.

5. Concluding Remarks

The ML-decoding performance of linear block codes in an erasure channel is an important problem from theoretical and practical points of view, since the decoding complexity is upper bounded by polynomial order $O(n^3)$ with length n, and because ereasure decoding forms the basis of information set and covering algorithms for error corection. In this paper the erasure decoding error probability is estimated via the weight spectrum. The estimates are exponentially tight for virtually all codes and meet the random coding exponent of the erasure channel. It is shown that long BCH codes with fixed or slowly growing distance and rate $R \rightarrow 1$, and their duals with $R \rightarrow 0$, achieve asymptotically the maximal possible capacity of erasure correction. The same holds for properly constructed concatenated codes of arbitrary rate R, with polynomial complexity.

Finally, we state two open problems, that stem from our considerations:

1. What are the erasure capacities of important algebraic (n,R)-codes of rate $0<R<1$, such as BCH codes and Reed-Muller codes?

2. Does the (n,R)-family $0<R<1$, achieve the maximal possible capacity 1-R of erasure correction, if the dual family achieves the capacity R?

Acknowledgement: The authors are grateful to the Royal Society, UK, for the financial support that made this research possible.

References

1. R. G. Gallager : Information Theory and Reliable Communications; New York:Wiley, 1968.

2. G. D. Forney : Concatenated codes; Cambridge, MA : MIT, 1966.

3. G. Zemor and G. Cohen : The threshold probability of a code (submitted to IEEE Trans IT).

4. F. J. MacWilliams and N. J. A. Sloane: The Theory of Error-Correcting Codes; North-Holland, 1977.

5. V. M. Sidelnikov : Weight Spectrum of Binary BCH Codes, Problemi Peredachi Informatsii, Vol 7, No 1, 1971, pp 14-22.

6. V. V. Zyablov : On estimation of complexity of construction of binary linear concatenated codes, Probl. Peredach. Inform., Vol 7, No 1, 1971, pp 5-13.

7. P. Delsarte and P. Piret : Algebraic Constructions of Shannon Codes for Regular Channels; IEEE-Trans IT, Vol. IT-28, No. 4, 1982, pp 593-599.

8. E. L. Blokh and V.V. Zyablov : Lineinie Kaskadnie Kodi, Moskva, Nauka, 1982 (in Russian).

9. Y. Xu : Maximum Likelihood Erasure Decoding Scheme for Concatenated Codes; IEE Proceedings-Part I, Vol 139, No 3, 1992, pp 336-339.

10. V.K. Wei: Generalised Hamming weights for linear codes; IEEE-Trans IT, Vol IT-37, No 5, 1991, pp 1412-1418.

Lecture Notes in Computer Science

For information about Vols. 1–709
please contact your bookseller or Springer-Verlag

Vol. 710: Z. Ésik (Ed.), Fundamentals of Computation Theory. Proceedings, 1993. IX, 471 pages. 1993.

Vol. 711: A. M. Borzyszkowski, S. Sokołowski (Eds.), Mathematical Foundations of Computer Science 1993. Proceedings, 1993. XIII, 782 pages. 1993.

Vol. 712: P. V. Rangan (Ed.), Network and Operating System Support for Digital Audio and Video. Proceedings, 1992. X, 416 pages. 1993.

Vol. 713: G. Gottlob, A. Leitsch, D. Mundici (Eds.), Computational Logic and Proof Theory. Proceedings, 1993. XI, 348 pages. 1993.

Vol. 714: M. Bruynooghe, J. Penjam (Eds.), Programming Language Implementation and Logic Programming. Proceedings, 1993. XI, 421 pages. 1993.

Vol. 715: E. Best (Ed.), CONCUR'93. Proceedings, 1993. IX, 541 pages. 1993.

Vol. 716: A. U. Frank, I. Campari (Eds.), Spatial Information Theory. Proceedings, 1993. XI, 478 pages. 1993.

Vol. 717: I. Sommerville, M. Paul (Eds.), Software Engineering – ESEC '93. Proceedings, 1993. XII, 516 pages. 1993.

Vol. 718: J. Seberry, Y. Zheng (Eds.), Advances in Cryptology – AUSCRYPT '92. Proceedings, 1992. XIII, 543 pages. 1993.

Vol. 719: D. Chetverikov, W.G. Kropatsch (Eds.), Computer Analysis of Images and Patterns. Proceedings, 1993. XVI, 857 pages. 1993.

Vol. 720: V.Mařík, J. Lažansky, R.R. Wagner (Eds.), Database and Expert Systems Applications. Proceedings, 1993. XV, 768 pages. 1993.

Vol. 721: J. Fitch (Ed.), Design and Implementation of Symbolic Computation Systems. Proceedings, 1992. VIII, 215 pages. 1993.

Vol. 722: A. Miola (Ed.), Design and Implementation of Symbolic Computation Systems. Proceedings, 1993. XII, 384 pages. 1993.

Vol. 723: N. Aussenac, G. Boy, B. Gaines, M. Linster, J.-G. Ganascia, Y. Kodratoff (Eds.), Knowledge Acquisition for Knowledge-Based Systems. Proceedings, 1993. XIII, 446 pages. 1993. (Subseries LNAI).

Vol. 724: P. Cousot, M. Falaschi, G. Filè, A. Rauzy (Eds.), Static Analysis. Proceedings, 1993. IX, 283 pages. 1993.

Vol. 725: A. Schiper (Ed.), Distributed Algorithms. Proceedings, 1993. VIII, 325 pages. 1993.

Vol. 726: T. Lengauer (Ed.), Algorithms – ESA '93. Proceedings, 1993. IX, 419 pages. 1993

Vol. 727: M. Filgueiras, L. Damas (Eds.), Progress in Artificial Intelligence. Proceedings, 1993. X, 362 pages. 1993. (Subseries LNAI).

Vol. 728: P. Torasso (Ed.), Advances in Artificial Intelligence. Proceedings, 1993. XI, 336 pages. 1993. (Subseries LNAI).

Vol. 729: L. Donatiello, R. Nelson (Eds.), Performance Evaluation of Computer and Communication Systems. Proceedings, 1993. VIII, 675 pages. 1993.

Vol. 730: D. B. Lomet (Ed.), Foundations of Data Organization and Algorithms. Proceedings, 1993. XII, 412 pages. 1993.

Vol. 731: A. Schill (Ed.), DCE – The OSF Distributed Computing Environment. Proceedings, 1993. VIII, 285 pages. 1993.

Vol. 732: A. Bode, M. Dal Cin (Eds.), Parallel Computer Architectures. IX, 311 pages. 1993.

Vol. 733: Th. Grechenig, M. Tscheligi (Eds.), Human Computer Interaction. Proceedings, 1993. XIV, 450 pages. 1993.

Vol. 734: J. Volkert (Ed.), Parallel Computation. Proceedings, 1993. VIII, 248 pages. 1993.

Vol. 735: D. Bjørner, M. Broy, I. V. Pottosin (Eds.), Formal Methods in Programming and Their Applications. Proceedings, 1993. IX, 434 pages. 1993.

Vol. 736: R. L. Grossman, A. Nerode, A. P. Ravn, H. Rischel (Eds.), Hybrid Systems. VIII, 474 pages. 1993.

Vol. 737: J. Calmet, J. A. Campbell (Eds.), Artificial Intelligence and Symbolic Mathematical Computing. Proceedings, 1992. VIII, 305 pages. 1993.

Vol. 738: M. Weber, M. Simons, Ch. Lafontaine, The Generic Development Language Deva. XI, 246 pages. 1993.

Vol. 739: H. Imai, R. L. Rivest, T. Matsumoto (Eds.), Advances in Cryptology – ASIACRYPT '91. X, 499 pages. 1993.

Vol. 740: E. F. Brickell (Ed.), Advances in Cryptology – CRYPTO '92. Proceedings, 1992. X, 593 pages. 1993.

Vol. 741: B. Preneel, R. Govaerts, J. Vandewalle (Eds.), Computer Security and Industrial Cryptography. Proceedings, 1991. VIII, 275 pages. 1993.

Vol. 742: S. Nishio, A. Yonezawa (Eds.), Object Technologies for Advanced Software. Proceedings, 1993. X, 543 pages. 1993.

Vol. 743: S. Doshita, K. Furukawa, K. P. Jantke, T. Nishida (Eds.), Algorithmic Learning Theory. Proceedings, 1992. X, 260 pages. 1993. (Subseries LNAI)

Vol. 744: K. P. Jantke, T. Yokomori, S. Kobayashi, E. Tomita (Eds.), Algorithmic Learning Theory. Proceedings, 1993. XI, 423 pages. 1993. (Subseries LNAI)

Vol. 745: V. Roberto (Ed.), Intelligent Perceptual Systems. VIII, 378 pages. 1993. (Subseries LNAI)

Vol. 746: A. S. Tanguiane, Artificial Perception and Music Recognition. XV, 210 pages. 1993. (Subseries LNAI).

Vol. 747: M. Clarke, R. Kruse, S. Moral (Eds.), Symbolic and Quantitative Approaches to Reasoning and Uncertainty. Proceedings, 1993. X, 390 pages. 1993.

Vol. 748: R. H. Halstead Jr., T. Ito (Eds.), Parallel Symbolic Computing: Languages, Systems, and Applications. Proceedings, 1992. X, 419 pages. 1993.

Vol. 749: P. A. Fritzson (Ed.), Automated and Algorithmic Debugging. Proceedings, 1993. VIII, 369 pages. 1993.

Vol. 750: J. L. Díaz-Herrera (Ed.), Software Engineering Education. Proceedings, 1994. XII, 601 pages. 1994.

Vol. 751: B. Jähne, Spatio-Temporal Image Processing. XII, 208 pages. 1993.

Vol. 752: T. W. Finin, C. K. Nicholas, Y. Yesha (Eds.), Information and Knowledge Management. Proceedings, 1992. VII, 142 pages. 1993.

Vol. 753: L. J. Bass, J. Gornostaev, C. Unger (Eds.), Human-Computer Interaction. Proceedings, 1993. X, 388 pages. 1993.

Vol. 754: H. D. Pfeiffer, T. E. Nagle (Eds.), Conceptual Structures: Theory and Implementation. Proceedings, 1992. IX, 327 pages. 1993. (Subseries LNAI).

Vol. 755: B. Möller, H. Partsch, S. Schuman (Eds.), Formal Program Development. Proceedings. VII, 371 pages. 1993.

Vol. 756: J. Pieprzyk, B. Sadeghiyan, Design of Hashing Algorithms. XV, 194 pages. 1993.

Vol. 757: U. Banerjee, D. Gelernter, A. Nicolau, D. Padua (Eds.), Languages and Compilers for Parallel Computing. Proceedings, 1992. X, 576 pages. 1993.

Vol. 758: M. Teillaud, Towards Dynamic Randomized Algorithms in Computational Geometry. IX, 157 pages. 1993.

Vol. 759: N. R. Adam, B. K. Bhargava (Eds.), Advanced Database Systems. XV, 451 pages. 1993.

Vol. 760: S. Ceri, K. Tanaka, S. Tsur (Eds.), Deductive and Object-Oriented Databases. Proceedings, 1993. XII, 488 pages. 1993.

Vol. 761: R. K. Shyamasundar (Ed.), Foundations of Software Technology and Theoretical Computer Science. Proceedings, 1993. XIV, 456 pages. 1993.

Vol. 762: K. W. Ng, P. Raghavan, N. V. Balasubramanian, F. Y. L. Chin (Eds.), Algorithms and Computation. Proceedings, 1993. XIII, 542 pages. 1993.

Vol. 763: F. Pichler, R. Moreno Díaz (Eds.), Computer Aided Systems Theory – EUROCAST '93. Proceedings, 1993. IX, 451 pages. 1994.

Vol. 764: G. Wagner, Vivid Logic. XII, 148 pages. 1994. (Subseries LNAI).

Vol. 765: T. Helleseth (Ed.), Advances in Cryptology – EUROCRYPT '93. Proceedings, 1993. X, 467 pages. 1994.

Vol. 766: P. R. Van Loocke, The Dynamics of Concepts. XI, 340 pages. 1994. (Subseries LNAI).

Vol. 767: M. Gogolla, An Extended Entity-Relationship Model. X, 136 pages. 1994.

Vol. 768: U. Banerjee, D. Gelernter, A. Nicolau, D. Padua (Eds.), Languages and Compilers for Parallel Computing. Proceedings, 1993. XI, 655 pages. 1994.

Vol. 769: J. L. Nazareth, The Newton-Cauchy Framework. XII, 101 pages. 1994.

Vol. 770: P. Haddawy (Representing Plans Under Uncertainty. X, 129 pages. 1994. (Subseries LNAI).

Vol. 771: G. Tomas, C. W. Ueberhuber, Visualization of Scientific Parallel Programs. XI, 310 pages. 1994.

Vol. 772: B. C. Warboys (Ed.),Software Process Technology. Proceedings, 1994. IX, 275 pages. 1994.

Vol. 773: D. R. Stinson (Ed.), Advances in Cryptology – CRYPTO '93. Proceedings, 1993. X, 492 pages. 1994.

Vol. 774: M. Banâtre, P. A. Lee (Eds.), Hardware and Software Architectures for Fault Tolerance. XIII, 311 pages. 1994.

Vol. 775: P. Enjalbert, E. W. Mayr, K. W. Wagner (Eds.), STACS 94. Proceedings, 1994. XIV, 782 pages. 1994.

Vol. 776: H. J. Schneider, H. Ehrig (Eds.), Graph Transformations in Computer Science. Proceedings, 1993. VIII, 395 pages. 1994.

Vol. 777: K. von Luck, H. Marburger (Eds.), Management and Processing of Complex Data Structures. Proceedings, 1994. VII, 220 pages. 1994.

Vol. 778: M. Bonuccelli, P. Crescenzi, R. Petreschi (Eds.), Algorithms and Complexity. Proceedings, 1994. VIII, 222 pages. 1994.

Vol. 779: M. Jarke, J. Bubenko, K. Jeffery (Eds.), Advances in Database Technology — EDBT '94. Proceedings, 1994. XII, 406 pages. 1994.

Vol. 780: J. J. Joyce, C.-J. H. Seger (Eds.), Higher Order Logic Theorem Proving and Its Applications. Proceedings, 1993. X, 518 pages. 1994.

Vol. 781: G. Cohen, S. Litsyn, A. Lobstein, G. Zémor (Eds.), Algebraic Coding. Proceedings, 1993. XII, 326 pages. 1994.

Vol. 782: J. Gutknecht (Ed.), Programming Languages and System Architectures. Proceedings, 1994. X, 344 pages. 1994.

Vol. 783: C. G. Günther (Ed.), Mobile Communications. Proceedings, 1994. XVI, 564 pages. 1994.

Vol. 784: F. Bergadano, L. De Raedt (Eds.), Machine Learning: ECML-94. Proceedings, 1994. XI, 439 pages. 1994. (Subseries LNAI).

Vol. 785: H. Ehrig, F. Orejas (Eds.), Recent Trends in Data Type Specification. Proceedings, 1992. VIII, 350 pages. 1994.

Vol. 786: P. A. Fritzson (Ed.), Compiler Construction. Proceedings, 1994. XI, 451 pages. 1994.

Vol. 787: S. Tison (Ed.), Trees in Algebra and Programming – CAAP '94. Proceedings, 1994. X, 351 pages. 1994.

Vol. 788: D. Sannella (Ed.), Programming Languages and Systems – ESOP '94. Proceedings, 1994. VIII, 516 pages. 1994.

Vol. 789: M. Hagiya, J. C. Mitchell (Eds.), Theoretical Aspects of Computer Software. Proceedings, 1994. XI, 887 pages. 1994.